# Theoretical Modeling of Organohalide Perovskites for Photovoltaic Applications

# Theoretical Modeling of Organohalide Perovskites for Photovoltaic Applications

Edited by

## Giacomo Giorgi
Department of Civil and Environmental Engineering (DICA)
University of Perugia, Italy

## Koichi Yamashita
Department of Chemical System Engineering
University of Tokyo, Japan

**CRC Press**
Taylor & Francis Group
Boca Raton London New York

CRC Press is an imprint of the
Taylor & Francis Group, an **informa** business

Cover: A cluster of methylammonium lead iodide. Figure drawn with the program VESTA, K. Momma and F. Izumi, *J. Appl. Cryst.*, 2008, 41, 653–658.

This manuscript has been authored by UT-Battelle, LLC under Contract No. DE-AC05-00OR22725 with the U.S. Department of Energy. The United States Government retains and the publisher, by accepting the article for publication, acknowledges that the United States Government retains a nonexclusive, paid-up, irrevocable, worldwide license to publish or reproduce the published form of this manuscript, or allow others to do so, for United States Government purposes. The Department of Energy (DOE) will provide public access to these results of federally sponsored research in accordance with the DOE Public Access Plan (http://energy.gov/downloads/doe-public-access-plan).

CRC Press
Taylor & Francis Group
6000 Broken Sound Parkway NW, Suite 300
Boca Raton, FL 33487-2742

First issued in paperback 2019

© 2017 by Taylor & Francis Group, LLC
CRC Press is an imprint of Taylor & Francis Group, an Informa business

No claim to original U.S. Government works

ISBN-13: 978-1-4987-5078-3 (hbk)
ISBN-13: 978-0-367-87574-9 (pbk)

---

### Library of Congress Cataloging-in-Publication Data

---

Names: Giorgi, Giacomo, 1976- editor. | Yamashita, Koichi (Professor of chemical engineering), editor.
Title: Theoretical Modeling of Organohalide Perovskites for Photovoltaic Applications / edited by Giacomo Giorgi, Koichi Yamashita.
Description: Boca Raton, FL : CRC Press, Taylor & Francis Group, [2017] | Includes bibliographical references and index.
Identifiers: LCCN 2016058861| ISBN 9781498750783 (hardback ; alk. paper) | ISBN 1498750788 (hardback ; alk. paper) | ISBN 9781498750790 (e-book) | ISBN 1498750796 (e-book)
Subjects: LCSH: Photovoltaic cells--Materials. | Solar cells--Materials. | Hybrid solar cells--Materials. | Perovskite. | Molecular dynamics.
Classification: LCC QC715.4 .T47 2017 | DDC 621.3815/420284--dc23
LC record available at https://lccn.loc.gov/2016058861

---

**Visit the Taylor & Francis Web site at**
**http://www.taylorandfrancis.com**

**and the CRC Press Web site at**
**http://www.crcpress.com**

# Contents

# Preface

RESEARCH OF ORGANIC–INORGANIC HALIDE PEROVSKITE materials achieved spectacular progress not only in the high efficiency of photovoltaic devices but also in the fundamental characterization of these unique hybrid materials. On our discovery of the perovskite materials in 2006–2009 as a light absorber of photoelectrochemical cells, we have met difficulty in stabilizing the material at the solid–liquid interface. This hybrid material is an ionic crystal with sufficient solubility in polar solvents, similar to silver halide crystals in photographic materials. This chemical property, however, enables a low-cost solution process for film preparation. Combined with solid-state hole transport materials, the quality of the perovskite absorbers has rapidly evolved to develop a new class of "printable" solar cell capable of high conversion efficiency beyond 22%. Simultaneously with device fabrication, investigation on the electronic structures and solid-state physics of the halide perovskite, as a narrow gap semiconductor, has unveiled many unique properties of this soft ionic crystal, which are rarely found in existing inorganic semiconductors. With its ambipolar nature in free carrier transport, the hybrid perovskite has large carrier lifetimes and diffusion lengths, allowance of ionic migration, and robust electronic properties with resistance against carrier recombination. These properties are involved in production of high voltage in photoelectric response, which makes a key contribution to the high efficiency of solar cells. When the efficiency is approaching near Shockley–Queisser limit (around 25% with a bandgap of 1.6 eV), tremendous efforts shall be directed to ensure high stability of the material against heat, moisture, and light soaking. A solution to this issue is to change the composition of this hybrid crystal, either by stoichiometric or non-stoichiometric approaches, for example, using doped and/or double perovskites. Besides conducting extremely time-consuming experiments, theoretical modeling of the potential structure makes significant contribution to propose candidate materials and methods for device engineers. Density Functional Theory–based computational simulation to describe optoelectronic properties has been already vigorously done for methylammonium and mix cation perovskites to elucidate influences of dielectric properties, influence of local defects, effect of ion migration, and so on. The state-of-the-art approaches to revealing inherent properties of mixed perovskites and their potential functionalities in enhancing efficiency and stability are reported in the chapters of this book. Versatile theoretical models and discussion across the chapters will be valuable gifts for readers who study this highly interdisciplinary field of perovskite-based optoelectronics and explore next-generation photovoltaic devices.

**Tsutomu Miyasaka**
*Toin University of Yokohama, Yokohama, Japan*

# Editors

**Giacomo Giorgi** is an associate professor at the Department of Civil and Environmental Engineering of the University of Perugia, Italy. He has worked at the Department of Chemical System Engineering at the University of Tokyo, where he was a postdoc and later a senior researcher in the group led by Prof. Koichi Yamashita. He was formerly an assistant professor at the Research Centre for Advanced Science and Technology (RCAST) at the University of Tokyo. His scientific interests focus on the theoretical analysis of materials for solar-to-energy conversion.

**Koichi Yamashita** has been a full professor at the Department of Chemical System Engineering at the University of Tokyo since 1994. He obtained both his undergraduate and graduate degrees at Kyoto University, supervised by Prof. Kenichi Fukui. He worked as a postdoctoral fellow with Prof. William Miller at the University of California, Berkeley. He has published more than 250 refereed journal articles in the fields of theoretical and computational chemistry.

# Contributors

**Fahhad H. Alharbi**
Qatar Environment and Energy Research
    Institute
and
College of Science and Engineering
Hamad Bin Khalifa University
Doha, Qatar

**Paolo Barone**
Consiglio Nazionale delle Ricerche
    (CNR-SPIN)
L'Aquila, Italy

and

Graphene Labs, Istituto Italiano di
    Tecnologia
Genova, Italy

**Claudia Caddeo**
Istituto Officina dei Materiali
Consiglio Nazionale delle Ricerche
CNR-IOM SLACS Cagliari
Monserrato, Italy

**Ivano E. Castelli**
Center for Atomic-Scale Materials Design
Department of Physics
Technical University of Denmark
Kongens Lyngby, Denmark

and

Department of Chemistry
University of Copenhagen
Copenhagen Ø, Denmark

**Filippo De Angelis**
Computational Laboratory for Hybrid/
    Organic Photovoltaics
CNR-ISTM
Perugia, Italy

and

CompuNet
Istituto Italiano di Tecnologia
Genova, Italy

**Domenico Di Sante**
Institut für Theoretische Physik und
  Astrophysik
Universität Würzburg, Am Hubland
  Campus Süd
Würzburg, Germany

and

Consiglio Nazionale delle Ricerche
  (CNR-SPIN)
L'Aquila, Italy

**Mao-Hua Du**
Materials Science and Technology Division
Oak Ridge National Laboratory
Oak Ridge, Tennessee, USA

**Fedwa El Mellouhi**
Qatar Environment and Energy Research
  Institute
Hamad Bin Khalifa University
Doha, Qatar

**Jacky Even**
Fonctions optiques pour les technologies
  de l'information, UMR 6082
CNRS, INSA Rennes
Rennes, France

**Alessio Filippetti**
Istituto Officina dei Materiali
Consiglio Nazionale delle Ricerche
CNR-IOM SLACS Cagliari
Monserrato, Italy

**Giacomo Giorgi**
Department of Civil and Environmental
  Engineering (DICA)
University of Perugia
Perugia, Italy

**Giulia Grancini**
Group for Molecular Engineering
  of Functional Materials
Institute of Chemical Sciences
  and Engineering
École Polytechnique Fédérale
  de Lausanne
Sion, Switzerland

**Jun Haruyama**
Global Research Center for Environment
  and Energy Nanoscience (GREEN)
and
International Center for Materials
Nanoarchitectonics (WPI-MANA)
National Institute for Materials Science
  (NIMS)
Tsukuba, Ibaraki, Japan

**Karsten W. Jacobsen**
Center for Atomic-Scale Materials
  Design
Department of Physics
Technical University of Denmark
Kongens Lyngby, Denmark

**Sabre Kais**
Qatar Environment and Energy Research
  Institute
and
College of Science and Engineering
Hamad Bin Khalifa University
Doha, Qatar

and

Department of Chemistry and Physics
Birck Nanotechnology Center
Purdue University
West Lafayette, Indiana, USA

**Claudine Katan**
Institut des Sciences Chimiques de Rennes
    UMR 6226, CNRS
Université de Rennes 1
Rennes, France

**Hiroki Kawai**
Department of Chemical System
    Engineering
School of Engineering
University of Tokyo
Bunkyo-ku, Tokyo, Japan

and

CREST-JST
Chiyoda-ku, Tokyo, Japan

**Alessandro Mattoni**
Istituto Officina dei Materiali
Consiglio Nazionale delle Ricerche
CNR-IOM SLACS Cagliari
Monserrato, Italy

**Edoardo Mosconi**
Computational Laboratory for Hybrid/
    Organic Photovoltaics
CNR-ISTM
Perugia, Italy

**Carlo Motta**
School of Physics and Centre for
    Research on Adaptive Nanostructures
    and Nanodevices
Trinity College
Dublin, Ireland

**Silvia Picozzi**
Consiglio Nazionale delle Ricerche
    (CNR-SPIN)
L'Aquila, Italy

**Claudio Quarti**
Laboratory for Chemistry of Novel
    Materials
Universite de Mons
Mons, Belgium

**Andrew M. Rappe**
Makineni Theoretical Laboratories
Department of Chemistry
University of Pennsylvania
Philadelphia, Pennsylvania, USA

**Sergey Rashkeev**
Qatar Environment and Energy Research
    Institute
Hamad Bin Khalifa University
Doha, Qatar

**Stefano Sanvito**
School of Physics and Centre for
    Research on Adaptive Nanostructures
    and Nanodevices
Trinity College
Dublin, Ireland

**Keitaro Sodeyama**
Center for Materials Research by
    Information Integration (cMI$^2$)
National Institute for Materials Science
    (NIMS)
Tsukuba, Ibaraki, Japan

and

PRESTO
Japan Science and Technology Agency
    (JST)
Kawaguchi, Saitama, Japan

**Alessandro Stroppa**
Consiglio Nazionale delle Ricerche
    (CNR-SPIN)
L'Aquila, Italy

and

International Centre for Quantum and
    Molecular Structures and Physics
    Department
Shanghai University
Shanghai, China

**Liang Z. Tan**
Makineni Theoretical Laboratories
Department of Chemistry
University of Pennsylvania
Philadelphia, Pennsylvania, USA

**Yoshitaka Tateyama**
Global Research Center for Environment
    and Energy Nanoscience (GREEN)
and
International Center for Materials
Nanoarchitectonics (WPI-MANA)
National Institute for Materials Science
    (NIMS)
Tsukuba, Ibaraki, Japan

and

Elements Strategy Initiative for Catalysts
    and Batteries
Kyoto University
Nishikyo-ku, Kyoto, Japan

**Kristian S. Thygesen**
Center for Atomic-Scale Materials Design
Department of Physics
Technical University of Denmark
Kongens Lyngby, Denmark

**Koichi Yamashita**
Department of Chemical System
    Engineering
School of Engineering
University of Tokyo
Bunkyo-ku, Tokyo, Japan

and

CREST-JST
Chiyoda-ku, Tokyo, Japan

# Structure and Thermodynamic Properties of Hybrid Perovskites by Classical Molecular Dynamics

Alessandro Mattoni, Alessio Filippetti, and Claudia Caddeo

## CONTENTS

IN THIS CHAPTER, THE STRUCTURAL AND THERMODYNAMIC PROPERTIES of hybrid perovskites are reviewed within the theoretical perspective of classical molecular dynamics applied to a material with an ionic character. An overview of the crystalline structural properties of both inorganic and hybrid perovskites is given using the concepts of tolerance factor and effective ionic radii for organic and inorganic ions. The simple ionic description of the material provides an effective representation of interatomic forces reproducing the fundamental finite-temperature properties of the hybrid perovskites including orthorhombic-to-tetragonal transitions, cation dynamics, molecular entropy, vibrational properties, and thermal conductivity.

## 1.1 INTRODUCTION TO HYBRID PEROVSKITES

Perovskites of the general formula $ABX_3$ have played a central role in the evolution of materials chemistry and condensed matter physics over the last 70 years (Kieslich et al. 2014). This family of solid-state materials covers a wide range of intriguing properties, including both application-oriented phenomena and fundamental physics and chemistry (Kieslich et al. 2015).

By definition, a perovskite is any material with the general chemical formula $ABX_3$ and crystal structure type such as calcium titanium oxide ($CaTiO_3$) (Wenk and Bulakh 2004); perovskites take their name from the natural mineral, which was discovered in 1839 and is named after mineralogist L. A. Perovski (1792–1856) (De Graef and McHenry 2007). Victor Goldschmidt was the first to describe the perovskite crystal structure in 1926, in his work on tolerance factors (Goldschmidt 1926). In the ideal cubic-symmetry perovskite the B cation is in 6-fold coordination, surrounded by an octahedron of anions, and the A cation is in 12-fold cuboctahedral coordination (Figure 1.1a). A and B are cations with different positive charges compensated by the negative X anion. The size of the ions must obey the Goldschmidt criterion (Filippetti and Mattoni 2014; Goldschmidt 1926), which requires that the sum of the radii of B and X ions must be in relation to the sum of A and X ones like the lengths of the edge and the diagonal of a square (see Figure 1.1b). The Goldschmidt condition is discussed further in Section 1.2.

Small departures from the ideal cubic ratio of perovskites are tolerated, which leads to buckling and distortions of the perovskite lattice (Glazer 1972). Distortions include tilting of the octahedra, displacements of the cations out of the centers of their coordination polyhedra, and distortions of the octahedra driven by electronic factors (Jahn–Teller distortions) (Jahn and Teller 1937). Distortions have great relevance in this class of crystal structures. For example, they are at the basis of the ferroelectric behavior of some perovskites such as $BaTiO_3$ (Pytte 1972; Resta 2003). The orthorhombic, tetragonal, and rhombohedral phases are the most common noncubic lower-symmetry distorted variants.

Materials with perovskite structure are very common because of the adaptability of this structure toward A, B, or X site substitution, which allows for a great variability in materials and properties (Kieslich et al. 2014). The most abundant inorganic perovskites are oxides ($ABO_3$). For example, the dense silicate $MgSiO_3$ perovskite-structured polymorph named bridgmanite is considered the most abundant mineral on Earth (Tschauner et al. 2014).

Many perovskite compounds can also form without oxygen. Some superperovskite structures can form with sulfur (or other group VI elements) (Snyder et al. 1992). More common alternatives to oxides are obtained by replacing oxygen atoms with atoms of group VII. Notable examples are the fluorides (such as $NaMgF_3$, $KMgF_3$, and $KZnF_3$) (Andersen et al. 1985). Particularly relevant is also the group of inorganic lead halide perovskites with cesium (e.g., $CsPbCl_3$, $CsPbBr_3$, and $CsPbI_3$).

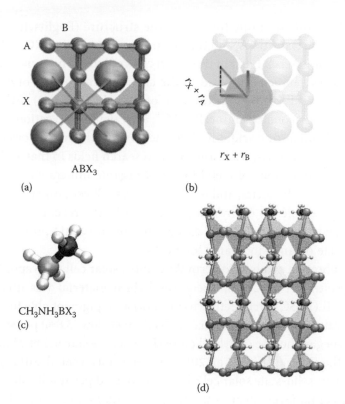

FIGURE 1.1    (a) Representation of the $ABX_3$ cubic perovskite; A and B are positively charged cations, and X the negatively charged anion. (b) Graphical representation of the Goldschmidt criterion for the corresponding atomic radii ($r_A$, $r_B$, and $r_X$, respectively). An organic positively charged molecule such as methylammonium (c) is used to replace the A cation in the inorganic structure to obtain a hybrid perovskite (d); B is a +2 atom (such as Pb or Sn) and X a negatively charged halide ion (such as I, Br, Cl); due to the ion size mismatch the hybrid crystals easily form a network of distorted $BX_3$ octahedra with the molecules providing a rich internal configurational space of different orientations.

One of the first works on inorganic lead halide perovskites was published in 1893 by Wells, but the perovskite crystal structure was characterized only much later in 1958 by Moller. This class of inorganic metal halides is particularly interesting. The large ionic radius of the lead ion is reflected in a relatively soft material and a large lattice spacing that is almost doubled with respect to perovskites such as $MgSiO_3$: the pseudo-cubic lattice parameters $CsPbI_3$ and $MgSiO_3$ are 6.17 Å (Møller 1958; Stoumpos et al. 2013) and 3.43 Å (Matsui 1988), respectively.

The large interatomic distances of lead halides require extremely large A cations. The largest group I element is cesium, which is the most common choice in inorganic lead halide perovskites. Of note, it is also possible to use small organic molecules such as methylammonium (MA) or formamidinium (FA) (Eperon et al. 2014; Jeon et al. 2015; Stoumpos et al. 2013), giving rise to hybrid organic inorganic perovskites.

The first synthesis of hybrid perovskites containing organic molecules was demonstrated in 1978 by Weber, who reported the formation, among the others, of MA lead iodide (MAPI) that is the archetypical hybrid perovskite. These works were followed by a number of studies aimed at understanding the crystalline structure (Poglitsch and Weber 1987), the nature of the phase transitions, and the dynamics of MA cations (Onoda-Yamamuro et al. 1990; Poglitsch and Weber 1987) in this novel class of hybrid materials. Exotic examples of hybrid perovskites can be also found in the class of dense metal-organic frameworks (MOFs) (Furukawa et al. 2013), as is the case of $[AmH]M(HCOO)_3$ hybrid perovskites, where A = $[AmH]^+$, a protonated amine; B = $M_2^{2+}$, a divalent metal cation; and X = $HCOO^-$, the anion derived from formic acid. MOFs formed by metal ions and organic linkers have evolved over the last 15 years into a fast-growing research fields in materials science owing to the unlimited combinations of metal ions and organic linkers with potential applications for optical devices, batteries, and semiconductors (Cheetham and Rao 2007).

However, the great interest on hybrid perovskites in the recent years is certainly due to hybrid lead halides because of their exceptional photovoltaic properties for low-cost, solution-processable organic–inorganic solar cells.

The first incorporation of a hybrid perovskite into a solar cell was reported by Miyasaka et al. in 2009 (Kojima et al. 2009). There, the hybrid material was used within a dye-sensitized solar cell (DSSC) architecture as an inorganic pigment able to absorb light and inject photogenerated carriers into titania, with relatively low (3.8%) power conversion efficiencies. In 2012 began the rise of hybrid perovskites as the star material for low-cost photovoltaics when the Grätzel (Kim et al. 2012) and Snaith (Lee et al. 2012) groups reported examples of efficient solid-state solar cells based on hybrid perovskite absorbers interfaced to an electron acceptor layer (such as titania or alumina) and an organic solid-state hole transporter. Solar cell efficiencies based on hybrid perovskites increased from 3.8% in 2009 (Kojima et al. 2009) to 22.1% in early 2016 (NREL 2016; Polman et al. 2016), making hybrid perovskites the fastest-advancing solar technology to date.

Sustained by a continuous progress of power conversion efficiencies and reduction of production costs, perovskite solar cells have become a commercially attractive solution for low-cost photovoltaics. Though there are a number of drawbacks affecting the material (e.g., toxicity of lead or fast degradation under thermal stress or humidity exposure) (Deretzis et al. 2015; Dualeh et al. 2014; Hailegnaw et al. 2015), the relevance of hybrid perovskites is impressive and extends far beyond photovoltaics, with applications in energy harvesting and sensing and potential impact in the fields of optoelectronics, thermoelectricity, photocatalysis, and many others (Boix et al. 2014; Chin et al. 2015; He and Galli 2014; Liu and Cohen 2016; Polman et al. 2016; Tan et al. 2014). Furthermore, recent progress indicates mixed monovalent cation $AA'BX_3$ (Saliba et al. 2016; Yi et al. 2016) as efficient photovoltaic systems able to tackle the stability problem.

## 1.2 CRYSTAL STRUCTURE OF HYBRID PEROVSKITES

The tolerance factor mentioned previously is a geometrical parameter introduced in 1926 by V. M. Goldschmidt to evaluate the maximum ionic size mismatches between ions that can be sustained by a perovskite structure. The tolerance factor $t$ is defined as

$$t = \frac{1}{\sqrt{2}} \frac{(r_A + r_X)}{(r_B + r_X)} \tag{1.1}$$

where $r_i$ ($i$ = A, B, and X) are the radii of ions in the perovskite $ABX_3$. Under ideal conditions, to maintain a high-symmetry cubic structure, the ionic radii of A, B, and X should satisfy the requirement that the tolerance factor $t$ should be close to 1 (see Figure 1.1b). Otherwise, the cubic structure will be distorted and crystal symmetry lowered. Of note, to satisfy $t \sim 1$, the A ion must be much larger than the B ion.

Equation 1.1 has been a guiding principle in the study of oxide perovskites that combines the idea of dense ionic packing with early estimates of ionic radii and that continues to be widely used. For values of $t$ in the range 0.9–1.0, cubic perovskites are expected, whereas values of 0.80–0.89 predominantly give rise to distorted perovskites that can be further classified by using Glazer's concept of octahedral tilting (Glazer 1972). Below 0.80, other structures such as the ilmenite type ($FeTiO_3$) are more stable owing to the similar sizes of cations A and B. Values of $t$ larger than 1 lead to hexagonal structures where layers of face-sharing octahedra are introduced. Accordingly, a strong correlation exists between tolerance factors and underlying crystal structures. For example, all currently known perovskite MOFs exhibit tolerance factors between 0.81 and 1.01. For the case of perovskite formates it is found that with decreasing size of cation A, the relative density steadily decreases until around $r \sim 0.81$, an unusual chiral structure becomes energetically comparable to the perovskite-type architecture (Kieslich et al. 2014).

In halide perovskites, the B site is usually occupied by a large Pb or Sn atom, so A must be extremely large. The most common example is Cs, that is, the largest group I element in the periodic table (Yin et al. 2014). However, Cs is still not large enough to hold a stable cubic or tetragonal perovskite structure in a wide range of temperatures, and different phases can form (e.g., the yellow phase, see Figure 1.2e). Improvement of stability is obtained by replacing A with a larger molecule such as MA. This can explain the better photovoltaic performance of hybrid $CH_3NH_3PbI_3$ with respect to $CsPbI_3$. For the MA lead halide systems, a cubic structure (see Figure 1.2a) is expected when $t$ lies between 0.89 and 1.0. Generally, smaller $t$ could lead to lower-symmetry tetragonal (Figure 1.2b) or orthorhombic (Figure 1.2b and c) structures, whereas larger $t$ ($t > 1$) could destabilize the three-dimensional (3D) B–X network, leading to a two-dimensional (2D) layer structure. Importantly, by increasing the temperature it is possible to increase the effective size of ions, thus changing the effective tolerance factor. This makes it possible to activate transformations thermally from lower to higher symmetry crystals. Accordingly, atomistic calculations at zero temperature of hybrid metal halides show that the orthorhombic γ phase is always the most stable. Most data identify the *Pnma* or *Pna2* structure as the most stable. In Figure 1.2c the phase is reported for two different molecular patterns.

The cubic phase is higher in energy with respect to γ. The stability of the orthorhombic phase of MAPI at low temperature is explained by the fact that for typical ions of hybrid perovskites it is difficult to satisfy $t \sim 1$ and the Pb–X–Pb distortions become energetically favorable. This is confirmed by the occurrence of imaginary frequencies in the phonon dispersion of the MAPI cubic phase (Mattoni et al. 2016) associated with a soft transverse acoustic phonon mode involving the B–X–B tilting. This mode makes anion X easily displaced from the B–B midpoint of the ideal cubic phase (Brivio et al. 2015; Mattoni et al. 2016)

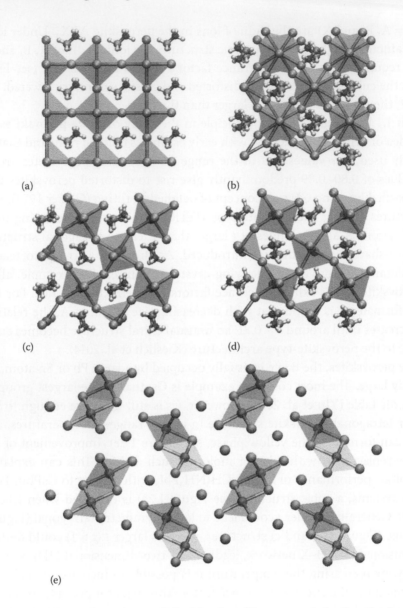

FIGURE 1.2 Atomistic representations of the (a) cubic, (b) tetragonal, (c, d) orthorhombic phases of hybrid MAPI and (e) yellow phase of inorganic $CsPbI_3$.

showing that the cubic phase at zero temperature is unstable and tends to transform spontaneously in a distorted orthorhombic crystal.

However, the crystals are observed to transform to the tetragonal and cubic phases at increasing temperature for most perovskites (Baikie et al. 2013; Stoumpos et al. 2013), indicating that such finite-temperature transformations are dynamic in nature (Baikie et al. 2013; Mattoni et al. 2015, 2016) and can be properly described only within a dynamic picture of the crystals, as discussed in Section 1.3. Depending on the specific ion sizes and the type of chemical elements and in turn on the hybrid interactions, the structural stabilities and the transition temperatures of different perovskites vary (see Table 1.1, second column).

TABLE 1.1 Structural Parameters of Hybrid Methylammonium Lead Halide Perovskites $CH_3NH_3PbI_3$ (MAPI), $CH_3NH_3PbBr_3$ (MAPBr), and $CH_3NH_3PbCl_3$ (MAPCl)

| Phase | T(K) | Sym | Exp a | b | c | $V^{1/3}$ | DFT $V^{1/3}$ | MYP[g] $V^{1/3}$ |
|---|---|---|---|---|---|---|---|---|
| **$CH_3NH_3PbI_3$** | | | | | | | | |
| α (cub) | >330 | Pm3m | 6.33[a] | | | | 6.34[f] PBE-vdW^TS | 6.31 |
| | | | 6.31[b] | | | 6.30 | | |
| | | | 6.28[c] | | | | | |
| β (tet) | 160–330 | I4/mcm | 8.85[a] | | 12.66 | | | 6.28 |
| | | | 8.85[b] | | 12.64 | 6.26 | | |
| | | | 8.85[c] | | 12.44 | | | |
| γ (orth) | <160 | Pna2 | 8.86[a] | 8.58 | 12.62 | 6.20 | 6.39, 6.19[d] PBE (DF2, DF2B88); 6.31[d] PBE[a]; 6.06[c] LDA | 6.23 |
| | | Pnma | 8.84[b] | 8.55 | 12.58 | | | |
| **$CH_3NH_3PbBr_3$** | | | | | | | | |
| α (cub) | >237 | Pm3m | 5.90[a] | | | 5.90 | | |
| β (tet) | 155–237 | I4/mcm | 8.32[a] | | | 5.89 | 5.97[f] PBE-vdW^TS | |
| γ (orth) | <144 | Pna2 | 7.98[a] | | | 5.88 | | |
| **$CH_3NH_3PbCl_3$** | | | | | | | | |
| α (cub) | >179 | Pm3m | 5.67[a] | | | 5.67 | | |
| β (tet) | 173–179 | P4/mmm | 5.66[a] | 5.63 | | 5.65 | 5.71[f] PBE-vdW^TS | |
| γ (orth) | <173 | P222 | 5.67[a] | 5.63 | 11.18 | 5.62 | | |

Sources: [a]Poglitsch and Weber (1987); [b]Stoumpos et al. (2013); [c]Baikie et al. (2013); [d]Menéndez-Proupin et al. (2014); [e]Wang et al. (2014); [f]Egger and Kronik (2014); [g]Mattoni et al. (2015).

Note: V here represents the volume per unit formula.

For $CH_3NH_3PbI_3$, the most important hybrid perovskite, the α to β and β to γ phase transitions occur at about 330 K and 160 K, respectively (Poglitsch and Weber 1987). Interestingly, a nonperovskite phase was found for perovskites such as $HC(NH_2)_2PbI_3$, $FAPbI_3$, $CsPbI_3$, and $CsSnI_3$ (Baikie et al. 2013; Chung et al. 2012; Stoumpos et al. 2013). Unlike tetragonal and orthorhombic phases, the phase cannot be derived from the cubic phase by B–X–B bond angle distortions. Instead, the B–X bond is broken (see Figure 1.2e).

Hybrid organic–inorganic perovskites are characterized by a much higher configurational complexity than inorganic perovskites, because noncentrosymmetric organic molecules have many possible orientations inducing a sizable molecular entropy. For MAPI the $CH_3NH_3$ molecules in the high-temperature phase are randomly oriented (Chen et al. 2015; Mattoni et al. 2015; Poglitsch and Weber 1987) and the overall crystals preserve $O_h$ symmetry. In the medium temperature (tetragonal) phase, the freedom of MA molecules is somewhat reduced but their orientations are still disordered (Mattoni et al. 2015; Poglitsch and Weber 1987). In the low-temperature (orthorhombic) phase, most of theoretical and experimental results indicate that the $CH_3NH_3$ molecular rotations are almost frozen (Chen et al. 2015; Mattoni et al. 2015) with a local dynamics around fixed alignments (Baikie et al. 2013), though the problem is still debated likely because of metastable ordering. The molecular disorder in a hybrid crystal is accordingly a relevant factor affecting hybrid perovskites that can be properly understood only within a statistical or dynamical modeling of the material at finite temperature, as discussed in Section 1.5. For example, the use of single static configurations gives conflicting results (Yin et al. 2014) concerning the structural parameters of the different crystalline phases of MAPI.

From the theoretical point of view, Density Functional Theory (DFT) is the most used method to predict the structural properties of solids. Hybrid perovskites are challenging because of the existence of a plethora of quasi-isoenergetic molecular configurations. DFT energies of $CH_3NH_3PbI_3$ show a slight dependence of the results on the orientations of the molecules and, for example, the most stable molecular direction in the cubic phase has been found to be different ([001] or [111]) (Yin et al. 2014) depending on the methodology or optimization details. As a result, the lattice parameters and energies calculated in the literature vary sizably (see Table 1.1). Differences can be due, for example, to the exchange-correlation functional. Consistent with the results for conventional semiconductors, local density approximation (LDA) underestimates and generalized gradient approximation (GGA) overestimates the lattice parameters of hybrid perovskites. The overestimation of GGA (PBE)-calculated lattice parameters was attributed to the neglect of van der Waals interactions between the $CH_3NH_3$ and Pb–I frameworks (Egger and Kronik 2014). When dispersive interactions are considered, it is possible (only for specific type of dispersive correction) to improve the agreement of the calculated lattice parameters with experiments (Egger and Kronik 2014; Menéndez-Proupin et al. 2014; Wang et al. 2014) (see Table 1.1).

Based on this, the van der Waals interactions are claimed to be an important factor in hybrid halide perovskites (Egger and Kronik 2014). Besides the importance of the theoretical method, we point out the importance of including the entropy associated with the molecular configurations when interested in structural properties at finite temperatures. Even considering only the high-symmetry directions ([001], [110], and [111]), the number of configurations rapidly grows with the number of molecules. The use of single specific configurations is justified for practical purposes but it is in principle inappropriate. Alternative strategies based on molecular dynamics are discussed in Section 1.5.

### 1.2.1 Readdressing the Tolerance Factor for Hybrid Perovskites

The concept of tolerance factor was originally developed for inorganic perovskites. The advent of hybrid perovskites has made it necessary to readdress and to extend the tolerance factor definition from isotropic ions to anisotropic molecules. The challenge in determining tolerance factors for organic–inorganic compounds lies in estimating the ionic radii of molecular cations. A set of thermochemical radii for molecular anions was proposed by Kapustinskii and Yatsimirskii in the 1940s (Mitzi 2001). Even highly symmetrical cations, such as $[NH_4]^+$ and $[(CH_3)_4N]^+$, exhibit radii that depend significantly on their anionic counterparts (Shannon 1976), making tolerance factors for hybrid perovskites a challenging problem. The $t$ factor was applied to hybrid lead iodides by Mitzi (2001) and recently the concept has been expanded also to the MOFs such as formate perovskites (Kieslich et al. 2014). In particular, assuming free rotational freedom around the molecule center of mass, a rigid sphere model is applicable to organic cations and leads to a consistent set of effective ionic radii $r_{Aeff} = r_{mass} + r_{ion}$, with $r_{mass}$ being the distance between the center of mass of the molecule and the atom with the largest distance to the center of mass, excluding hydrogen atoms; $r_{ion}$ is the corresponding ionic radius of this atom (Kieslich et al. 2014).

By using crystallographic data from known perovskite-like hybrid frameworks Kieslich et al. (2014) estimated a consistent set of effective radii for different organic cations, for example, 2.17 Å for MA $[(CH_3)NH_3]^+$ and 2.53 Å for FA cations $[NH_2(CH)NH_2]^+$, respectively. For the case of lead iodide (PI) this approach gives a tolerance factor for MAPI and FAPI of 0.9 and 1.0, confirming the stability of the perovskite structure. Conversely, nonperovskite structures are expected, for example, for ammonium $(NH_4^+)$ and guanidinium $C[(NH_2)_3]^+$, with $t$ values higher than 1 and smaller than 0.8, respectively. It was also possible to extend the concept to molecular anions such as $HCOO^-$, $CN^-$, and $N^{3-}$ (for which the situation is complicated by the high anisotropy of the anion) by treating all molecular anions as rigid cylinders, with effective radius $r_{Xeff}$ and an effective height $h_{Xeff}$ (Kieslich et al. 2014). By using the tolerance factor criterion and the extended definition for hybrid perovskites it was possible to propose potential existence of more than 600 undiscovered hybrid perovskites including alkaline earth metal and lanthanide-based materials (Kieslich et al. 2015). The tolerance factor has been revisited recently in order to tune the band gap of hybrid perovskites by steric engineering of ionic constituents (Filip et al. 2014).

The aforementioned works show the importance of the tolerance factor to predict the stability and occurrence of the perovskite phase. However, the importance of this concept goes beyond the stability; it can also be used to understand in detail the distortions within a given orthorhombic or tetragonal structure. An example was reported in Filippetti and Mattoni (2014). There the tolerance factor was applied to hybrid tetragonal crystals such as MAPI, MAPbICl$_2$, and MaPbI$_2$Cl by introducing a generalization in which the tolerance factor is related to the molecular orientations. The crystals are reported in Figure 1.3, showing MAPbI$_3$ and MAPbI$_2$Cl with [110]-oriented molecules, both within tetragonal *I4/mcm* symmetry. The same symmetry was used to describe the MAPbI$_2$Cl structure with [100]-oriented molecules; *I4/mcm* symmetry is characterized by octahedral rotations labeled $a^0a^0c^-$ (in Glazer notation), which indicates no rotation along $a = b$ axes, and antiphase rotations along $c$ (notice that $a$ and $b$ here refer to the conventional cubic cell, which is 45°-rotated with respect to the tetragonal unit cells). Thus, once the tetragonal cell parameters are fixed, only one more internal degree of freedom, namely the octahedral rotation in the ($a$, $b$) plane, is left to be further optimized.

In Figure 1.3 we report the relaxed structures for these three perovskites, with both cubic and rotated reference system indicated. At the end of relaxation, the [100]-oriented molecules are actually slightly tilted in the (001) plane with respect to the [100] direction, and the [110]-oriented molecules also tilted out the (001) plane. We label them [100] and [110] according to their starting molecule orientation. We see in the Figure 1.3 that the three structures are characterized by quite different Pb–I–Pb angles: close to (170°) for the MAPbI$_2$Cl [100], remarkably rotated for the two [110] systems (153° and 147° for MAPbI$_2$Cl and MAPbI$_3$, respectively).

Typically, a tolerance factor lower than 0.9 indicates the tendency to lose the ideal cubic symmetry and move toward low-symmetry tetragonal, orthorhombic, or rhombohedral structures characterized by cooperative octahedral rotations. In the present case of hybrid perovskites, however, two additional ingredients complicate the picture: (1) the chemical mixture of Cl and I ions; (2) the nonsphericity of the molecule; clearly, factor (1) is the determinant distinguishing the tolerance factors of [110] MAPbI$_3$ and MAPbI$_2$Cl systems, while (2) distinguishes [100] and [110] MAPbI$_2$Cl.

To include these ingredients in the tolerance factors of the hybrid perovskites, we considered the following generalization: for each structure, three inequivalent planes [($a$, $b$), ($a$, $c$), and ($b$, $c$)] and three corresponding tolerance factors ($t_{ab}$, $t_{ac}$, $t_{bc}$, respectively) are defined as follows (now A = M and B = Pb, X = I or Cl):

$$t_{ab} = \frac{r_M^{ab} + r_{Cl}}{r_{Pb} + r_I} \cos \vartheta_{ab}$$

$$t_{ac} = \frac{r_M^{ac} + r_{Cl}}{r_{Pb} + r_I} \cos \vartheta_{ac} \tag{1.2}$$

$$t_{bc} = \frac{r_M^{bc} + r_{Cl}}{r_{Pb} + r_I} \cos \vartheta_{ab}$$

where $\theta_{ab}$ $\theta_{bc}$ are the face-diagonal angle projections expressed in terms of ionic radii: $r_{Pb} = 1.33$ Å for Pb$^{2+}$, $r_I = 2.06$ Å for I$^-$, and $r_{Cl} = 1.67$ Å for Cl$^-$. The geometrical factor $\cos \vartheta$ keeps

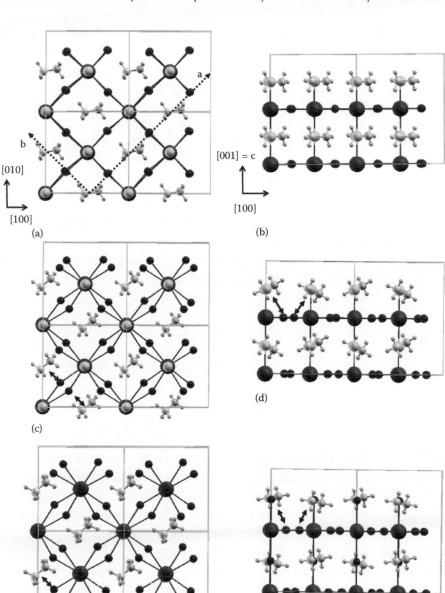

FIGURE 1.3 Structure of the $2\sqrt{2} \times 2\sqrt{2} \times 2$ supercell for [100]-oriented MAPbI$_2$Cl (a, b), [110]-oriented MAPbI$_2$Cl (c, d), and [110]-oriented MAPbI$_3$ (e, f). The reference system indicated by black solid arrows refers to the $\sqrt{2} \times \sqrt{2} \times 2$ tetragonal unit cell, while $a$ and $b$ are 45°-rotated axes of the conventional cubic cell. Small arrows indicate the important M–I distances shortened by the octahedral rotations (see text). Atomic colors are yellow (C), gray (N), cyan (H), red (I), violet (Pb), green (Cl). (Reprinted with permission from Filippetti, A., and A. Mattoni. 2014. Hybrid perovskites for photovoltaics: Insights from first principles. *Phys. Rev. B.* 89; 125203–125208. Copyright 2014 American Physical Society.)

into account the fact that the face may not be squared, due to the presence of two different anions (I$^-$ and Cl$^-$) in the same face:

$$\vartheta_{ab} = \tan^{-1} \frac{(r_{Pb} + r_I)}{(r_{Pb} + r_I)}$$

$$\vartheta_{ac} = \tan^{-1} \frac{(r_{Pb} + r_{Cl})}{(r_{Pb} + r_I)} \qquad (1.3)$$

$$\vartheta_{bc} = \vartheta_{ac}$$

Thus for $(a, b)$ or any face of the MAPbI$_3$ perovskite cos $\vartheta = 1/\sqrt{2}$ and the tolerance factor goes back to the standard expression.

Furthermore, in Equation 1.2 we introduced a face-dependent effective ionic radius for the molecule, to include the effect of different orientations:

$$r_M^{ab} = \frac{d}{2} \cos\alpha + r_N$$

$$r_M^{ac} = \frac{d}{2} \cos\beta + r_N \qquad (1.4)$$

$$r_M^{bc} = \frac{d}{2} \cos\gamma + r_N$$

where $d = 2.47$ Å is the molecule length, $r_N = 1.32$ Å for N$^{3-}$, and the cosines project the dimer direction onto the molecule–anion distance on each face. Clearly, for spherically averaged molecules the effective radii are all equal and the standard definition is recovered.

In Table 1.2 we report the calculated parameters for each considered structure. The key quantity to explain the different rotation amplitudes of the examined perovskites is $t_{bc}$, very small for the two [110] structures and fairly large for the [100]. The reason can be easily understood in looking at Figure 1.3. For the [100] orientation (Figure 1.3a and b), planes $(a, c)$ and $(b, c)$ are nearly equivalent; on the contrary, for the [110] orientation, $(a, c)$ and $(b, c)$ are quite different, due to the asymmetric orientation of the molecules along axes $a$ and $b$; the visible emptiness present in the $(b, c)$ planes between ions M and I produces very small $t_{bc}$; this emptiness is compensated by octahedral rotations parallel to $(a, b)$ that reduce the M–I distances (see the small black arrows in the figure). Nicely, the model also distinguishes between [110] MAPbI$_2$Cl and MAPbI$_3$ systems: while they have the same molecule effective radius, $t_{bc}$ is slightly smaller than the latter, due to the fact that $r_I > r_{Cl}$.

Finally, it is interesting to notice that even $t_{ab}$ is rather small, albeit larger than $t_{bc}$, in all the considered systems. This signals the tendency of developing octahedral rotations in order to close the space between M and Cl (or I for MPbI$_3$) in the $(a, b)$ planes. These rotations around $a$ and $b$ axes are ruled out in tetragonal symmetry, but can be activated by a structural transition to orthorhombic symmetry, which is known to occur for these hybrid perovskites at lower temperature.

TABLE 1.2 Generalized Tolerance Factors $t$, Angles, and Effective Molecular Radius $r_M^{AB}$ (Å) for the Three Perovskites Considered in the Chapter

| | MPbI$_2$Cl [100] | MPbI$_2$Cl [110] | MPbI$_3$ [110] |
|---|---|---|---|
| $t_{ab}$ | 0.88 | 0.81 | 0.89 |
| $t_{ac}$ | 0.94 | 1.02 | 0.96 |
| $t_{bc}$ | 0.94 | 0.75 | 0.71 |
| $\vartheta_{ab}$ | 45° | 45° | 45° |
| $\vartheta_{ac}$ | 41.5° | 41.5° | 45° |
| $\vartheta_{bc}$ | 41.5° | 41.5° | 45° |
| $\alpha$ | 0 | 45° | 45° |
| $\beta$ | 45° | 0 | 0 |
| $\gamma$ | 45° | 90° | 90° |
| $r_M^{ab}$ | 2.55 | 2.19 | 2.19 |
| $r_M^{ac}$ | 2.19 | 2.55 | 2.55 |
| $r_M^{bc}$ | 2.19 | 1.32 | 1.32 |

*Note:* The definitions are given in the text.

## 1.3 BEYOND THE STATIC MODELING: IMPORTANCE OF ORGANIC–INORGANIC INTERACTIONS IN HYBRID PEROVSKITES

After having described the crystalline structure of hybrid perovskites it is important to discuss the nature of the interatomic forces. This is a necessary step to go beyond the static modeling of crystals. Since the first theoretical studies of the electronic structure of the hybrid perovskites (Chang et al. 2004; Umebayashi et al. 2003), it was recognized that the molecules do not participate in the optoelectronic activity taking place at the electronic gap; rather, the orbitals of organic cations are separated in energy and located deep within the valence and conduction bands (Chang et al. 2004; Umebayashi et al. 2003). This is a quite different situation with respect to standard hybrid organic–inorganic materials such as polymer metaloxides (Canesi et al. 2012; Oosterhout et al. 2010) or molecule–oxide systems (Mattioli et al. 2012, 2014) used in DSSCs (O'Reagan and Grätzel 1991). The current understanding is that hybrid perovskites behave as inorganic ionic semiconductors based on excellent optoelectronic properties [e.g., absorption (Filippetti and Mattoni 2014), weakly bound hole–electron pairs, radiative recombination (Filippetti et al. 2014)] but with a hybrid and solution-processable body (Filippetti et al. 2014). Also, the electron transport properties of the material confirm the behavior of an ionic semiconductor with the electron–polar optical phonon interaction being the relevant mechanism of scattering at room temperature (Filippetti et al. 2016). In this scenario, the molecules behave as electronically separated units interacting electrostatically through its atomic partial charges.

The electronic separation in energy between organic and inorganic levels does not imply that the molecular configurations are not relevant for optoelectronic properties (Filippetti et al. 2014). We have already discussed how critical the molecular disorder is on the actual lattice parameters of the hybrid crystals. Moreover, there exists an interplay between the molecule orientation and the tilting of inorganic octahedra that in turn affects the opto-electronic and photovoltaic properties through a change of the electronic orbital overlap. A clear demonstration of this effect is shown in Figure 1.4 (Filippetti et al. 2014). Also, the effect of the local electric field induced by ordered molecular configurations can have important electronic effects [Stark effects (Dar et al. 2016; Roiati et al. 2014)]. The organic–inorganic interplay can also contribute to localize electron or holes within the material as a result of local PbI distortions induced by molecular disorder (Giorgi and Yamashita 2016; Ma and Wang 2015), similarly to localization phenomena occurring in disordered semiconductors (Bagolini et al. 2010, 2014) and possibly contributing to the formation of polarons (Neukirch et al. 2016; Zhu et al. 2016). Much progress has been made in recent years in understanding such dynamic organic–inorganic interactions in hybrid perovskites. However, many questions are still open and are the object of an intense ongoing research (Egger et al. 2016; Frost and Walsh 2016).

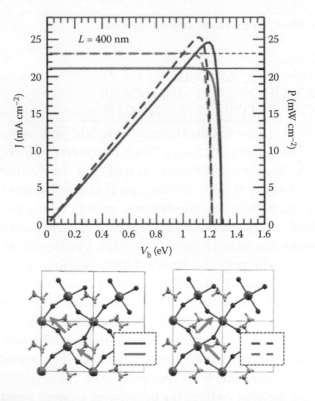

FIGURE 1.4 The interplay between photovoltaic properties and molecules is demonstrated by sizable variations in the ideal photovoltaic properties (top) calculated for two crystalline structures with different molecular ordering (bottom panels). Light (dark) line corresponds to output current (power); dashed (continuous) line corresponds to left (right) molecular pattern. (Adapted from Filippetti, A., P. Delugas, and A. Mattoni. 2014. Radiative recombination and photoconversion of methylammonium lead iodide perovskite by first principles: Properties of an inorganic semi-conductor within a hybrid body. *J. Phys. Chem. C* 118; 24843–24853. Copyright © 2014 American Chemical Society.)

In conclusion, hybrid perovskites behave as ionic crystal with a soft body that requires taking into account the large structural fluctuations (e.g., the rich variety of octahedral distortions) and the molecular entropy due to directional disorder. The most suitable perspective is, accordingly, a thermodynamic one.

## 1.3.1 The Molecular Dynamics Approach

Molecular dynamics (MD) is a well-established method (Allen and Tildesley 1989; Marx and Hutter 2000; Tuckerman 2010) to study the thermodynamics and statistical mechanics of complex materials. MD is particularly appropriate for soft crystals such as lead halide perovskites. A considerable number of MD studies of hybrid perovskites have in fact recently appeared in the literature (see later).

MD consists in calculating numerically the time evolution of a set of particles (considering in principle both electronic and ionic degrees of freedom) representing the material of interest in the appropriate state of aggregation. Within the Born–Oppenheimer approximation the nuclei can be treated as classical point-like masses whose trajectories $X(t)$ are obtained by solving the classical Newton's equation of motion:

$$M\ddot{X} = F(X) \tag{1.5}$$

The interatomic forces $F(X)$ can be calculated by the numerical derivative of the total energy $F(X) = -\nabla_X \langle \psi, X|H(X)| \psi, X \rangle$ with respect to nuclei displacements $X$; the total energy $\langle \psi, X|H(X)|h, X \rangle$ in principle requires taking into account the electronic states $|\psi, X \rangle$ (or the electronic density) by solving, for the instantaneous nuclei positions, the corresponding Schrödinger equation $H|\psi, X \rangle = E|\psi, X \rangle$. The details depend on the different levels of electronic theory (Marx and Hutter 2000), on the different electron–nuclei types of coupling [Car-Parrinello (CP) (Car and Parrinello 1985; Hutter 2012) or Born–Oppenheimer (BO) (Marx and Hutter 2000)], and on technical details that are discussed extensively elsewhere. Under the ergodic hypothesis, the ensemble-averaged properties $\langle O \rangle$ of the system at the physical temperature can be calculated by time-averaging the corresponding instantaneous values $O(t) = \langle \psi(t), X(t)|\hat{O}(X)|\psi(t), X(t) \rangle$ during MD

$$\langle O \rangle = \frac{1}{t} \int_0^t O(\tau) d\tau \tag{1.6}$$

In this way, MD not only gives the evolution of the system but it also provides, at least in principle, the whole set of thermodynamic functions (Allen and Tildesley 1989). *Ab initio* MD refers to the cases in which the forces $F(X)$ are calculated from *ab initio* methods. In particular, DFT-based molecular dynamics provides both accuracy and relatively low computational cost and it has become a common tool of modern computational physics (Marx and Hutter 2000).

Though a large majority of *ab initio* theoretical works on hybrid perovskites are based on local structural optimization and static electronic calculations (Yin et al. 2014), several *ab initio* MD studies have been recently published on hybrid perovskite dynamical properties.

Since the first application *ab initio* CP molecular dynamics (CPMD) to hybrid perovskites attributable to Mosconi et al. (2014), several authors have investigated the molecular rotations at finite temperature (Carignano et al. 2015; Frost et al. 2014; Goehry et al. 2015; Lahnsteiner et al. 2016; Meloni et al. 2016), polarization and possible ferroelectric molecular domains (Frost et al. 2014; Quarti et al. 2014a), the interplay between structural and electronic properties (Carignano et al. 2015; Frost et al. 2014; Mosconi et al. 2014), structural transformation [e.g., the tetragonal-to-cubic transition (Quarti et al. 2016)], the onset of degradation of MAPI in water (Mosconi et al. 2015), the diffusivity and dynamics of defects (Mosconi and De Angelis 2016), and many others.

### 1.3.2 Classical Molecular Dynamics

The preceding discussion notwithstanding, many important thermally activated phenomena in hybrid perovskites are out of reach of *ab initio* MD methods. For example, defects clustering or diffusion, degradation kinetics, mixed-phases with coexistence of order–disorder domains, grain boundaries, nanostructures, and thermal conductivity are only a few examples of problems not easily amenable to *ab initio* methods and requiring thousands to millions atoms and nanosecond to microsecond simulation times (Mattoni et al. 2017).

To overcome the computational limitations of *ab initio* MD it is possible to approximate the total energy by model potentials that depend only on the positions of the nuclei (Marx and Hutter 2009):

$$\langle \psi, X | H(X) | \psi, X \rangle \sim U(X) \qquad (1.7)$$

The electronic degrees of freedom are not explicitly included in the calculations but are implicitly taken into account as effective classical interactions between nuclei (i.e., by setting empirical parameters).

The first model potential for hybrid perovskites (MYP) has been developed recently by Mattoni et al. (2015). By classical MD it has been possible to perform a comprehensive study of the orientational dynamics of cations (see Section 1.5) in a wide range of temperatures. Several applications of MYP have been reported to study temperature evolution of vibrations (Mattoni et al. 2016), diffusion of point defects (Delugas et al. 2016), and thermal transport (Caddeo et al. 2016; Wang and Lin 2016). Examples are discussed more extensively in the next sections. Recently the MYP potential has been used to investigate various caloric effects in MAPI (Liu and Cohen 2016), as well as fracture.

Classical models different from the MYP have also been proposed to study organohalide perovskite precursors in solvents (Gutierrez-Sevillano et al. 2015). Other classical models based on bonded interactions have been proposed by Qian et al. (2016) and by Hata et al. (2016) to study thermal transport. The models are based on the fitting of trajectories and forces during *ab initio* MD. They are expected to be accurate around a local region in the configurational space of the material but cannot be easily applied, for example, to systems containing defects.

## 1.4 UNDERSTANDING THE THERMODYNAMIC PROPERTIES OF HYBRID PEROVSKITES BY CLASSICAL FORCE FIELDS

The development of simple classical force fields makes it possible to perform long simulations that are necessary to fully take into account volume fluctuations and temperature effects and study comprehensively the thermodynamics of the hybrid perovskites at finite temperature. There is also another important benefit: the reduction of the complexity of the material to a simple classical model provides physical insight into the nature and properties of hybrid perovoskites. Before studying the evolution with temperature of the crystal structure and lattice parameters of hybrid perovskites we describe the functional form of the MYP potential and its physical implications.

Classical force fields for inorganic perovskites have been known since the 1980s as a result of the pioneering work of Matsui (1988). The simplest interatomic model for inorganic perovskites is the Buckingham–Coulomb (BC) potential:

$$\Phi = \sum_{ij} A_{ij} \exp(-r_{ij}/\rho_{ij}) - \frac{\sigma_{ij}}{r_{ij}^6} + \frac{q_i q_j}{4\pi\varepsilon_0 r_{ij}} \qquad (1.8)$$

developed to describe ionic crystals (Matsui 1988). The model explicitly includes the long-range electrostatic interactions $\frac{q_i q_j}{4\pi\varepsilon_r}$ between atomic charges $q_i q_j$. $r_{ij}$ is the distance between atoms $i$ and $j$, $\varepsilon$ is the dielectric constant. The other terms

$$V_{\text{Buck}} = \sum_{ij} A_{ij} \exp(-r_{ij}/\rho_{ij}) - \frac{\sigma_{ij}}{r^6} \qquad (1.9)$$

are known as Buckingham potential (Buckingham 1938) that was originally introduced for gaseous helium (Buckingham 1938). The attractive term $\sigma_{ij} r_{ij}^{-6}$ (with coefficients $\sigma_{ij}$ for each pair of $i, j$ atoms) is included to account for the dispersive forces due to instantaneous dipole–dipole interactions. The repulsive term is controlled by the distance $\rho_{ij}$ and prefactor $A_{ij}$. Overall the BC model contains four types of parameters—A, $\rho$, $\sigma$, and $q$—that can be calibrated on the material properties. BC models have been developed for oxides (Saba and Mattoni 2015) [e.g., $MgSiO_3$ (Matsui 1988), ZnO (Kulkarni et al. 2005), and $TiO_2$ (Matsui and Akaogi 1991)].

There exist also several force fields for the study of organic molecules (Saba and Mattoni 2015). One of the most widely adopted is the GAFF force field (Wang et al. 2004) obtained as a generalization of the previous AMBER model specifically designed for proteins and nucleic acids (Ponder and Case 2003).

The MYP model for hybrid perovskites (Mattoni et al. 2015) is derived from the afore-mentioned models. In fact, it describes the atomic interactions as the sum of organic-organic $U_{OO}$ (intra- and inter-MA interactions), inorganic–inorganic $U_{II}$ (Pb–I interactions), and organic–inorganic $U_{OI}$ interactions (MAPI). $U_{OO}$ and $U_{II}$ are of type GAFF and BC, respectively.

Taking into account the fact that the molecules and the PbI orbitals are electronically separated and that they are positively charged cations, it is natural to include electrostatic interactions through BC terms. Furthermore, dispersive interactions are taken into account through Lennard–Jones 12-6 (LJ) terms, in the spirit of previous works for hybrids (Caddeo et al. 2011; Saba and Mattoni 2014), obtaining hybrid interactions of the general form:

$$U_{IO} = \sum_{ij}\left[ A_{ij}e^{-r_{ij}/\rho_{ij}} - \frac{\sigma_{ij}}{r_{ij}^6}\right] + \sum_{ij}\frac{1}{4\pi\varepsilon_0}\frac{q_i q_j}{r_{ij}}$$
$$+ \sum_{ij}\varepsilon_{ij}\left[\left(\frac{\sigma_{ij}}{r_{ij}^{12}}\right)^{12} - \left(\frac{\sigma_{ij}}{r_{ij}^6}\right)^6\right] \tag{1.10}$$

The parameters of the model can be fitted on a dataset obtained by first-principles calculations.

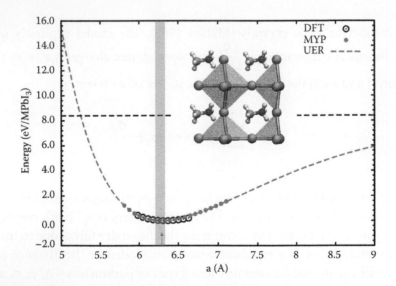

FIGURE 1.5 (Left) Energy as a function of lattice spacing of the cubic $CH_3NH_3PbI_3$ bulk during a hydrostatic deformation for the MYP model potential and for *ab initio* DFT-GGA (shifted to the MYP minimum). Atomistic data are fit by the universal energy relation (Mattoni et al. 2007). (Reprinted with permission from Mattoni, A., A. Filippetti, M. I. Saba, and P. Delugas. 2015. Methylammonium rotational dynamics in lead halide perovskite by classical molecular dynamics: The role of temperature. *J. Phys. Chem. C* 119; 17421–17428. Copyright © 2015 American Chemical Society.)

The fitting procedure and a detailed description of the parameters can be found elsewhere (Mattoni et al. 2015, 2017).

The MYP cohesive energy of MAPI as a function of lattice spacing is reported in Figure 1.5 together with experimental values and DFT (Mattoni et al. 2015). The DFT curve is shifted to the MYP minimum, and this latter falls nicely within the experimental range (dashed region).

In addition, MYP gives a bulk modulus (0.18 Mbar) and cohesive energy (7.5 eV/ stoichiometric unit) in reasonable agreement with DFT GGA. This represents a first indication of the ability of an ionic interatomic scheme to reproduce the material properties.

It has been shown in the literature (Rose et al. 1984) that most solids follow the simple Universal Energy Relation (UER) between energy $U(r)$ and lattice spacing $r$:

$$U(r) = -E_0\left(\frac{r-r_0}{s}+1\right)e^{-\frac{r-r_0}{s}} \quad (1.11)$$

where $r = r_0$ is the minimum energy distance where $U(r_0) = -E_0$, that is, the cohesive energy $E_0$; at long distances the energy vanishes $U(r \gg s) \to 0$ and under compression the energy increases to positive repulsive values. At distance $r_0 - s$ the energy is $U(r_0 - s) = 0$. The three parameters of UER ($r_0$, $s$, $E_0$) set equilibrium distance, anharmonicity parameter, and the cohesive energy (against dissociation of the solid into ions) (Mattoni et al. 2007). The energy derivatives are

$$U' = \frac{E_0}{s}\frac{r-r_0}{s}e - \frac{r-r_0}{s}$$

$$U'' = \frac{E_0}{s^2}\left(-\frac{r-r_0}{s}+1\right)e^{-\frac{r-r_0}{s}} \quad (1.12)$$

The universal relation makes use of the scaling invariant quantity $(r - r_0)/s$. The observed universality implies that the anharmonic parameter $s$ of most materials scales together with the lattice spacing (i.e., $s \to s\lambda$ when $r_0 \to r_0\lambda$). This is relevant for the crystal stiffness. At the equilibrium distance, $U''(r_0) = \frac{E_0}{s^2}$. Within an ionic force field the cohesive energy is due to the coulomb energy ($E_0 = e^2/r_0$) so that $U''(r_0) = \frac{e^2}{r_0 s^2}$. By scaling $s \to s\lambda$ and $r_0 \to r_0\lambda$ we finally have $U''(r_0) = \frac{e^2}{r_0 s^2} = \beta \to \frac{1}{\lambda^3}\frac{e^2}{r_0 s^2} = \frac{1}{\lambda^3}\beta$. In summary, under the assumption of constant charges the bulk modulus decreases by a factor $\frac{1}{\lambda^3}$ for a scaling $\lambda$ of the lattice parameters. This result can be obtained also within the BC model (Mattoni et al. 2017) and it is consistent with the fact that the large lattice spacing of lead halides is associated with lower bulk modulus and softer crystals. As observed by Wang and Lin (2016), the low bulk modulus is, in turn, at the origin of the low sound velocity of the material. This point is discussed further in the text that follows.

It is important to stress here that, according to the above analysis, the soft nature of metal halides is not related to the hybrid interactions and to the nature of the organic cations. The soft nature of the crystals ultimately derives from the inorganic Pb–I bonds as is confirmed by first-principles calculations (Feng 2014). Here we emphasize the ability of the simple electrostatic-dispersive scheme (as is MYP potential) to reproduce the overall structural properties of the material, supporting the description of hybrid perovskites as ionic semiconductors.

Though the details of interactions between the molecule and the inorganic PbX lattice are not crucial for the elastic properties of the crystal (controlled mainly by PbX lattice and charges) they nevertheless determine the molecular orientations and the Pb–X–Pb tilting (important for the overall structural parameters).

Under this respect, a fundamental physical ingredient of hybrid perovskites is the energy cost associated with molecular rotations. The energy for the rotation of the molecules within the (001) symmetry plane is reported in Figure 1.6a, showing a very good agreement between DFT and MYP (the cubic periodicity is imposed during the rotation). This static barrier calculated for the collective molecular rotation in a cubic crystal represents an upper limit of the real activation energy for molecular rotations. The real activation is lower because of atomic relaxations, entropic contributions, and more favorable rotational paths that are relevant during finite temperature dynamics.

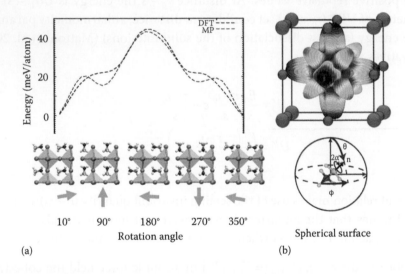

(a)                                                          (b)

FIGURE 1.6   (a) Total energy of the cubic MAPI bulk (cubic periodicity imposed) during a rotation (see bottom insets) of the molecular cation within a (001) crystallographic plane; curves are DFT and MYP. (Reprinted with permission from Mattoni, A., A. Filippetti, M. I. Saba, and P. Delugas. 2015. Methylammonium rotational dynamics in lead halide perovskite by classical molecular dynamics: The role of temperature. *J. Phys. Chem. C* 119; 17421–17428. Copyright © 2015 American Chemical Society.) (b) Radial representation of DFT energy as a function molecule direction (bottom inset) with respect to cubic PbI cage. (Reprinted with permission from Bechtel, J. S., R. Seshadri, and A. Van der Ven. 2016. Energy landscape of molecular motion in cubic methylammonium lead iodide from first-principles. *J. Phys. Chem. C* 120; 12403–12410. Copyright © 2015 American Chemical Society.)

These effects are naturally included in finite-temperature MD simulations and reduce the aforementioned static barrier by almost one order of magnitude, as discussed in Section 1.5. The static energy landscape of the molecular motion in cubic MAPI has been calculated recently by using first-principles methods (Bechtel et al. 2016) and is reported in Figure 1.6b. Also the MAPbBr (Motta et al. 2016) has been investigated from first principles.

### 1.4.1 MAPI Thermodynamic Transformation Studied Using the MYP Potential

Notably, the simple MYP ionic model reproduces also the thermodynamical features of the material. This was shown in Mattoni et al. (2015) by a series of constant-temperature constant-stress N$\sigma$T simulations under zero stress performed on orthorhombic crystals (4 × 4 × 4 cells × 48 atoms = 3072 atoms). At variance with most *ab initio* simulations of hybrid perovskites in which the volume is kept fixed for computational convenience, in classical dynamics the metric tensor was allowed to fluctuate without constraints on volume and symmetry of the simulation cell. Each system was equilibrated at temperatures in the range of 0–400 K.

Snapshots of the crystals [obtained by using the VMD package (Humphrey et al. 1996)] at $T < 100$ K (Figure 1.7, left) and 300 K (right) clearly show the stability of the perovskite crystal structure. The $a^-a^-c^+$ tilting pattern (in Glazer notation) (Glazer 1972) proper of the orthorhombic *Pnma* structure can be recognized at low temperatures, while at room temperature the $a^0a^0c^+$ tilting pattern of the tetragonal phase is clearly visible. Experiments show that MA lead halide perovskite is orthorhombic below 160 K, with lattice parameters different along the three crystal axes. As for the density, experiments show that it increases with temperature with discontinuities and large fluctuations at the phase transition around 160–200 K. In Figure 1.7, middle and bottom panels, the MD calculated volume per formula unit and the anisotropy ratio, respectively, as a function of temperature are reported. Two sets of experimental data (Baikie et al. 2013; Poglitsch and Weber 1987) are reported for comparison. The calculated anisotropy (i.e., the ratio between the largest and the minimum parameters, indicated by $c = b$) is sizable at low temperatures (100–150 K). As the temperature increases, the anisotropy decreases with a jump at ~160 K followed by a continuous decrease of the anisotropy down to $c/b$ ~1 (above 300 K) when the cubic phase is reached.

In summary, it can be concluded that the MYP calculations compare well with experiments: (1) at low temperature the crystal is orthorhombic with sizable anisotropy with a jump around 160–180 K; (2) the anisotropy decreases with temperature and tends to one (cubic phase) above 300 K; (3) the volume evolution with temperature has a change of slope (though smaller than experimental) at the orthorhombic-to-tetragonal phase transition reported experimentally at 160 K; (4) the volume increases with temperature and it is close to the experimental value in the high-temperature phase 160–350 K, where the molecular rotations are thermally activated. The aforementioned analysis shows an overall good agreement between MYP and the main relevant properties of the material, confirming the dynamic nature of the phase transitions.

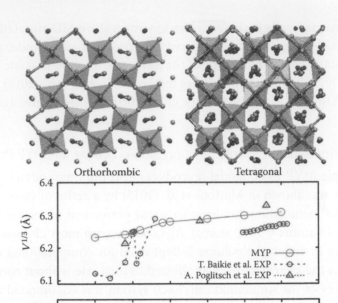

FIGURE 1.7 (Top) Snapshots of the MAPI crystal at $T < 100$ K (left) and 300 K (right) during zero stress NσT run. Pseudo-cubic lattice (i.e., $V^{1/3}$) (middle) and anisotropy factor (bottom) of MPbI$_3$ as a function of temperature calculated using MYP (circles) compared to experiments [small circles (Baikie et al. 2013); triangles (Poglitsch and Weber 1987)]. (Reprinted with permission from Mattoni, A., A. Filippetti, M. I. Saba, and P. Delugas. 2015. Methylammonium rotational dynamics in lead halide perovskite by classical molecular dynamics: The role of temperature. *J. Phys. Chem. C* 119; 17421–17428. Copyright © 2015 American Chemical Society.)

## 1.5 MOLECULAR ENTROPY

The incorporation of organic molecules in hybrid perovskites increases the configurational space of the crystals by adding the rotational and internal modes of the molecules. The associated configurational entropy and dynamical disorder have been recognized and studied since the first synthesis of hybrid perovskites.

Wasylishen et al. (1985) used nuclear magnetic resonance to characterize the rotational relaxation time of $N$-deuterated samples, concluding that, at room temperature, the cations undergo rapid overall reorientation and a complete orientational disorder within the material.

In 1987 Poglitsch and Weber further analyzed the evolution of cation disorder with temperature by millimeter-wave spectroscopy and complex permittivity. They assumed that cations explore the high-symmetry directions of the cubic crystal and predicted the relaxation time according to a Debye model in which the cations have a permanent dipole and the polarization results from their orientations. It was possible in this way to fit the experiments and estimate the temperature-dependent relaxation (i.e., correlation) time $\tau = \tau_0 e^{-E_a/k_B T}$, where $E_a$ is the activation energy. At room temperature, the correlation time was found to be 5.37, 2.73, and 5.63 ps in MA lead iodide, bromide, and chloride, respectively.

An alternative analysis was reported a few years later by Onoda-Yamamuro et al. (1990). The authors found a correlation between the dynamics of the organic cations and the temperature-dependent infrared (IR) spectra of MAPI. In particular, the IR peak at about 910 $cm^{-1}$ was attributed to the rocking modes of the cations, and its evolution with temperature was used to study the ability of molecules to rotate. An activation energy of 2.6 kJ $mol^{-1}$ with rotational time at room temperature as fast as ~1 ps was found. It was reported that at low temperatures, the IR feature consists of two sharp peaks that broaden into a unique structure with temperature. This evolution was associated with the hindered rotational motion in the orthorhombic phase.

The aforementioned pioneering experiments show that the correlation time $\tau$ and the activation energy $E_a$ are suitable quantities to characterize the cation dynamics at different temperatures. Further experimental investigations have been performed recently, motivated by the renewed interest in hybrid perovskites. Ultrafast vibrational spectroscopy (Bakulin et al. 2015) has revealed two characteristic time constants: fast "wobbling-in-a-cone" dynamics (0.3 ps) around the crystal axis; and a relatively slow (~3 ps) jump-like reorientation of the molecular dipole with respect to the iodide lattice.

Two different neutron diffraction measurements of MAPI have been performed recently (Chen et al. 2015; Leguy et al. 2015). Both studies found reorientational times in the picosecond time scale but conclusions were different. Chen et al. (2015) analyzed the neutron data in terms of molecular symmetries, finding relaxation times of 5 and 1 ps for $C_4$ (reorientation of the CN axis within the molecular cage) and $C_3$ (rotation around the CN axis) rotations at room temperature, respectively. A relatively large activation energy (68 meV) was found in the tetragonal phase for the $C_4$ rotations and a small 6 meV value for the $C_3$ rotations. Interestingly, the $C_4$ rotations were found to disappear at $T < 160$ K in the orthorhombic phase. Conversely, Leguy et al. did not find freezing of the molecules in the orthorhombic phase. Moreover, they reported larger relaxation times (14 ps) and lower activation energy (about 10 meV). Chen et al. (2015) attributed the difference to the richer statistics of their experiments (in energy and Q-range) with respect to the work of Leguy et al. (2015).

The observed freezing of the cation rotations in the genuine orthorhombic phase requires some additional understanding. It has been shown that molecular dynamics simulations are appropriate to address these problems by directly studying the dynamics of the cations with temperature. Most of early *ab initio* MD simulations showed the possibility of molecules to rotate at room temperature inducing molecular disorder (Carignano et al. 2015; Goehry et al. 2015; Meloni et al. 2016; Mosconi et al. 2014). However, discordant results were found, and in a few the possibility of collective molecular ordering and residual polarization was reported (Goehry et al. 2015), probably due to the small size and cubic symmetry of the simulated systems. Classical MD recently made it possible to study large crystals as large as $4 \times 4 \times 4$ cells $\times 48$ atoms = 3072 atoms (Mattoni et al. 2015). Crystals were fully equilibrated at zero pressure and the cation dynamics was analyzed in a dense grid of temperatures within 10–350 K. We notice that a full equilibration of the crystal volume for several nanoseconds is critical to explore correctly the dynamics of cations. The orientations of MAPI molecules during MD (Mattoni et al. 2015) at three different temperatures are reported in Figure 1.8. The instantaneous molecular orientation $\hat{n}_i(t) = (\sin\vartheta\cos\varphi, \sin\varphi\cos\vartheta, \cos\vartheta)$ with respect to the crystallographic orthorhombic axes ($\varphi$ angle lies in the $ab$ plane and $\vartheta$ is the polar angle with respect to the $c$-axis) can be represented in the spherical plane ($\varphi$, $\cos\vartheta$) as points. Collecting data during the dynamics, the directional map reported in Figure 1.8 is obtained. The cations give different patterns at different temperatures. At 150 K (top) the pattern is symmetric with respect to $\varphi$ transformation into $\varphi + 180°$ corresponding to the inversion of the C–N molecular axis (i.e., an antiferroelectric-like compensation of molecular dipole orientations).

At higher temperatures, the distribution tends to assume a $C_4$ symmetry $\varphi \rightarrow \varphi + \pi/2$ with a quasirandom exploration of the spherical space (Figure 1.8, bottom). By analyzing the temperature dependence, Mattoni et al. concluded that the molecular trajectories exhibit a transition from a dynamics dominated by high-symmetry directions at low temperatures to a fast dynamics at room temperature in which the molecule can reorient quasirandomly. The change in the dynamics occurs at the orthorhombic-to-tetragonal transition temperature, a result that is consistent with the freezing of $C_4$ rotations found in the experiments of Chen et al. (2015).

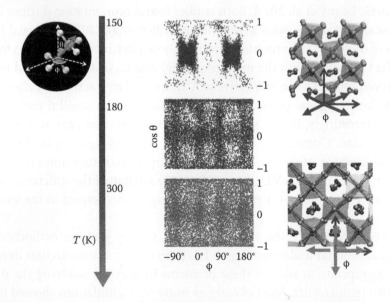

FIGURE 1.8 Directional maps ($\phi$, $z = \cos\theta$) as a function of temperature. The maps are obtained by considering the directions of molecules of a perfect monocrystal formed by 256 formula units (f.u.), sampled every 30 fs during 10 ps of annealing at constant temperature N$\sigma$T.

Beside the directional distribution of the cation dynamics, MD provides directly the molecular relaxation time from the calculated autocorrelation functions $\langle \hat{n}_i(t) \cdot \hat{n}_i(0) \rangle$ at the desired thermodynamic conditions. The results are reported in Figure 1.9a. A detailed fitting of the curves can be found in the literature (Mattoni et al. 2015). Here we notice that the correlation times confirm the temperature dependence observed in the directional patterns; at low temperatures ($T < 160$ K) in correspondence with the nonuniform map, the molecules do not decorrelate in time and the order is preserved for long times (see arrows in right top panel). It is important to notice that the orthorhombic configuration has no net molecular dipole because of the observed inversion symmetry φ +180° (planes of opposite dipole are found along the c direction).

Conversely, at $T > 160$ K corresponding to quasi-uniform maps the molecular ordering is lost in the picosecond scale and the autocorrelation decreases to zero with faster decay at higher temperatures. The molecular order is depicted schematically as arrows in Figure 1.9c. In Figure 1.9b we report the PbI pattern obtained through *ab initio* MD by Meloni et al. (2016). The results are in qualitative agreement.

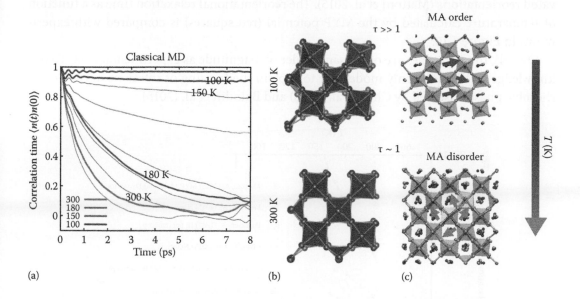

(a)  (b)  (c)

FIGURE 1.9 Methylammonium directional correlation function of $\langle \hat{n}_i(t) \cdot \hat{n}_i(0) \rangle$ crystalline MAPI calculated using classical MD MYP potential (a). Stick-and-ball representation of crystal structures obtained by *ab initio* CPMD (b) and classical MD MYP (c) at different temperatures. Arrows in c indicate the molecular order (top) and disorder (bottom). ([b] adapted from Meloni, S., T. Moehl, W. Tress, M. Franckevicius, M. Saliba, Y. H. Lee, P. Gao, M. K. Nazeeruddin, S. M. Zakeeruddin, U. Röthlisberger, and M. Grätzel. 2016. Ionic polarization-induced current voltage hysteresis in $CH_3NH_3PbX_3$ perovskite solar cells. *Nat. Commun.* 7; 10334–10339. License CC BY 4.0; (a, c) adapted with permission from Mattoni, A., A. Filippetti, M. I. Saba, and P. Delugas. 2015. Methylammonium rotational dynamics in lead halide perovskite by classical molecular dynamics: The role of temperature. *J. Phys. Chem. C* 119; 17421–17428. Copyright © 2015 American Chemical Society.)

Concerning *ab initio* correlation time at 100 K, the molecules tend to decorrelate in the *ab initio* simulations more than in the MYP studies. However, we observe that the limited equilibration (a few picoseconds) before calculating the correlation functions (sampled over 30 ps), and the constant volume adopted in the *ab initio* calculations could possibly explain the quantitative differences with MYP. Another possible explanation of the discrepancy is related to the relative small size of the *ab initio* MD computational sample as suggested by recent large-scale *ab initio* calculations (Lahnsteiner et al. 2016). The stochastic reorientation of a limited number of MA cations [see Supporting Information in Meloni et al. (2016)] can lead to an overestimation of the correlation in this smaller sample. In agreement with Bakulin et al. (2015), from the reorientational correlation time, a fast (subpicosecond) and a longer (picosecond) dynamics can be distinguished corresponding to local vibration and jump-like reorientations, respectively.

Classical MD results provides a sound validation of the aforementioned results because the subpicosecond and picosecond dynamics are characterized by different temperature dependence (power $1/\sqrt{T}$ and exponential Arrhenius-like $e^{-1/T}$, respectively) standing for the attribution of the former to local angular fluctuations and the latter to thermally activated reorientations (Mattoni et al. 2015). The reorientational relaxation time as a function of temperature calculated by the MYP potential (red squares) is compared with experiments in Figure 1.10.

Overall, the MD times are of the same order of magnitude as experimental data. To our knowledge, MYP is the only model able to explain the freezing of rotations in the orthorhombic phase observed by Chen et al. (2016) and Bakulin et al. (2015).

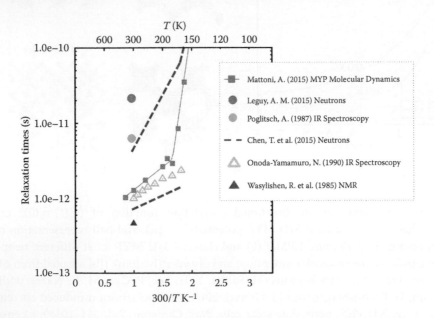

FIGURE 1.10 Reorientational correlation times as a function of temperature calculated by classical MYP MD (squares) compared to experiments (other symbols). Data are taken from the literature (see text).

## 1.6 VIBRATIONAL PROPERTIES OF HYBRID PEROVSKITES

Phonons characterize the dynamical properties of crystalline solids controlling important physical properties such as sound velocity, elastic constants, and thermal transport. Furthermore, a good understanding of the vibrational modes is relevant for the optoelectronic properties because of the importance of electron-phonon scattering that affects the carrier mobilities and broadening of the photoluminescence (PL) emission lines in semiconductors (Wright et al. 2016). For example, Wehrenfennig et al. (2014) recently showed that luminescence from hybrid perovskites exhibits significant homogeneous broadening as a result of strong electron–phonon coupling.

The interaction of charge carriers with lattice vibrations (phonons) has been largely discussed (Brenner et al. 2016; Zhu and Podzorov 2015) to explain the apparently low mobility of hybrid perovskites compared to other semiconductors (Brenner et al. 2016). Several groups have assumed only acoustic phonon scattering based on the mobility temperature dependence that is found experimentally as $T^{-1.3} - T^{-1.6}$ (Brenner et al. 2016), that is, close to the ideal electron-acoustic case $T^{-1.5}$. Recent studies based on *ab initio* calculations (Filippetti et al. 2016; Wright et al. 2016) have clarified that the scattering from longitudinal optical phonons via the Frohlich interaction is the dominant source of electron–phonon coupling near room temperature (Filippetti et al. 2016; Wright et al. 2016). This result is consistent with the polar nature of lead–halide interactions. The electron–polar optical phonon scattering can explain (together with low electron masses) the observed values of mobility and its temperature dependence (Filippetti et al. 2016), making it possible to classify MAPI as a polar semiconductor (Filippetti et al. 2016).

The characterization of acoustic and optical phonons is accordingly very important and has been the objects of several recent studies. The vibrational properties of MAPI have been characterized by IR (Glaser et al. 2015; Mosconi et al. 2014; Onoda-Yamamuro et al. 1990), Raman (Ledinský et al. 2015; Quarti et al. 2014b), and terahertz time-domain spectroscopy (La-o-vorakiat et al. 2015). As for theoretical studies, several first-principles analyses were published (Brivio et al. 2015; Leguy et al. 2016; Mosconi et al. 2014; Pérez-Osorio et al. 2015; Quarti et al. 2014b). Most studies adopted the harmonic approximation to calculate the vibrational density of states (VDOS) and IR spectra of the tetragonal, orthorhombic (Mattoni et al. 2016; Pérez-Osorio et al. 2015; Quarti et al. 2014b), and cubic phases (Brivio et al. 2015; Mattoni et al. 2016). Raman spectra were also simulated by DFT (Leguy et al. 2016; Quarti et al. 2014b).

The vibrational structure of the orthorhombic MAPI crystal calculated using DFT is reported in Figure 1.11d. For clarity, only the frequencies below 1000 cm$^{-1}$ are reported. The vibrational modes of the isolated MA molecules are reported in Figure 1.11c for comparison. These modes consist of singlets or doublets corresponding to the representations of the $C_{3v}$ molecular symmetry occurring above 290 cm$^{-1}$ (the twisting $T_{\omega}$ around the C–N axis) up to 3000 cm$^{-1}$ (rocking and bending of the NH$_3$ group). The full spectrum can be found elsewhere (Mattoni et al. 2016; Pérez-Osorio et al. 2015). In addition to the internal modes, the isolated molecule is characterized by six roto-translations (R-T) of zero frequency corresponding to energy invariance under rigid transformation of the molecule.

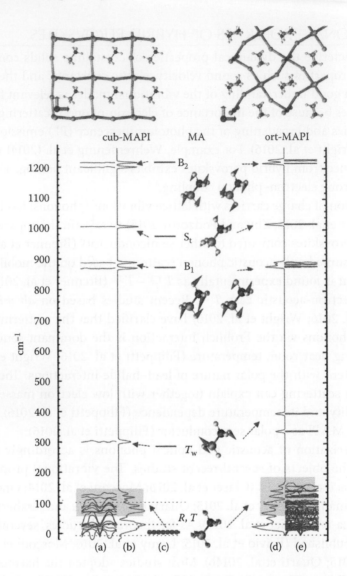

FIGURE 1.11   Vibrational DFT spectrum of cubic (b) and orthorhombic (d) crystalline MAPI hybrid perovskite compared to the MA molecular levels (c). The phonon dispersions of cubic and orthorhombic crystals are reported in a and e, respectively. (Reprinted with permission from Mattoni, A., A. Filippetti, M. Saba, C. Caddeo, and P. Delugas. 2016. Temperature evolution of methylammonium trihalide vibrations at the atomic scale. *J. Phys. Chem. Lett.* 7; 529–535. Copyright © 2016 American Chemical Society.)

The zero R-T modes gain finite frequencies when the molecules are embedded in the lead halide cage (confining the molecules and breaking the roto-translation invariance). The resulting vibrational structure is formed by a distribution of mixed modes located below 200 cm⁻¹ (shadowed regions), spurring from the relative motions of molecules and octahedra (Brivio et al. 2015; Pérez-Osorio et al. 2015). Above this threshold, only flat molecular modes appear, slightly perturbed with respect to the isolated molecule (energy-shifted and degeneracy-broken).

In Figure 1.11e, the whole phonon dispersion for the orthorhombic phonons calculated in Mattoni et al. (2016) is reported [most works study only the $\Gamma$ point (Pérez-Osorio et al. 2015) or focus on the cubic phase (Brivio et al. 2015)]. A sizable dispersion of molecular levels is found in the 100–200 $cm^{-1}$ vibrational band. Conversely, the absence of dispersion confirms the localization of molecular levels above 300 $cm^{-1}$.

The MAPI spectrum, and in particular, the molecular band below 200 $cm^{-1}$, slightly changes depending on the crystal structure (compare, for example, Figure 1.11b and e) and molecular order (Mattoni et al. 2016) but it preserves the general features and the mixed continuum-discrete nature that is peculiar to this kind of hybrid crystal.

It is an important result that the main properties of vibrational structure of hybrid perovskites can be reproduced by the simple classical MYP potential (Mattoni et al. 2016; Wang and Lin 2016). Consequently, the ionic nature of the organic–inorganic interactions and the mass difference between organic and inorganic ions [actually the ratio between the bulk modulus and the density (Wang and Lin 2016)] are the key ingredients to explain the progressive separation in frequency between the organic–inorganic modes, the phonon dispersion, and the evolution of the vibrational peaks with temperature.

The reduced computational cost of the classical MYP interatomic potential with respect to *ab initio* methods makes it possible to easily calculate the phonon dispersion curves of the different phases (see Figure 1.12, right panels corresponding to the static analysis). Considering the *Pnma* orthorhombic phase (right) the lowest internal molecular mode can be observed at about 400 $cm^{-1}$ [twisting (Mattoni et al. 2016)]; at variance in the tetragonal and cubic phases the formation of a molecular mode (labeled by X) is observed at lower frequencies (~250 $cm^{-1}$ in MD and ~300 $cm^{-1}$ in DFT).

Concerning roto-translational modes below 200 $cm^{-1}$, a progressive reduction of the S peak can be observed at about 150 $cm^{-1}$ when passing from the orthorhombic to tetragonal and cubic phases. Temperature effects on the vibrational properties are relevant for electronic transport. The phonon population changes with temperature. As optical phonons in semiconductors typically have energies of the order of tens of milli-electron volts, their population at low temperatures ($T \sim 100$ K) is very small whereas it becomes important at room temperature. This effect is used to explain the electron–optical phonons scattering in MAPI at room temperature (Filippetti et al. 2016; Wright et al. 2016).

Furthermore, temperature controls the structural evolution and the phase transitions of the material, as well as the cation dynamics (Carignano et al. 2015; Chen et al. 2015; Goehry et al. 2015; Leguy et al. 2015; Wasylishen et al. 1985). As discussed in the preceding text, at low temperatures the molecules are constrained in the orthorhombic crystal and harmonically vibrate around local equilibrium configurations; as the temperature increases, the organic cations progressively explore a larger set of configurations and, beyond the orthorhombic-to-tetragonal transition ($T = 160$ K), the temperature gives rise to a complex anharmonic dynamics with fast reorientations.

Concerning the vibrational density of states, it is important to observe that the phonon spectrum at zero temperature refers to a single molecular configuration, while the low-temperature phase of this material is characterized by a plethora of structures with different MA orientations (Filippetti and Mattoni 2014). The tetragonal and cubic structures

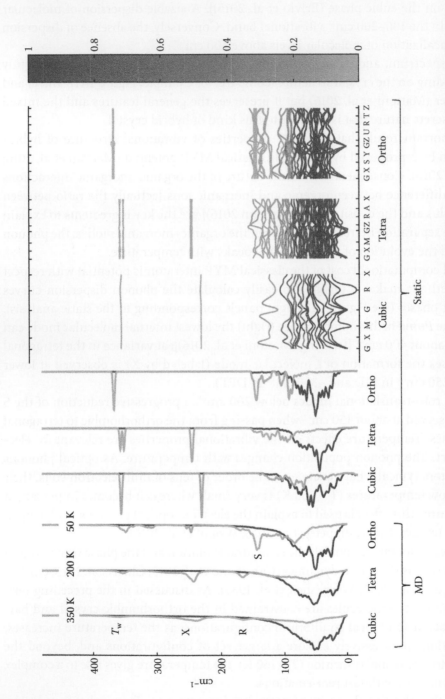

FIGURE 1.12  Phonon dispersion and vibrational density of states of cubic, tetragonal, and orthorhombic MAPI crystals (right) obtained by static calculations compared with MD vibrational density of states calculated using the MYP interatomic potential. (Reprinted with permission from Mattoni, A., A. Filippetti, M. Saba, C. Caddeo, and P. Delugas. 2016. Temperature evolution of methylammonium trihalide vibrations at the atomic scale. *J. Phys. Chem. Lett.* 7; 529–535. Copyright © 2016 American Chemical Society.)

in particular have a large number of quasi-isoenergetic configurations that give molecular modes with imaginary frequencies. In this sense, the $T$-dependent MD simulation (Mattoni et al. 2017) is a more appropriate treatment of the atomic vibrations than $T = 0$ K DFT results.

A comparison between static (right panels) and dynamic MD approach is reported in Figure 1.12 (Mattoni et al. 2016). Dynamic effects include phonon–phonon scattering that reduces the phonon lifetime, giving rise to the broadening and slightly shifting the molecular peaks at zero temperature (compare the MD and static vibrational density of states for the $T_\omega$ and X internal molecular peaks at ~400 cm$^{-1}$ and ~300 cm$^{-1}$). Sizable dynamical effects can be observed also in the 100–150 cm$^{-1}$ region, affecting, for example, the S peak that tends to disappears and giving rise to the formation of the R feature that is absent in the static VDOS (Mattoni et al. 2016).

Several spectroscopic measurements have been used to study the temperature evolution of vibrations. Beside PL broadening that gives information on electron–phonon scattering (Wehrenfennig et al. 2014; Wright et al. 2016), important information can be extracted from IR, Raman, and terahertz time-domain spectroscopy (La-o-vorakiat et al. 2015). In the high-energy range of the vibrational spectra the evolution of the molecular peaks at 900–1000 cm$^{-1}$ has been correlated to the orthorhombic-tetragonal phase transition (Mosconi et al. 2014; Onoda-Yamamuro et al. 1990). At medium frequencies, temperature-dependent bands in the Raman and IR spectra are observed in the tetragonal phase of MAPI at 200–300 cm$^{-1}$ and attributed to the molecular order (Quarti et al. 2014b) or degradation phenomena (Deretzis et al. 2015; Ledinský et al. 2015; Pérez-Osorio et al. 2015). In the low-frequency range (below 100 cm$^{-1}$) the temperature evolution of the MAPI vibrations has been characterized also by terahertz time-domain (La-o-vorakiat et al. 2015) in addition to Raman and IR spectroscopy. The IR experimental spectra (Mosconi et al. 2014) are reported in Figure 1.13, top (Mattoni et al. 2016).

Both *ab initio* and classical MD are able to reproduce the main experimental features (see Figure 1.13b, c and d, e); the spectrum of MAPI at low temperature shows a peak at 40–60 cm$^{-1}$ (Brivio et al. 2015; Mosconi et al. 2014; Pérez-Osorio et al. 2015; Quarti et al. 2014b). At increasing temperature the peak slightly blue-shifts, forming at 300 K a shoulder above 60–70 cm$^{-1}$. This trend is reproduced by the simulated phonons and IR spectrum [see also (Mattoni et al. 2016) for a comparison with terahertz time-domain spectroscopy (La-o-vorakiat et al. 2015)]. Concerning higher frequencies (>100 cm$^{-1}$), there is a contribution from molecular levels that extend up to 300 cm$^{-1}$ at room temperatures. MD makes it possible to analyze the evolution of such modes that are reproduced both classically and *ab initio*. Detailed classical analysis (Mattoni et al. 2016) predicts the formation of two dynamic features referred to as R and X (see bottom panels of Figure 1.13) that can be compared with similar features in Raman (Quarti et al. 2014b) *ab initio* (middle panels) and IR measurements (Mosconi et al. 2014) (top panels) (though the X peak resulting from MD at 250 cm$^{-1}$ is likely underestimated, corresponding to a DFT peak at 300 cm$^{-1}$).

Further experimental characterization, possibly based on neutrons, is necessary to better characterize the molecular peak in this region. We further notice that this frequency

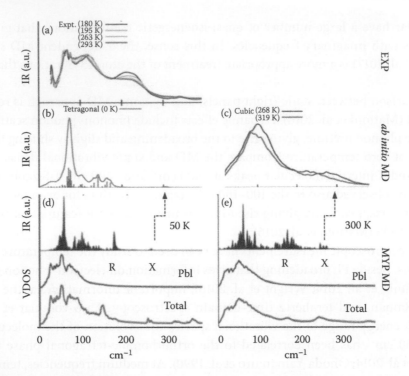

**FIGURE 1.13** (d, e) Vibrational density of states and simulated IR spectra by MYP potential at temperatures 50 K and 300 K. (Adapted with permission from Mattoni, A., A. Filippetti, M. Saba, C. Caddeo, and P. Delugas. 2016. Temperature evolution of methylammonium trihalide vibrations at the atomic scale. *J. Phys. Chem. Lett.* 7; 529–535. Copyright © 2016 American Chemical Society.) (b, c) IR spectra by DFT at zero and close to room temperature. (a) IR experimental data at different temperatures. (Adapted with permission from Mosconi, E., C. Quarti, T. Ivanovska, G. Ruani, and F. De Angelis. 2014. Structural and electronic properties of organo-halide lead perovskites: A combined IR-spectroscopy and *ab initio* molecular dynamics investigation. *Phys. Chem. Chem. Phys.* 16; 16137–16144. Copyright © 2016 American Chemical Society.)

range is also affected by degradation of the material during measurements (Ledinský et al. 2015), making a conclusive attribution difficult based on the experimental data available in the literature.

## 1.7 THERMAL CONDUCTIVITY OF HYBRID PEROVSKITES INVESTIGATED BY MOLECULAR DYNAMICS

One of the major drawbacks of hybrid perovskites and in particular MAPI is the instability of the material to thermal stress. This material is characterized by a low sublimation temperature (Dualeh et al. 2014) and by degradation reactions (e.g., $CH_3NH_3PbI_3 \rightarrow CH_3NH_3I + PbI_2$) that can be easily activated with small thermal budgets (Eperon et al. 2014; Stranks and Snaith 2015).

Accordingly, understanding the thermal properties of hybrid perovskites is of great interest. In particular, the heat control in crystalline layers of MAPI is technologically relevant, requiring a precise determination of the thermal conductivity κ. Moreover, it is important to understand the dependence of κ on parameters such as temperature and crystal size and to find approaches to tune the thermal conductivity. Increasing κ can be beneficial for the efficient heat dissipation and for preventing the device overheating.

On the other hand, lowering thermal conductivity could increase the thermoelectric figure of merit for applications of hybrid perovskites to thermoelectric devices, as suggested in the literature (He and Galli 2014; Mettan et al. 2015). In general, thermal conductivity has electronic and lattice contributions. However, experiments (Pisoni et al. 2014) and theoretical analysis suggest that, in the absence of significant doping, the electronic contribution to thermal conductivity is discardable at any temperature, and only the lattice contribution matters. In the range of temperatures 80 K ≤ T ≤ 350 K higher or comparable to the Debye temperature of MAPI (~120 K) (Pisoni et al. 2014), the thermal conductivity is only marginally affected by quantum features.

Calorimetric measurements (molar heat capacity) have been reported since 1990 (Onoda-Yamamuro et al. 1990). However, thermal conductivity κ has been reported only recently (Chen et al. 2016; Guo et al. 2016; Pisoni et al. 2014). The results for monocrystalline MAPI are shown in Figure 1.14. Experimental data are conflicting. In most cases, ultralow thermal conductivity dominated by the lattice contribution is reported (Guo et al. 2016, Pisoni et al. 2014); however, also higher values of κ (corresponding to efficient thermal conductance) have been reported (Chen et al. 2016). This likely can be attributed to the different synthesis methods and sample quality, and further experimental investigations will be necessary to clarify the discrepancies.

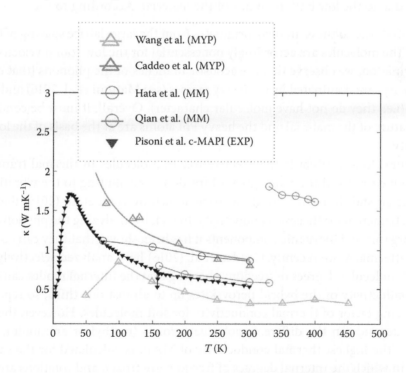

FIGURE 1.14  Lattice thermal conductivity of crystalline MAPI obtained from experiments (Pisoni et al.) and MD calculations. Data were taken from the literature (see text).

As under intrinsic conditions κ is due to the lattice, we observe that classical MD is an ideal method to predict the actual κ of perfect crystals and clarify the experimental discrepancies. Several classical molecular dynamics simulations of thermal conductivity have been published recently (Caddeo et al. 2016; Hata et al. 2016; Qian et al. 2016; Wang and Lin 2016), by using the MYP potential (Caddeo et al. 2016; Wang and Lin 2016) or other local potentials directly derived from *ab initio* MD.

The thermal conductivity κ as a function of temperature calculated by the different theoretical works is reported in Figure 1.14 compared to the data of Pisoni et al. (2014). The results depend both on the interatomic model and on the methods used to calculate κ. For example, the two sets of data using the same MYP potentials (Caddeo et al. 2016; Wang and Lin 2016) only differ by the use of non-equilibrium molecular dynamics (NEMD) and approach-to-equilibrium molecular dynamics (AEMD) methods (Caddeo et al. 2016).

Despite the differences in the calculated data, all the κ values at room temperature fall within the range 0.3–1 W mK$^{-1}$, supporting the low thermal conductivity of MAPI measured by Pisoni et al. Furthermore, data are consistent with the reported $1/T$ temperature dependence that can be attributed to phonon–phonon scattering. It is interesting to understand the origin of the low value of κ of hybrid perovskite. It would be tempting to attribute the low κ value to the hybrid nature of the material and to the role of molecules. However, the origin of the low κ must be searched mainly in the inorganic PbI lattice. κ depends on the sound velocity of the material. As observed by Wang and Lin (2016), the sound velocity $\sqrt{\dfrac{B}{\rho}}$ is low due to the low bulk modulus of the material. According to the scaling rule for ionic potentials the softness, in turn, originates from the large lattice spacing of lead halide sublattice. The molecules are accordingly not essential for the low sound velocity. As a further confirmation, we observe that the acoustic branches of the phonons (that control the sound velocity) are dominated by the heavy masses (see Mattoni et al. 2016) and not by the molecules (i.e., they do not have molecular character). Overall, it must be concluded that the ionic nature of the material and the heavy PbI atoms are at the basis of the low thermal conductivity.

It is natural to ask, What is the effective role of molecules in thermal transport? The molecules give rise to additional degrees of freedom contributing to the vibrational spectrum of the crystal. In particular, it has been shown by Hata et al. (2016) that the rotations are detrimental for thermal conductivity. In fact, by analyzing the phonon dispersion curves of organic and inorganic components it has been shown that they can contribute to phonon scattering. More recently, Caddeo et al. (2016) have analyzed selectively the effect of different molecular degrees of freedom, showing that the internal modes can reduce the thermal conductivity of the hybrid perovskites up to almost one third, so representing a major reducing factor of thermal conductivity for soft molecules. However, the substructure effect depends on the different molecular groups (methyl or ammonium) (Caddeo et al. 2016). The highest thermal conductivity of MAPI is calculated for the case of rigid molecules in which the internal degrees of freedom are frozen and rotations are left as the leading mechanism of thermal transport perturbation.

## 1.8 CONCLUSIONS

Hybrid perovskites are a class of crystalline solids of tremendous importance for photovoltaics, optoelectronics, photocatalysis, and many other applications. These materials are also the archetypal hybrid crystalline materials formed by organic moieties distributed periodically within an inorganic metal halide network and bound together by ionic interactions. It has been shown that the tolerance factor is the key concept to understand the stability of inorganic perovskite structure. In the presence of organic molecules it is possible to generalize the tolerance factor concept by including the anisotropy of the organic components. The molecules introduce a sizable configurational disorder in the hybrid crystals that affects both structural and electronic properties of the material. The appropriate perspective to model and fully understand hybrid perovskites at finite temperature is dynamic, as it necessary to include the relevant molecular entropy to understand the material properties (from ferroelectricity to optoelectronic ones). By molecular dynamics it is possible to average over the relevant molecular configurations taking into account the large fluctuations of the soft body of lead halides. It has been described in this chapter how a simple ionic interatomic force field (MYP) also including dispersive interactions can reproduce the main structural parameters of MAPI, that is, the prototype of hybrid perovskites.

Notably also the main thermodynamic features of MAPI at finite temperatures are well described. This is a strong validation of the dynamic nature of phase transitions during which the temperature and molecular thermal vibrations induce transformation from a low-symmetry distorted orthorhombic crystal at low temperature to the cubic high-symmetry phase above room temperature. Furthermore, such a simple interatomic view of the material is able to reproduce the vibrational properties as well as the thermal properties of the material, providing physical insight into the ionic and dispersive nature of interatomic forces.

The emerging scenario is that hybrid perovskites, such as MAPI, can be described as ionic inorganic materials in terms of optoelectronic and transport properties but with a hybrid solution-processable and dynamic soft body. Accordingly, hybrid perovskites differ from standard hybrids and really combine the best of the organic and inorganic worlds in terms of both functionalities and easy processability. The subtle interplay between the organic and inorganic components, together with the flexible body, makes hybrid perovskites a unique class of crystalline solids with an intriguing dynamic character.

## ACKNOWLEDGMENT

We acknowledge computational support by the Italian supercomputing center CINECA (Casalecchio di Reno, Italy) through the ISCRA initiative (Projects MYPALLOY, THEHYPE, UNWRAPIT).

## REFERENCES

Allen, M. P., and D. J. Tildesley. 1989. *Computer simulation of liquids*. Oxford: Clarendon Press.
Andersen, N. H., J. Kjems, and W. Hayes. 1985. Ionic conductivity of the perovskites $NaMgF_3$, $KMgF_3$, $KMgK_3$ and $KZnF_3$ at high temperatures. *Solid State Ionics*. 17; 143–145.
Bagolini, L., A. Mattoni, R. T. Collins, and M. T. Lusk. 2014. Carrier localization in nanocrystalline silicon. *J. Phys. Chem. C* 118; 13417–13423.

Bagolini, L., A. Mattoni, G. Fugallo, L. Colombo, E. Poliani, S. Sanguinetti, and E. Grilli. 2010. Quantum confinement by an order-disorder boundary in nanocrystalline silicon. *Phys. Rev. Lett.* 104; 176803–176804.

Baikie, T., Y. Fang, J. M. Kadro, M. Schreyer, F. Wei, S. G. Mhaisalkar, M. Grätzel, and T. J. White. 2013. Synthesis and crystal chemistry of the hybrid perovskite $(CH_3NH_3)PbI_3$ for solid-state sensitised solar cell applications. *J. Mater. Chem. A* 1; 5628–5641.

Bakulin, A. A., O. Selig, H. J. Bakker, Y. L. Rezus, C. Müller, T. Glaser, R. Lovrincic, Z. Sun, Z. Chen, A. Walsh, J. M. Frost, and T. L. C. Jansen. 2015. Real-time observation of organic cation reorientation in methylammonium lead iodide perovskites. *J. Phys. Chem. Lett.* 6; 3663–3669.

Bechtel, J. S., R. Seshadri, and A. Van der Ven. 2016. Energy landscape of molecular motion in cubic methylammonium lead iodide from first-principles. *J. Phys. Chem. C* 120; 12403–12410.

Boix, P. P., K. Nonomura, N. Mathews, and S. G. Mhaisalkar. 2014. Current progress and future perspectives for organic/inorganic perovskite solar cells. *Mater. Today* 17; 16–23.

Brenner, T. M., D. M. Egger, L. Kronik, G. Hodes, and D. Cahen. 2016. Hybrid organic-inorganic perovskites: Low-cost semiconductors with intriguing charge-transport properties. *Nat. Rev. Mater.* 1; 15007–15016.

Brivio, F., J. M. Frost, J. M. Skelton, A. J. Jackson, O. J. Weber, M. T. Weller, A. R. Goñi, A. M. A. Leguy, P. R. F. Barnes, and A. Walsh. 2015. Lattice dynamics and vibrational spectra of the orthorhombic, tetragonal, and cubic phases of methylammonium lead iodide. *Phys. Rev. B* 92; 144308.

Buckingham, R. A. 1938. The classical equation of state of gaseous helium, neon and argon. *Proc. R. Soc. A Math. Phys. Eng. Sci.* 168; 264–283.

Caddeo, C., R. Dessi, C. Melis, L. Colombo, and A. Mattoni. 2011. Poly(3–hexylthiophene) adhesion on zinc oxide nanoneedles. *J. Phys. Chem. C* 115; 16833–16837.

Caddeo, C., C. Melis, M. I. Saba, A. Filippetti, L. Colombo, and A. Mattoni. 2016. Tuning the thermal conductivity of methylammonium lead halide by the molecular substructure. *Phys. Chem. Chem. Phys.* 18; 24318–24324.

Canesi, E. V., M. Binda, A. Abate, S. Guarnera, L. Moretti, V. D'Innocenzo, R. Sai Santosh Kumar, C. Bertarelli, A. Abrusci, H. Snaith, A. Calloni, A. et al. 2012. The effect of selective interactions at the interface of polymeroxide hybrid solar cells. *Energy Environ. Sci.* 5; 9068–9076.

Car, R., and M. Parrinello. 1985. Unified approach for molecular dynamics and density-functional theory. *Phys. Rev. Lett.* 55; 2471–2474.

Carignano, M. A., A. Kachmar, and J. Hutter. 2015. Thermal effects on $CH_3NH_3PbI_3$ perovskite from ab initio molecular dynamics simulations. *J. Phys. Chem. C* 119; 8991–8997.

Chang, Y. H., C. H. Park, and K. Matsuishi. 2004. First-principles study of the structural and the electronic properties of the lead-halide-based inorganic-organic perovskites $(CH_3NH_3)PbX_3$ and $CsPbX_3$ (X = Cl, Br, I). *J. Korean Phys. Soc.* 44; 889–893.

Cheetham, A. K., and C. N. R. Rao. 2007. There's room in the middle. *Science* 318; 58–59.

Chen, Q., C. Zhang, M. Zhu, S. Liu, M. E. Siemens, S. Gu, J. Zhu, J. Shen, X. Wu, C. Liao, J. Zhang, X. Wang, and M. Xiao. 2016. M Efficient thermal conductance in organometallic perovskite $CH_3NH_3PbI_3$ films. *Appl. Phys. Lett.* 108; 081902.

Chen, T., B. J. Foley, B. Ipek, M. Tyagi, J. R. D. Copley, C. M. Brown, J. J. Choi, and S. H. Lee. 2015. Rotational dynamics of organic cations in the $CH_3NH_3PbI_3$ perovskite. *Phys. Chem. Chem. Phys.* 17; 31278–31286.

Chin, X. Y., D. Cortecchia, J. Yin, A. Bruno, and C. Soci. 2015. Lead iodide perovskite light-emitting field-effect transistor. *Nat. Commun.* 6; 7383–7389.

Chung, I., J. H. Song, J. Im, J. Androulakis, C. D. Malliakas, H. Li, A. J. Freeman, J. T. Kenney, and M. G. Kanatzidis. 2012. $CsSnI_3$: Semiconductor or metal? High electrical conductivity and strong near-infrared photoluminescence from a single material. High hole mobility and phase-transitions. *J. Am. Chem. Soc.* 134; 8579–8587.

Dar, M. I., G. Jacopin, S. Meloni, A. Mattoni, N. Arora, A. Boziki, S. M. Zakeeruddin, U. Röthlisberger, and M. Grätzel. 2016. Origin of unusual bandgap shift and dual emission in organic-inorganic lead halide perovskites. *Sci. Adv.* 2; e1601156–1601159.

De Graef, M., and M. E. McHenry. 2007. *Structure of materials: An introduction to crystallography, diffraction, and symmetry.* Cambridge: Cambridge University Press.

Delugas, P., C. Caddeo, A. Filippetti, and A. Mattoni. 2016. Thermally activated point defect diffusion in methylammonium lead trihalide: Anisotropic and ultrahigh mobility of iodine. *J. Phys. Chem. Lett.* 7; 2356–2361.

Deretzis, I., A. Alberti, G. Pellegrino, E. Smecca, F. Giannazzo, N. Sakai, T. Miyasaka, and A. La Magna. 2015. Atomistic origins of $CH_3NH_3PbI_3$ degradation to $PbI_2$ in vacuum. *Appl. Phys. Lett.* 106; 131904.

Dualeh, A., P. Gao, S. I. Seok, M. K. Nazeeruddin, and M. Grätzel. 2014. Thermal behavior of methylammonium lead-trihalide perovskite photovoltaic light harvesters. *Chem. Mater.* 26; 6160–6164.

Egger, D. A., and L. Kronik. 2014. Role of dispersive interactions in determining structural properties of organic inorganic halide perovskites: Insights from first-principles calculations. *J. Phys. Chem. Lett.* 5; 2728–2733.

Egger, D. A., A. M. Rappe, and L. Kronik. 2016. Hybrid organic-inorganic perovskites on the move. *Acc. Chem. Res.* 49; 573–581.

Eperon, G. E., S. D. Stranks, C. Menelaou, M. B. Johnston, L. M. Herz, and H. J. Snaith. 2014. Formamidinium lead trihalide: A broadly tunable perovskite for efficient planar heterojunction solar cells. *Energy Environ. Sci.* 7; 982–988.

Feng, J. 2014. Mechanical properties of hybrid organic-inorganic $CH_3NH_3BX_3$ (B = Sn, Pb; X = Br, I) perovskites for solar cell absorbers. *APL Mater.* 2; 081801–081808.

Filip, M. R., G. E. Eperon, H. J. Snaith, and F. Giustino. 2014. Steric engineering of metal-halide perovskites with tunable optical band gaps. *Nat. Commun.* 5; 5757–5759.

Filippetti, A., P. Delugas, and A. Mattoni. 2014. Radiative recombination and photoconversion of methylammonium lead iodide perovskite by first principles: Properties of an inorganic semiconductor within a hybrid body. *J. Phys. Chem. C* 118; 24843–24853.

Filippetti, A., and A. Mattoni. 2014. Hybrid perovskites for photovoltaics: Insights from first principles. *Phys. Rev. B.* 89; 125203–125208.

Filippetti, A., A. Mattoni, C. Caddeo, M. I. Saba, and P. Delugas. 2016. Low electron-polar optical phonon scattering as a fundamental aspect of carrier mobility in methylammonium lead halide $CH_3NH_3PbI_3$ perovskites. *Phys. Chem. Chem. Phys.* 18; 15352–15362.

Frost, J. M., K. T. Butler, and A. Walsh. 2014. Molecular ferroelectric contributions to anomalous hysteresis in hybrid perovskite solar cells. *APL Mater.* 2; 081506–081510.

Frost, J. M., and A. Walsh. 2016. What is moving in hybrid halide perovskite solar cells? 2016. *Acc. Chem. Res.* 49; 528–535.

Furukawa, H., K. E. Cordova, M. O' Kee, and O. M. Yaghi. 2013. The chemistry and applications of metal-organic frameworks. *Science* 341; 974.

Giorgi, G., and K. Yamashita. 2016. Zero-dimensional hybrid organic inorganic halide perovskite modeling: Insights from first principles. *J. Phys. Chem. Lett.* 7; 888–899.

Glaser, T., C. Müller, M. Sendner, C. Krekeler, O. E. Semonin, T. D. Hull, O. Yaffe, J. S. Owen, W. Kowalsky, A. Pucci, and R. Lovrinčić. 2015. Infrared spectroscopic study of vibrational modes in methylammonium lead halide perovskites. *J. Phys. Chem. Lett.* 6; 2913–2918.

Glazer, A. M. 1972. The classification of tilted octahedra in perovskites. *Acta Cryst.* B28; 3384–3392.

Goehry, C., G. A. Nemnes, and A. Manolescu. 2015. Collective behavior of molecular dipoles in $CH_3NH_3PbI_3$. *J. Phys. Chem. C* 119; 19674–19680.

Goldschmidt, V. M. 1926. Die Gesetze der Krystallochemie. *Naturwissenschaften.* 14; 477–485.

Guo, Z., S. J. Yoon, J. S. Manser, P. V. Kamat, and T. Luo. 2016. Structural phase- and degradation-dependent thermal conductivity of $CH_3NH_3PbI_3$ perovskite thin films. *J. Phys. Chem. C* 120; 6394–6401.

Gutierrez-Sevillano, J. J., S. Ahmad, S. Calero, and J. A. Anta. 2015. Molecular dynamics simulations of organohalide perovskite precursors: Solvent effects in the formation of perovskite solar cells. *Phys. Chem. Chem. Phys.* 17; 22770–22777.

Hailegnaw, B., S. Kirmayer, E. Edri, G. Hodes, and D. Cahen. 2015. Rain on methylammonium lead iodide based perovskites: Possible environmental effects of perovskite solar cells. *J. Phys. Chem. Lett.* 6; 1543–1547.

Hata, T., G. Giorgi, and K. Yamashita. 2016. The effects of the organic-inorganic interactions on the thermal transport properties of $CH_3NH_3PbI_3$. *Nano Lett.* 16; 2749–2753.

He, Y., and G. Galli. 2014. Perovskites for solar thermoelectric applications: A first principle study of $CH_3NH_3AI_3$ (A = Pb and Sn). *Chem. Mater.* 26; 5394–5400.

Humphrey, W., A. Dalke, and K. Schulten. 1996. VMD: Visual molecular dynamics. *J. Mol. Graph.* 14; 33–38.

Hutter, J. 2012. Car-Parrinello molecular dynamics. *Wiley Interdiscip. Rev. Comput. Mol. Sci.* 2; 604–612.

Jahn, H. A., and E. Teller. 1937. Stability of polyatomic molecules in degenerate electronic states. I-orbital degeneracy. *Proc. R. Soc. A Math. Phys. Eng. Sci.* 161; 220–235.

Jeon, N. J., J. H. Noh, W. S. Yang, Y. C. Kim, S. Ryu, J. Seo, and S. I. Seok. 2015. Compositional engineering of perovskite materials for high-performance solar cells. *Nature.* 517; 476–480.

Kieslich, G., S. Sun, and A. K. Cheetham. 2014. Solid-state principles applied to organic–inorganic perovskites: New tricks for an old dog. *Chem. Sci.* 5; 4712–4715.

Kieslich, G., S. Sun, and A. K. Cheetham. 2015. An extended tolerance factor approach for organic–inorganic perovskites. *Chem. Sci.* 6; 3430–3433.

Kim, H. S., C. R. Lee, J. H. Im, K. B. Lee, T. Moehl, A. Marchioro, S. J. Moon, R. Humphry-Baker, J. H. Yum, J. E. Moser, M. Grätzel, and N. G. Park. 2012. Lead iodide perovskite sensitized all-solid state submicron thin film mesoscopic solar cell with efficiency exceeding 9%. *Sci. Rep.* 2; 591.

Kojima, A., K. Teshima, Y. Shirai, and T. Miyasaka. 2009. Organometal halide perovskites as visible-light sensitizers for photovoltaic cells. *J. Am. Chem. Soc.* 131; 6050–6051.

Kulkarni, A. J., M. Zhou, and F. J. Ke. 2005. Orientation and size dependence of the elastic properties of zinc oxide nanobelts. *Nanotechnology* 16; 2749–2756.

Lahnsteiner, J., G. Kresse, A. Kumar, D. D. Sarma, C. Franchini, and M. Bokdam. 2016. Room temperature dynamic correlation between methylammonium molecules in lead-iodine based perovskites: An ab-initio molecular dynamics perspective. *Phys. Rev. B* 94 (21); 214114.

La-o-vorakiat, C., J. M. Kadro, T. Salim, D. Zhao, T. Ahmed, Y. M. Lam, J. X. Zhu, R. A. Marcus, M. E. Michel-Beyerle, and E. E. Chia. 2015. Phonon mode transformation across the orthorhombic-tetragonal phase transition in a lead-iodide perovskite $CH_3NH_3PbI_3$: A terahertz time-domain spectroscopy approach. *J. Phys. Chem. Lett.* 7; 1–6.

Ledinský, M., P. Löper, B. Niesen, J. Holovský, S. J. Moon, J. H. Yum, S. De Wolf, A. Fejfar, and C. Ballif. 2015. Raman spectroscopy of organic-inorganic halide perovskites. *J. Phys. Chem. Lett.* 6; 401–406.

Lee, M. M., J. Teuscher, T. Miyasaka, T. N. Murakami, and H. J. Snaith. 2012. Efficient hybrid solar cells based on meso-superstructured organometal halide perovskites. *Science* 338; 643–647.

Leguy, A. M. A., J. M. Frost, A. P. McMahon, V. G. Sakai, W. Kochelmann, C. Law, X. Li, F. Foglia, A. Walsh, B. C. O'Reagan, J. Nelson, J. T. Cabral, and P. R. F. Barnes. 2015. The dynamics of methylammonium ions in hybrid organic-inorganic perovskite solar cells. *Nat. Commun.* 6; 7124, 10 pp.

Leguy, A. M. A., A. R. Goñi, J. M. Frost, J. Skelton, F. Brivio, X. Rodrìguez-Martìnez, O. J. Weber, A. Pallipurath, M. I. Alonso, M. Campoy-Quiles, M. T. Weller et al. 2016. Dynamic disorder, phonon lifetimes, and the assignment of modes to the vibrational spectra of methylammonium lead halide perovskites. *Phys. Chem. Chem. Phys.* 18; 27051–27066.

Liu, S., and R. E. Cohen. 2016. Response of methylammonium lead iodide to external stimuli and caloric effects from molecular dynamics simulations. *J. Phys. Chem. C* 120; 17274–17281.

Ma, J., and L.-W. Wang. 2015. Nanoscale charge localization induced by random orientations of organic molecules in hybrid perovskite $CH_3NH_3PbI_3$. *Nano Lett.* 15; 248–253.

Marx, D., and J. Hutter. 2000. Ab initio molecular dynamics: Theory and implementation. In J. Grotendorst (ed.), *Modern methods and algorithms of quantum chemistry* (pp. 301–449). NIC Series, Vol. 1. Jülich, Germany: Forschungszentrum Jülich.

Marx, D., and J. Hutter. 2009. *Ab initio molecular dynamics basic theory and advanced methods.* Cambridge: Cambridge University Press.

Matsui, M. 1988. Molecular dynamics study of $MgSiO_3$ perovskite. *Phys. Chem. Miner.* 16; 234–238.

Matsui, M., and M. Akaogi. 1991. Molecular dynamics simulation of the structural and physical properties of the four polymorphs of $TiO_2$. *Mol. Simul.* 6; 239–244.

Mattioli, G., S. B. Dkhil, M. I. Saba, G. Malloci, C. Melis, P. Alippi, F. Filippone, P. Giannozzi, A. K. Thakur, M. Gaceur, O. Margeat et al. 2014. Interfacial engineering of P3HT/ZnO hybrid solar cells using phthalocyanines: A joint theoretical and experimental investigation. *Adv. Energy Mater.* 4; 1301694, 11 pp.

Mattioli, G., C. Melis, G. Malloci, F. Filippone, P. Alippi, P. Giannozzi, A. Mattoni, and A. Amore Bonapasta. 2012. Zinc oxide zinc phthalocyanine interface for hybrid solar cells. *J. Phys. Chem. C* 116; 15439–15448.

Mattoni, A., A. Filippetti, and C. Caddeo. 2017. Modeling hybrid perovskites by molecular dynamics. *J. Phys.: Condens. Matter.* 29; 043001–043023.

Mattoni, A., A. Filippetti, M. Saba, C. Caddeo, and P. Delugas. 2016. Temperature evolution of methylammonium trihalide vibrations at the atomic scale. *J. Phys. Chem. Lett.* 7; 529–535.

Mattoni, A., A. Filippetti, M. I. Saba, and P. Delugas. 2015. Methylammonium rotational dynamics in lead halide perovskite by classical molecular dynamics: The role of temperature. *J. Phys. Chem. C* 119; 17421–17428.

Mattoni, A, M. Ippolito, and L. Colombo. 2007. Atomistic modeling of brittleness in covalent materials. *Phys. Rev. B* 76; 224103.

Meloni, S., T. Moehl, W. Tress, M. Franckevicius, M. Saliba, Y. H. Lee, P. Gao, M. K. Nazeeruddin, S. M. Zakeeruddin, U. Röthlisberger, and M. Grätzel. 2016. Ionic polarization-induced current voltage hysteresis in $CH_3NH_3PbX_3$ perovskite solar cells. *Nat. Commun.* 7; 10334–10339.

Menéndez-Proupin, E., P. Palacios, P. Wahnón, and J. C. Conesa. 2014. Self-consistent relativistic band structure of the $CH_3NH_3PbI_3$ perovskite. *Phys. Rev. B* 90; 045207.

Mettan, X., R. Pisoni, P. Matus, A. Pisoni, J. Jaćimović, B. Náfrádi, M. Spina, D. Pavuna, L. Forró, and E. Horváth. 2015. Tuning of the thermoelectric figure of merit of $CH_3NH_3MI_3$ (M = Pb, Sn) photovoltaic perovskites. *J. Phys. Chem. C* 119; 11506–11510.

Mitzi, D. B. 2001. Templating and structural engineering in organic–inorganic perovskites. *J. Chem. Soc. Dalt. Trans.* 1; 1–12.

Mosconi, E., J. M. Azpiroz, and F. De Angelis. 2015. Ab initio molecular dynamics simulations of methylammonium lead iodide perovskite degradation by water. *Chem. Mater.* 27; 4885–4892.

Mosconi, E., and F. De Angelis. 2016. Mobile ions in organohalide perovskites: Interplay of electronic structure and dynamics. *ACS Energy Lett.* 1; 182–188.

Mosconi, E., C. Quarti, T. Ivanovska, G. Ruani, and F. De Angelis. 2014. Structural and electronic properties of organo-halide lead perovskites: A combined IR-spectroscopy and *ab initio* molecular dynamics investigation. *Phys. Chem. Chem. Phys.* 16; 16137–16144.

Motta, C., F. El-Mellouhi, and S. Sanvito. 2016. Exploring the cation dynamics in lead-bromide hybrid perovskites. *Phys. Rev. B* 93; 235412.

Møller, C. K. 1958. Crystal structure and photoconductivity of caesium plumbohalides. *Nature* 182; 1436.

Neukirch, A. J., W. Nie, J. C. Blancon, K. Appavoo, H. Tsa, M. Y. Sfeir, C. Katan, L. Pedesseau, J. Even, J. J. Crochet, G. Gupta, A. D. Mohite, and S. Tretiak. 2016. Polaron stabilization by cooperative lattice distortion and cation rotations in hybrid perovskite materials. *Nano Lett.* 16; 3809–3816.

NREL (National Renewable Energy Laboratory). Research cell efficiency records. http://www.nrel.gov/ncpv/images/efficiency chart.jpg (Accessed February 1, 2016).

Onoda-Yamamuro, N., T. Matsuo, and H. Suga. 1990. Calorimetric and IR spectroscopic studies of phase transitions in methylammonium trihalogenoplumbates (II). *J. Phys. Chem. Solids.* 51; 1383–1395.

Oosterhout, S. D., M. M. Wienk, S. S. van Bavel, R. Thiedmann, L. J. A. Koster, J. Gilot, J. Loos, V. Schmidt, and R. A. J. Janssen. 2010. The effect of three-dimensional morphology on the efficiency of hybrid polymer solar cells. *Nat. Mater.* 8; 818–824.

O'Reagan, B., and M. Grätzel. 1991. A low-cost, high-efficiency solar cell based on dye-sensitized colloidal $TiO_2$ films. *Nature* 353; 737–740.

Pérez-Osorio, M. A., R. L. Milot, M. R. Filip, J. B. Patel, L. M. Herz, M. B. Johnston, and F. Giustino. 2015. Vibrational properties of the organic–inorganic halide perovskite $CH_3NH_3PbI_3$ from theory and experiment: Factor group analysis, first-principles calculations, and low-temperature infrared spectra. *J. Phys. Chem. C* 119; 25703–25718.

Pisoni, A., J. Jaćimović, O. S. Barišić, M. Spina, R. Gaál, L. Forró, and E. Horváth. 2014. Ultralow thermal conductivity in organic inorganic hybrid perovskite $CH_3NH_3PbI_3$. *J. Phys. Chem. Lett.* 5; 2488–2492.

Poglitsch, A., and D. Weber. 1987. Dynamic disorder in methylammoniumtrihalogenoplumbates (II) observed by millimeter-wave spectroscopy. *J. Chem. Phys.* 87; 6373–6378.

Polman, A, M. Knight, E. C. Garnett, B. Ehrler, and W. C. Sinke. 2016. Photovoltaic materials: Present efficiencies and future challenges. *Science* 352; aad4424–4410.

Ponder, J. W., and D. A. Case. 2003. Force fields for protein simulations. *Adv. Prot. Chem.* 66; 27–85.

Pytte, E. 1972. Theory of perovskite ferroelectrics. *Phys. Rev. B* 5: 3758–3769.

Qian, X., X. Gu, and R. Yang. 2016. Lattice thermal conductivity of organic-inorganic hybrid perovskite $CH_3NH_3PbI_3$. *Appl. Phys. Lett.* 108; 063902.

Quarti, C., G. Grancini, E. Mosconi, P. Bruno, J. M. Ball, M. M. Lee, H. J. Snaith, A. Petrozza, and F. De Angelis. 2014b. The Raman spectrum of the $CH_3NH_3PbI_3$ hybrid perovskite: Interplay of theory and experiment. *J. Phys. Chem. Lett.* 5; 279–284.

Quarti, C., E. Mosconi, J. M. Ball, V. D'Innocenzo, C. Tao, S. Pathak, H. J. Snaith, A. Petrozza, and F. De Angelis. 2016. Structural and optical properties of methylammonium lead iodide across the tetragonal to cubic phase transition: Implications for perovskite solar cells. *Energy Environ. Sci.* 9; 155–163.

Quarti, C., E. Mosconi, and F. De Angelis. 2014a. Interplay of orientational order and electronic structure in methylammonium lead iodide: Implications for solar cells operation. *Chem. Mater.* 26; 6557–6569.

Resta, R. 2003. Ab initio simulation of the properties of ferroelectric materials. *Model. Simul. Mater. Sci. Eng.* 11; R69–R96.

Roiati, V., S. Colella, G. Lerario, L. De Marco, A. Rizzo, A. Listorti, and G. Gigli. 2014. Investigating charge dynamics in halide perovskite-sensitized mesostructured solar cells. *Energy Environ. Sci.* 7; 1889–1894.

Rose J. H., J. R. Smith, F. Guinea, and J. Ferrante. 1984. Universal features of the equation of state of metals. *Phys. Rev. B* 29; 2963–2969.

Saba, M. I., and A. Mattoni. 2014. Effect of thermodynamics and curvature on the crystallinity of P3HT thin films on ZnO: Insights from atomistic simulations. *J. Phys. Chem. C* 118; 4687–4694.

Saba, M. I., and A. Mattoni. 2015. Simulations of oxide/polymer hybrids. *Encyclopedia of Nanotechnology*, pp. 1–13.

Saliba, M., T. Matsui, K. Domanski, J. Y. Seo, A. Ummadisingu, S. M. Zakeeruddin, J. P. Correa-Baena, W. R. Tress, A. Abate, A. Hagfeldt, and M. Grätzel. 2016. Incorporation of rubidium cations into perovskite solar cells improves photovoltaic performance. *Science* 354; 206–209.

Shannon, R. D. 1976. Revised effective ionic radii and systematic studies of interatomic distances in halides and chalcogenides. *Acta Crystallogr. Sect. A* 32; 751–767.

Snyder, G. J., M. E. Badding, and F. J. DiSalvo. 1992. Synthesis, structure, and properties of barium cobalt sulfide ($Ba_6Co_{25}S_{27}$) a perovskite-like superstructure of $Co_8S_6$ and $Ba_6S$ clusters. *Inorg. Chem.* 31; 2107–2110.

Stoumpos, C. C., C. D. Malliakas, and M. G. Kanatzidis. 2013. Semiconducting tin and lead iodide perovskites with organic cations: Phase transitions, high mobilities, and near-infrared photoluminescent properties. *Inorg. Chem.* 52 9019–9038.

Stranks, S. D., and H. J. Snaith. 2015. Metal-halide perovskites for photovoltaic and light-emitting devices. *Nat. Nanotechnol.* 10; 391–402.

Tan, Z. K., R. S. Moghaddam, M. L. Lai, P. Docampo, R. Higler, F. Deschler, M. Price, A. Sadhanala, L. M. Pazos, D. Credgington, F. Hanusch et al. 2014. Bright light-emitting diodes based on organometal halide perovskite. *Nat. Nanotechnol.* 9; 687–692.

Tschauner, O., C. Ma, J. R. Beckett, C. Prescher, V. B. Prakapenka, and G. R. Rossman. 2014. Discovery of bridgmanite, the most abundant mineral in Earth, in a shocked meteorite. *Science* 346; 1100–1102.

Tuckerman, M. 2010. *Statistical mechanics: Theory and molecular simulation.* New York: Oxford University Press.

Umebayashi, T., K. Asai, T. Kondo, and A. Nakao. 2003. Electronic structures of lead iodide based low-dimensional crystals. *Phys. Rev. B* 67; 155405.

Wang, J., R. M. Wolf, J. W. Caldwell, P. A. Kollman, and D. A. Case. 2004. Development and testing of a general amber force field. *J. Comput. Chem.* 25; 1157–1174.

Wang, Y., T. Gould, J. F. Dobson, H. Zhang, H. Yang, X. Yao, and H. Zhao. 2014. Density functional theory analysis of structural and electronic properties of orthorhombic perovskite $CH_3NH_3PbI_3$. *Phys. Chem. Chem. Phys.* 16; 1424–1429.

Wang, M., and S. Lin. 2016. Anisotropic and ultralow phonon thermal transport in organic-inorganic hybrid perovskites: Atomistic insights into solar cell thermal management and thermoelectric energy conversion efficiency. *Adv. Funct. Mater.* 26; 5297–5306.

Wasylishen, R., O. Knop, and J. Macdonald. 1985. Cation rotation in methylammonium lead halides. *Solid State Commun.* 56; 581–582.

Weber, D. 1978. $CH_3NH_3PbX_3$, a Pb(II)-system with cubic perovskite structure. *Zeitschrift für Naturforsch.* 33b; 1443–1445.

Wehrenfennig, C., M. Liu, H. J. Snaith, M. B. Johnston, and L. M. Herz. 2014. Homogeneous emission line broadening in the organo lead halide perovskite $CH_3NH_3PbI_{3-x}Cl_x$. *J. Phys. Chem. Lett.* 5; 1300–1306.

Wells, H. L. 1893. Über die Cäsium- und Kalium- Bleihalogenide. *Zeitschrift für Anorg. und Allg. Chemie.* 3; 195–210.

Wenk, H. R., and A. Bulakh. 2004. *Minerals: Their constitution and origin.* Cambridge: Cambridge University Press.

Wright, A. D., C. Verdi, R. L. Milot, G. E. Eperon, M. A. Perez-Osorio, H. J. Snaith, F. Giustino, M. B. Johnston, and L. M. Herz. 2016. Electron-phonon coupling in hybrid lead halide perovskites. *Nat. Commun.* 7; 11755, 9 pp.

Yi, C., J. Luo, S. Meloni, A. Boziki, N. Ashari-Astani, C. Grätzel, S. M. Zakeeruddin, U. Röthlisberger, and M. Grätzel. 2016. Entropic stabilization of mixed A-cation $ABX_3$ metal halide perovskites for high performance perovskite solar cells. *Energy Environ. Sci.* 9; 656–662.

Yin, W. J., J. H. Yang, J. Kang, Y. Yan, and S.-H. Wei. 2014. Halide perovskite materials for solar cells: A theoretical review. *J. Mater. Chem. A* 3; 8926–8942.

Zhu, H., K. Miyata, Y. Fu, J. Wang, P. P. Joshi, D. Niesner, K. W. Williams, S. Jin, and X. Y. Zhu. 2016. Screening in crystalline liquids protects energetic carriers in hybrid perovskites. *Science* 353; 1409–1413.

Zhu, X. Y., and V. Podzorov. 2015. Charge carriers in hybrid organic–inorganic lead halide perovskites might be protected as large polarons. *J. Phys. Chem. Lett.* 6; 4758–4761.

# CHAPTER 2

# Bulk Structural and Electronic Properties at the Density Functional Theory and Post-Density Functional Theory Level of Calculation

Claudia Caddeo, Alessandro Mattoni,
Alessio Filippetti, and Mao-Hua Du

## CONTENTS

## 2.1 INTRINSIC ELECTRONIC, OPTICAL, AND RECOMBINATION PROPERTIES DESCRIBED BY THE DENSITY FUNCTIONAL THEORY CALCULATIONS

*Claudia Caddeo, Alessandro Mattoni, and Alessio Filippetti*

Istituto Officina dei Materiali, Consiglio Nazionale delle Ricerche,
CNR-IOM SLACS Cagliari, Monserrato, Italy

### 2.1.1 Introduction

The class of hybrid (i.e., mixed organic–inorganic) $MAPbI_{3-x}Cl_x$ perovskites, where MA refers to the methylammonium molecule ($CH_3-NH_3$) located at the A site of the perovskite ($ABX_3$), is today the most promising alternative to Si-based solar cells as efficient and economically feasible photovoltaic devices (Norris and Aydill 2012; NREL 2016). These materials can be processed quite inexpensively from solution at low temperature, creating innovative opportunities for economically viable mass production solar cell technology.

Highly efficient sunlight-converting devices based on $MAPbI_3$ (Bi et al. 2013; Burschka et al. 2013; Cai et al. 2013; Chen et al. 2013; Etgar et al. 2012; Im et al. 2011; Kim et al. 2012; Kojima et al. 2009; Qiu et al. 2013), $MAPbBr_3$ (Burschka et al. 2013; Cai et al. 2013; Edri et al. 2013; Qiu et al. 2013), and $MAPbI_{3-x}Cl_x$ (Ball et al. 2013; Lee et al. 2012) as light absorber and carrier transporter have been massively realized and described in recent publications. All these studies indicate $MAPbI_{3-x}Cl_x$ perovskites as the first emerging material to possess all the required features for efficient solar cell applications: low-cost processing, capability of light absorption and carrier generation, and good ambipolar transport properties (Filippetti et al. 2016). It is thus of great importance to shed light on the reasons for these performances from a fundamental viewpoint.

Theoretical work based on *ab initio* calculations analyzed in detail the intrinsic fundamental characteristics of hybrid perovskites, pertaining to their structural (Ball et al. 2013; Brivio et al. 2013; Colella et al. 2013; Mosconi et al. 2013), electronic (Ball et al. 2013; Brivio et al. 2013; Even et al. 2013; Koutselas et al. 1996; Umari et al. 2014; Wang et al. 2014), and vibrational (Quarti et al. 2014) properties. It is remarkable that the simple Local Density Approximation–Density Functional Theory (LDA-DFT) used for all the results illustrated in the first part of the chapter is accurate enough to describe the fundamental optoelectronic properties of hybrid perovskites, and to capture all the important characteristics, which are the basis of their outstanding performances. More sophisticated aspects of the electronic structure such as spin–orbit coupling or many-body electronic correlation effects were reported by several authors (Brivio et al. 2013; Colella et al. 2013; Even et al. 2013; Wang et al. 2014). However, it was demonstrated that errors due to the lack of spin–orbit and correlation effects largely cancel out, so that the band gap reported by LDA is to date the most accurate achieveble by *ab initio* results, to our knowledge. Also, considering the large increase of computational effort required to include beyond-local contributions, we can conclude that neglecting these effects is a reasonable strategy for the scope of this analysis. A detailed methodological description is provided in Section 2.1.2.

This chapter focuses in particular on the description of the optical absorption and recombination properties of $MAPbI_3$ and $MAPbI_2Cl$. In Section 2.1.3.1 the mechanisms that relate the basic features of these systems to the observed optical properties are described from a fundamental viewpoint. It will be shown with clear evidence that the outstanding optoelectronic performance of these systems is related mainly to the orbital character of the band extrema, with valence band top (VBT) dominated by $Pb^{2+}$ 6$s$ states, and conduction band bottom almost purely of Pb 6$p$ character. These fairly extended orbitals give rise to broadly dispersed, light-mass bands, which clearly favor good electron and hole mobility, and absorbance comparable to that of the best semiconducting absorbers. It follows that, in practice, hybrid perovskites can be totally assimilated to inorganic semiconductors, at least with regard to photoabsorption. The role of the organic molecules is also well described: on the one hand, the molecule acts as an electron donor to the halogen ions, while its electronic levels do not directly couple with the bands derived from the inorganic side, remaining well separated in energy from the band-gap region where photoabsorption occurs. On the other hand, the orientation of the molecules is an important aspect of the resulting atomic structure, as it is tightly coupled with the type of pattern and amplitude of the $PbI_3$ octahedral rotations, which in turn affect electron hopping and bandwidth as well. Consistent with this analysis, the vast majority of optical measurements (D'Innocenzo et al. 2014; Stranks et al. 2013) consistently report that photoabsorption and photoemission are largely dominated by band-to-band recombination involving extended electronic states, while the role of the excitons that are typically dominant in organic absorbers is minor or negligible, at least at room temperature.

In Section 2.1.3.2, we recall the theoretical limit for ideal photoconversion. An analysis of radiative band-to-band recombination rate and minority carrier lifetime is presented, evidencing a remarkable similarity between $MAPbI_3$ and a standard semiconductor such as gallium arsenide (GaAs). The determination of recombination rate is also used to quantify the current-voltage characteristic, and power conversion efficiency in perovskite-based solar cells, in the ideal situation of lack of any current and/or potential loss due to shunt or parasitic resistances or band disalignment between absorber and transporting materials. The calculated efficiency as a function of the perovskite film thickness thus fixes the upper (i.e., intrinsic) limit of the attainable power conversion efficiency for this material. One paramount aspect of these materials is the apparent resilience to be severely affected by extrinsic factors such as as trapping defects or structural disorder, which can typically degrade solar cell performances. For reasons still to be entirely understood, trapping and disorder seem to have a relatively small impact on photoabsorption and recombination, which, at least in some specific carrier concentration range, appear instead determined solely by their basic structural and electronic characteristics.

## 2.1.2 Methodology

LDA is used, with plane-wave basis set and ultrasoft pseudopotentials (Vanderbilt 1992), as implemented in the PWSIC code (PWSIC 2016). This approach accurately describes the atomic structure of the isolated molecule. The variational pseudo-self-interaction correction (VPSIC) (Filippetti and Fiorentini 2009; Filippetti and Spaldin 2003; Filippetti et al. 2011) is also used as a check to quantify electronic correlation effects.

As well known from previous reports, LDA describes with good accuracy the band gap of the perovskites, although this agreement is fortuitous and related to the cancelation of correlation and relativistic (spin–orbit) effects, both large but opposite in sign (Brivio et al. 2014; Umari et al. 2014). However, the correct description of the band gap is a key aspect for the determination of radiative band-to-band recombination; thus the use of the LDA functional is considered the best option for the aims of the analysis. Atomic structures are calculated by force minimization below a threshold of $10^{-3}$ Ry Å$^{-1}$ on each atom. The electronic structure is self-consistently determined using a $6 \times 6 \times 6$ special $k$-point grid and 30 Ry cutoff energy. The Density of States (DOS) is calculated on an extremely dense $20 \times 20 \times 20$ $k$-point grid (critical for the precise determination of the intrinsic charge doping concentration). The optical functions (dielectric function, absorption, and refraction index) are calculated in random-phase approximation on an ultradense mesh of more than 5000 $k$-points and 150 bands. The optical properties of GaAs (for which there is abundance of experimental data) are also calculated as a test case to validate the accuracy of the chosen theoretical framework, and as a scientifically sound term of comparison for the perovskite. However, to treat both materials at the same level of accuracy, for GaAs (whose band gap is notoriously underestimated in LDA) the VPSIC is used, so that the calculated band gap could be fixed at the experimental value. In this way, a very good description of the optical properties of GaAs could be obtained.

The radiative recombination rate (number of events of electron–hole recombination per unit volume and time) can be calculated using the Van Roosbroek–Shockley formulation from the *ab initio* calculated absorption function $\alpha(E)$ and refraction index $n_r(E)$, as (Van Roosbroeck and Shockley 1954):

$$R_{\text{rad}}(T) = \int_{E_G}^{\infty} dE \rho_{\text{ph}}(E,T) \alpha(E) v_{\text{ph}}(E) \qquad (2.1.1)$$

where $v_{\text{ph}}(\varepsilon) = c/n_r(\varepsilon)$ is the photon velocity in the matter, and

$$\rho_{\text{ph}}(E,T) = \frac{8\pi E^2}{(v_{\text{ph}}(E)h)^3} \left( \frac{1}{e^{E/k_B T} - 1} \right) \qquad (2.1.2)$$

is the equilibrium photon distribution. Another important quantity is the intrinsic carrier concentration:

$$n_i(T) = \int_{E_{CBB}}^{\infty} dE n(E) \left( \frac{1}{1+e^{(E-\mu)/k_B T}} \right) = \int_{\infty}^{E_{VBT}} dE n(E) \left( \frac{1}{1+e^{(E-\mu)/k_B T}} \right) \tag{2.1.3}$$

where $n(\varepsilon)$ is the electronic DOS and $\mu = E_F(T)$ the chemical potential. From $n_i$ we can calculate the so-called $B$-factor $\left( B_{rad} = R_{rad}/n_i^2 \right)$, which is more commonly used in the literature than $R_{rad}$.

The calculated radiative recombination rate can be also used to make quantitative estimates of the power conversion efficiency for perovskite-based solar cells. According to the detailed balance principle first introduced by Shockley and Queisser (1961), the total current flowing through the absorber in a solar cell circuit under a bias potential $V_b$ is

$$J_{total} = J_{sc} - J_{dark}(V_b) \tag{2.1.4}$$

where $J_{sc}$ is the short-circuit current flowing through the circuit at zero bias, given by two terms:

$$J_{sc} = J_L - J_{rad}(0) = f \frac{e2\pi}{c^2 h^3} \int_{E_G}^{\infty} dE \left( \frac{E^2}{e^{E/k_B T_S} - 1} \right) P_L(E) - f \frac{e2\pi}{h^3} \int_{E_G}^{\infty} dE \left( \frac{E^2}{e^{E/k_B T} - 1} \right) \frac{P_L(E)}{v_{ph}^2(E)} \tag{2.1.5}$$

$J_L$ is the illumination current due to the solar flux (AM0 black-body radiation) incident on the surface of the absorber, while the second term is the current lost due to radiative recombination at zero bias. The geometric factor $f$ takes into account the solar cone incident on the surface; $T_s = 5759$ K is the sun temperature, and the absorption probability $P_L$ is: $P_L(\varepsilon) = 1 - e^{-\alpha(\varepsilon)L}$, where $L$ is the absorber thickness along the incident light direction. In practice, $J_L$ is largely dominant over the recombination term, which can thus be safely discarded. If zero-bias recombination is ignored, $P_L(E)$ corresponds to the so-called "external quantum efficiency" or "incident photon-to-current conversion efficiency" (IPCE, percentage of incident photons transformed to current) in the hypothesis that the "internal quantum efficiency" (IQE, percentage of electron–hole couples transformed to current) is set to 100% (IPCE = IQE · $P_L$).

The bias-induced "dark" current is given by

$$J_{dark}(V_b) = -\frac{e2\pi}{h^3} \int_{E_G}^{\infty} dE \left( \frac{E^2}{e^{E/k_B T} - 1} \right) \frac{P_L(E)}{v_{ph}(E)^2} + \frac{e2\pi}{h^3} \int_{E_G}^{\infty} dE \left( \frac{E^2}{e^{(E-eV_b)/k_B T} - 1} \right) \frac{P_L(E)}{v_{ph}(E)^2} \tag{2.1.6}$$

where the first term compensates the $V_b = 0$ contribution of the second ($J_{dark}(0) = 0$). The delivered power $P$ is calculated from the current as $P = J \cdot V_b$ and the power conversion efficiency (PCE) as the ratio between the maximum power $P_{max}$ and the incident power $P_{in}$ given by

$$P_{in} = \frac{2\pi f}{c^2 h^3} \int_0^\infty dE \left( \frac{E^3}{e^{E/k_B T_S} - 1} \right) = 135 \, \text{mW cm}^{-2} \qquad (2.1.7)$$

For a more accurate comparison with the experiments, $P_{in}$ and $J_L$ (Equation 2.1.5) are computed using, in place of the AM0 black-body radiation, the numerical AM1.5 solar illumination spectrum, which delivers a $P_{in} = 100 \, \text{mW cm}^{-2}$.

### 2.1.3 Results

#### 2.1.3.1 Structural, Electronic, and Optical Properties of MAPbI₃ and MAPbI₂Cl

2.1.3.1.1 Structure    For all calculations presented in this work, the adopted lattice parameters and symmetries are the reported experimental values (Baikie et al. 2013; Kawamura et al. 2002). At low temperature the structure is orthorhombic *Pnma*, and characterized by a pattern of Glazer octahedral rotations (tilting) of the kind $a^-a^-c^+$. At room temperature, MAPbI₃ is tetragonal *I4/mcm* with octahedral tilting pattern $a^0a^0c^-$. Lattice parameters are $a_0 = 8.80$ Å, $c_0 = 12.68$ Å (Baikie et al. 2013; Filippetti et al. 2015; Kawamura et al. 2002; Stoumpos et al. 2013).

Both symmetries are described by a $\sqrt{2} \times \sqrt{2} \times 2$ formula-formula unit supercell, whose $(a, b)$ plane is rotated by 45° with respect to the $(a_0, b_0)$ plane of the cubic cell (see Figure 2.1.1). A peculiar difficulty when dealing with this hybrid structure is the determination of the MA atomic positions: in absence of a well-defined high-symmetry configuration, the molecule can be seen as a C–N dimer whose center is located near the A-site of the perovskite; however, the dimer orientation remains substantially undetermined to x-ray diffraction. However, the most likely orientations are those with the C–N axis parallel to high-symmetry directions [110], [100], and [111], owing to their large configurational entropy. Furthermore, also the relative orientation of the molecules with respect to each other should be considered. For a complete analysis, all the possible molecular patterns compatible with tetragonal and orthorhombic supercells should be included. Having four molecules per unit cell, this greatly expands the number of nonequivalent molecular configurations to be examined. Because each configuration corresponds to a structural local minimum, the determination of the absolute ground state requires the structural optimization of a large number of configurations assumed as a possible starting point. After the full atomic relaxation, the ground state is selected as the structure with the lowest energy. The resulting structures with minimal energy among both tetragonal and orthorhombic symmetries are displayed in Figure 2.1.1.

These results draw a complex scenario formed by a large number of possible MA sublattice configurations in subtle competition with each other. The most stable molecule orientations are those with C–N dimers nearly parallel to the [110] axis of the supercell; in comparison, configurations with molecules parallel to [100] and [111] directions are

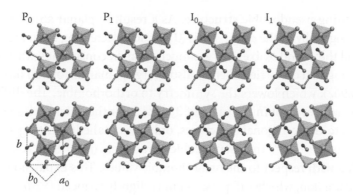

FIGURE 2.1.1 Top view of the four structures analyzed in the text, resulting from our *ab initio* total energy minimization. Those are $\sqrt{2} \times \sqrt{2} \times 2$ supercells with 4 formula units and 48 atoms per cell; the $(a, b)$ plane is 45°-rotated with respect to the $(a_0, b_0)$ plane of the cubic perovskite cell, while $c$ and $c_0$ axis coincide and are orthogonal to the page. $P_0$ and $P_1$ are the orthorhombic *Pnma* structures with the lowest (ground state) and second-lowest (first excited state) energies: $I_0$ and $I_1$ are the corresponding states for the tetragonal *I4/mcm* symmetry. The two (001) layers of the supercell are shown separately, to better distinguish the molecular orientation in the two layers. Top panels: $z = 0$; bottom panels: $z = c/2$. Atomic code: ball-and-stick dimers represent MA molecules (hydrogens are not shown for clarity), while octahedra plus ball-and-stick represent the inorganic frame; I atoms are at the octahedra vertices and center, while Pb atoms are hidden by the I atoms placed at the center of the octahedra.

highly unfavorable and can be discarded. Within the [110] molecular orientation there is still a large variety of possible structures. Two main typologies can be distinguished: parallel ("chain-wise") and orthogonal ("checkboard-wise") MA orientations. In the former the dimers align in parallel forming [110]-oriented parallel chains. $P_0$, the orthorhombic ground state, belongs to this class (see Figure 2.1.1). Note also that the molecules in two consecutive (001) monolayers are flipped by 90° with respect to each other. The first "structurally excited" state $P_1$ belongs instead to the class of orthogonal orientation: now any two adjacent dimers are tilted by 90° along [110], giving rise to a checkboard molecular ordering in the $(a, b)$ plane. Again, the molecules in two adjacent planes are tilted by 90° with respect to each other. For the tetragonal symmetry the situation is reversed: the lowest-energy structure $I_0$ displays a checkboard molecular sublattice, while in the first excited state $I_1$ the dimers are oriented in [110]-parallel chains. At variance with the other cases, for $I_1$ the molecules in two consecutive (001) monolayers of the cell are also parallel to each other.

The difference in energy per formula unit is rather small, on the order of a few tenths of milli-electron volts; thus it is reasonable to assume that the actual equilibrium structure could be best represented as a thermodynamic average [as, e.g., in Filippetti et al. (2015)] over molecular orientations [at finite temperature dynamical approaches could be even more suitable (Filippetti et al. 2015; Mattoni et al. 2015, 2016)]. Consider first the orthorhombic *Pnma* structure. It is useful to introduce the pseudo-cubic lattice parameters $a_{pc} = \sqrt{2}(\Omega/4)^{1/3} = 8.76$ Å and $c_{pc} = 2(\Omega/4)^{1/3} = 12.39$ Å, that is, the lattice parameters of a cubic cell with equivalent volume $\Omega$. As the $(a, b)$ average is 8.70 Å and $c = 12.58$ Å, it follows that the *Pnma* structure is squeezed in the $(a, b)$ plane, and stretched along $c$ with respect to the corresponding pseudocubic structure. As a result of planar shrinking, the Pb–I–Pb angles rotate orthogonally to the $c$ (or $c_0$) axis ($\theta_c$), sizably decreasing from the ideal 180° value. Rotations around the $a_0$ and $b_0$ axes ($\theta_a$) are much smaller, consistent with $c$-axis elongation. On top of the tetragonal distortion there is also a large planar distortion ($a/b = 1.03$, $\sqrt{2}a/c = 0.993$, $\sqrt{2}b/c = 0.962$), which lowers the symmetry to orthorhombic. The difference between $P_0$ and $P_1$ is especially evident in the value of $\theta_a$, which for the latter is close to the ideal 180° value. In relation to the tetragonal structures, only the octahedral tilting around the $c$-axis is present, whereas $\theta_a = 180°$. The pseudocubic parameters ($a_{pc} = 8.85$ Å, $c_{pc} = 12.52$ Å) are slightly expanded with respect to the orthorhombic values. This expansion tends to inhibit the tetragonal distortion, which still persists but is slightly reduced ($\sqrt{2}a/c = 0.98$). The complete list of all the important structural parameters can be found in Filippetti et al. (2014).

To summarize, for both symmetries the octahedra are strongly tilted around the $c$-axis, and weakly tilted (for P states) or not tilted at all (for I states) around the $a_0$ and $b_0$ axes. It follows that the Pb–I–Pb bonds are nearly straight segments along $c_0$, and strongly bent in the ($a_0$, $b_0$) plane. This Pb–I–Pb bending crucially affects the overlap of Pb($6s$, $6p$) and I($5p$) states, and in turn, electron hopping and bandwidth in the plane: of course the larger the tilting, the smaller the Pb–I electron hopping and bandwidth. From this analysis one should expect the Pb-derived bands to be remarkably anisotropic, that is, broader and with lighter masses along $k_z$ than in the ($k_x$, $k_y$) plane, as shown in the next section. Finally, the driving force of the octahedral tilting and its relation to molecular patterns leads back to the concept of the tolerance factor, for which a detailed discussion can be found in Filippetti and Mattoni (2014).

For the Cl-doped perovskite, only the room temperature phase is considered here. Again, the tetragonal $I4/mcm$ symmetry observed by x-ray diffraction experiments is assumed ($\sqrt{2} \times \sqrt{2} \times 2$ 48-atom unit cells), with cell parameters taken from experiments (Lee et al. 2012) ($a_0 = 8.83$ Å, $c_0 = 11.24$ Å). Our optimized structure is shown in Figure 2.1.2.

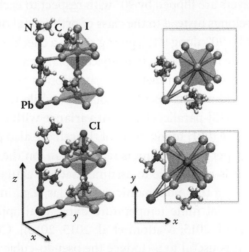

FIGURE 2.1.2 (Top) Three-dimensional view (left) and top view (right) of the unit cell for MAPbI$_3$. (Bottom) Same for MAPbI$_2$Cl. Atomic names are shown in the figure, with small white balls representing hydrogens.

Concerning the difference from MAPbI$_3$, it is apparent that MAPbI$_2$Cl presents smaller octahedral rotations in the (001) plane (see top view), that is, larger planar Pb–I–Pb angles (153°) than MAPbI$_3$ (147°). This can be understood in terms of the different ionic radii for I$^-$ (2.06) and Cl$^-$ (1.67), which is reflected in different tolerance factors (Goldschmidt 1926) (see Chapter 1 for a detailed description) and $c_0/a_0$ ratios (1.02 and 0.9, respectively). Thus, whereas the former is close to cubic, MAPbI$_2$Cl is highly anisotropic, and remarkably squeezed along [001], whereas $a_0$ is similar for the two compounds. Furthermore, in MAPbI$_2$Cl the molecule is also slightly tilted out-of-plane, due to the stronger tetragonal anisotropy.

2.1.3.1.2 Electronic Properties    The band energies for the orthorhombic and tetragonal MAPbI$_3$ perovskite structures illustrated in Figure 2.1.1 are reported in Figure 2.1.3. For clarity, only a small energy interval surrounding the band gap is shown. The conduction

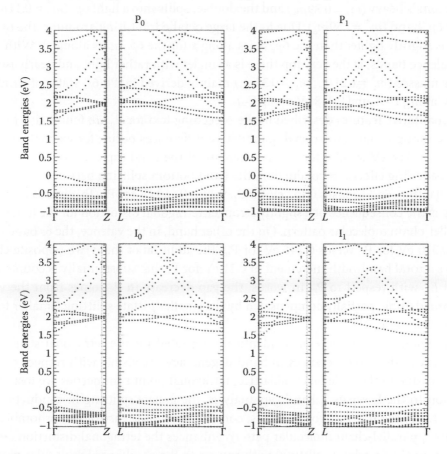

FIGURE 2.1.3    *Ab initio* calculated band energies for MAPbI$_3$ in four different structures (described in the text). The k-points are $Z = [0, 0, \pi/c]$, $L = [\pi/a, \pi/a, 0]$ in the $(a, b, c)$ Cartesian reference of the $\sqrt{2} \times \sqrt{2} \times 2$ supercell; $Z = [0, 0, \pi/(2a_0)]$ and $L = [\pi/(2a_0), 0, 0]$ in the $(1 \times 1 \times 1)$ reference system. Energy zero is fixed at the VBT. (Reprinted with permission from Filippetti, A., P. Delugas, and A. Mattoni. 2014. Radiative recombination and photoconversion of methylammonium lead iodide perovskite by first principles: Properties of an inorganic semiconductor within a hybrid body. *J. Phys. Chem. C* 118; 24843–24853. Copyright © 2014 American Chemical Society.)

bands are dominated by Pb (6$p$) states in the 2 eV–wide region above the conduction band bottom (CBB). Due to the tetragonal distortion, at the $\Gamma$-point they split into a singlet and a doublet; being antibonding, the stretching along the $c$-axis favors the stabilization of $6p_z$ with respect to ($6p_x$, $6p_y$) states. Consider first the $P_0$ band structure: along $\Gamma$–$Z$ (i.e., [001]) the splitting between a dispersed, highly parabolic band derived from the ligand $6p_z$ orbital, of energy 1.58 eV at $\Gamma$, and two flat bands located 163 meV above, and derived from $6p_x$, $6p_y$ states can be observed. The doublet actually splits (by ~50 meV) because of the orthorhombic distortion. Moving to higher energy, another doublet is visible just above 2 eV, which is nothing but the downfolding of the same bands due to the cell doubling along the $c$-axis. Finally, the other dispersed band of $6p_z$ orbital character (there are two Pb per plane in the supercell) is present at 2.3 eV. Looking now at the $\Gamma$–$L$ direction (the [110] direction of the rotated supercell, corresponding to [100] of the cubic lattice), the lowest $6p_z$ band becomes heavy ($m_z^* = 0.89m_e$) and the doublet splits into a light $6p_x$ ($m_x^* = 0.11m_e$) and a heavy $6p_y$ band ($m_y^* = 1.59m_e$). Due to the large octahedral tilting around $c$, the $6p_z$ mass along $z$ is slightly lighter than the $6p_x$ mass along $x$ (or the $6p_y$ mass along $y$). With regard to the valence bands, at the very top there is a singlet with rather light and nearly isotropic effective masses ($m_z^* = 0.21m_e$, $m_x^* = 0.33m_e$) mainly derived by Pb(6$s$) states of bandwidth 0.246 eV along $\Gamma$–$Z$, and 0.318 eV along $\Gamma$–$L$.

The qualitative features described in the preceding text for $P_0$ are found also for the $P_1$ structure though small but crucial quantitative differences occur; for example, the band gap is about 0.08 eV smaller for $P_1$. The decrease of the band gap comes from very subtle band broadening effects: in conduction, the band bottom splitting between the broad $6p_z$ and the flat ($6p_x$, $6p_y$) bands is 0.152 eV for $P_1$, versus 0.163 eV for $P_0$. This increase of $p$-state splitting is the consequence of a slightly stronger tetragonal distortion for the latter, due to its parallel-chain molecular pattern. On the other hand, in the valence, the 6$s$ band dispersion is 0.370 eV for $P_1$, versus 0.246 eV for $P_0$. The addition of these two opposite changes produces a total bandwidth increment of 0.11 eV for $P_1$ that substantially accounts for the reduced $E_G$ with respect to $P_0$: in words, the gap decrease in $P_1$ comes from the valence $s$-state broadening, partially compensated by the smaller $p$-state splitting (smaller tetragonal distortion) of $P_1$.

The previous description can be identically repeated for the tetragonal structures as well. Consider that no differences in band degeneracy are visible with respect to $P_0$ and $P_1$ because, due to the molecular sublattice, tetragonal point symmetries are lost anyway. In relation to the $I_0$ versus $I_1$ band gap analysis, the $p$-state splitting in conduction gives 0.188 eV for $I_0$ and 0.254 eV for $I_1$; that is, consistent with the case of orthorhombic structures, the $I_1$ parallel-chain molecular pattern enhances the tetragonal distortion (and as a consequence the 6$p$ orbital splitting) with respect to the checkboard molecular pattern of $I_0$. The $s$-state bandwidth in the valence is 0.266 eV for $I_0$ and 0.327 eV for $I_1$. The addition of these two band broadening effects (0.127 eV) accounts for the decrease of the $I_1$ gap with respect to $I_0$.

A comparison of band structure calculated for the tetragonal phases of MAPbI$_3$ and MAPbI$_2$Cl is reported in Figure 2.1.4. In the figure, the atomic character for each group of bands is also indicated. The same comparison concerning the calculated orbital-projected

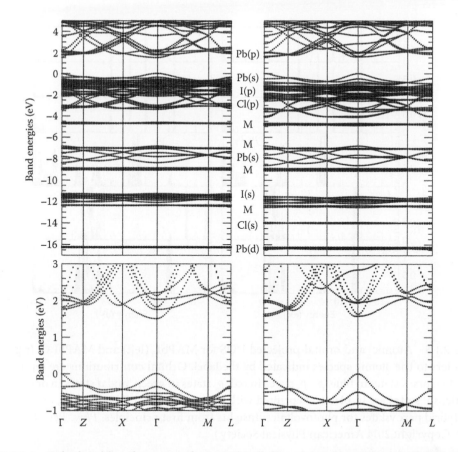

FIGURE 2.1.4   Calculated band energies for MAPbI$_3$ (left) and MAPbI$_2$Cl (right). Labels indicate the atomic and orbital character for each group of bands (M labels molecular states). In the bottom panels the enlargement around the band gap is shown. The *k*-point coordinates (unit of $2\pi/a$) are $X = (1/2, 0, 0)$, $L = (1/2, 1/2, 0)$, $M = (1/2, 1/2, a/2c)$, $Z = (0, 0, a/2c)$. (Reprinted with permission from Filippetti, A., and A. Mattoni. 2014. Hybrid perovskites for photovoltaics: Insights from first principles. *Phys. Rev. B* 89; 125203–125208. Copyright 2014 American Physical Society.)

DOS is reported in Figure 2.1.5; the integrals of the DOS components show that the ionic charges are close to the expected nominal ionic picture: MA$^+$, Pb$^{2+}$, I$^-$, and Cl$^-$. The occupied molecular states are easily recognizable as four perfectly flat horizontal lines in the band structure and four corresponding vertical lines in the DOS; this indicates that the electronic states of the molecule remain well localized in space, without substantial hopping with the surrounding inorganic environment. Moreover, these states are far below the energy gap region that controls the photoconversion properties: the highest occupied MA$^+$ level (HOMO) lies about 5 eV below the VBT. Concerning the empty MA states, they appear 2 eV above the CBB. From these results, a paramount characteristic of these hybrid perovskites emerges: organic and inorganic sides are basically decoupled from the electronic viewpoint; thus the molecules do not interfere with the active region of the perovskite. Its only role, in terms of the electronic properties, is to donate one electron to the surrounding environment.

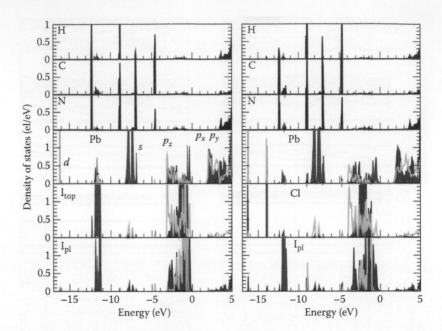

FIGURE 2.1.5   Atomic- and orbital-projected DOS for MAPbI$_3$ (left) and MAPbI$_2$Cl (right). Each panel refers to the atomic species indicated by the label. Orbital contributions are represented by different colors: $s$ states in blue, $p_x$–$p_y$ states in red, $p_z$ states in green, and $d$ states in orange. For the molecule, $sp^3$ states are in black. (Reprinted with permission from Filippetti, A., and A. Mattoni. 2014. Hybrid perovskites for photovoltaics: Insights from first principles. *Phys. Rev. B* 89; 125203–125208. Copyright 2014 American Physical Society.)

The HOMO of the neutral MA is located about 10 eV above the MA$^+$ HOMO, thus 4 eV above the I and Cl $p$-type VBT (see the DOS in Figure 2.1.5). This large difference between the MA and MA$^+$ HOMO levels ultimately induces the electron transfer from the molecule to the inorganic side, and in turn the insulating character of the perovskite. This brings us to the second key element of these results: Pb$^{2+}$ states govern the energy-conversion properties of these systems. The VBT is characterized by hybridized Pb(6$s$)–I/Cl(5$p$) states, while the CBB is of pure Pb(6$p$) character; thus the low-energy absorption region is Pb dominated. The large bandwidth of these states is consistent with the good transport properties argued for both types of carriers.

It is important to emphasize the general validity of this picture: (1) it holds equally for MAPbI$_3$ and MAPbI$_2$Cl; (2) it does not depend on the specific MA orientation or on the details of the atomic structure (other structural local minima with different MA orientations held a similar DOS); and (3) it is not an artifact of LDA (calculations done by the VPSIC approach obtain similar results). Notice, however, that the role of MA is not irrelevant in photoconversion efficiency, as it directly influences the structural properties, which in turn determine the electronic properties as well. For example, a different MA orientation with the dimers parallel to the [100] direction may reduce the octahedral rotations, and in turn increase the Pb($s$)–I($p$) hopping and bandwidth. Despite those large structural effects, the MA$^+$ HOMO energy remains equally far below the band-gap region.

Concerning the calculated band masses, we have seen that $MAPbI_3$ displays light electron mass along $c$ and heavier masses in the $(a, b)$ plane, as a consequence of the anisotropic $a^0a^0c^-$ octahedral tilting pattern. In $MAPbI_2Cl$, on the other hand, the $c/a$ shrinking and the smaller octahedral rotations stabilize the planar Pb $p_x$ and $p_y$ states over the $p_z$ state. Thus the former dominate the energy region between 1.5 and 3 eV (see the DOS in Figure 2.1.5). Now the mass is heavy along $c$ ($m_z = 2.71$), while in the $(a, b)$ plane two light masses (0.21 and 0.59) are found, relative to two Pb $p_x$–$p_y$ bands separated by 0.1 eV. Concerning the valence bands (i.e., $p$-type transport), both compounds show large bandwidth, in turn due to the large $Pb(s)$–$I(p)$ orbital hybridization that characterizes the VBT. Due to the smaller octahedral rotations, $MAPbI_2Cl$ is the system with the lightest masses ($m_x = m_y = 0.51$, $m_z = 0.67$).

In summary, relevant structural differences occur between $MAPbI_3$ and $MAPbI_2Cl$ in turn causing sizable differences in electronic and transport properties. For $n$-type transport, $MAPbI_3$ and $MAPbI_2Cl$ are expected to have good mobility in one and two dimensions, respectively. This result indicates that $MAPbI_2Cl$ may be best suited as an $n$-type conductor in layered heterostructures. On the other hand, both perovskites are expected to be good hole conductors, although $MAPbI_2Cl$ may arguably display higher hole mobility due to slightly lower effective masses.

2.1.3.1.3 Optical Properties    The optical properties of the ground-state orthorhombic $P_0$ structure of $MAPbI_3$ are reported in Figure 2.1.6. The difference among the three Cartesian components indicates a remarkable tetragonal anisotropy. Interestingly, the absorption along $c$ is visibly higher than along the in-plane directions. This may appear in contrast to the results for the effective masses, lower along $c$ and higher in the plane. However, it can be understood considering that, due to the tetragonal distortion (elongation along $c$-axis), the dominant dipole transition at $\Gamma$-point involves $Pb(6s)$ and $Pb(6p_z)$ orbitals as initial and final states. By symmetry considerations, it follows that only the $c$-component of the dipole does not vanish. Clearly, outside $\Gamma$-point, the symmetry cancellation does not apply. According to these results, the perovskite is characterized by high-frequency electronic dielectric constant ~5–5.2 and refraction index ~2.2–2.4.

In Figure 2.1.7 we focus attention on the absorption properties of $MAPbI_3$, showing the calculated 3D-averaged absorption in comparison with the analogous quantity for GaAs, which is a standard reference material for excellent photoabsorption. Notice that at room temperature the excitonic peak is smoothed away [see, e.g., the optical spectra measured in (D'Innocenzo et al. 2014)], thus discarding the excitonic contribution to the absorption is legitimate in the calculation. For $MAPbI_3$ we consider the four previously described structures, that is, $P_0$ and $P_1$ with orthorhombic $Pnma$ symmetry and $I_0$ and $I_1$ with tetragonal $I4/mcm$ symmetry. Notice that their different band gap $E_G$ (reported in Table 2.1.1) produces visible differences in the absorption onset, and as we will see in the next section, in the corresponding recombination properties as well. In particular, the band gap of $P_0$ is very close to the measured value 1.57 eV; thus this structure is likely the closest reference to the actual experimental determination of the absorption. GaAs is also included in the analysis as a relevant term of comparison.

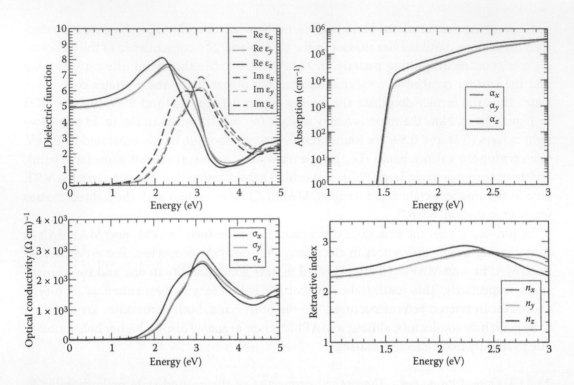

FIGURE 2.1.6 *Ab initio* calculated optical properties for MAPbI$_3$ (P$_0$ structure). (Reprinted with permission from Filippetti, A., P. Delugas, and A. Mattoni. 2014. Radiative recombination and photoconversion of methylammonium lead iodide perovskite by first principles: Properties of an inorganic semiconductor within a hybrid body. *J. Phys. Chem. C* 118; 24843–24853. Copyright © 2014 American Chemical Society.)

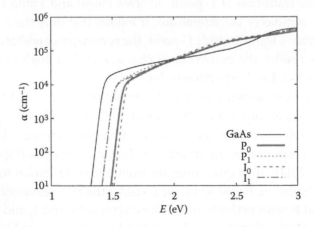

FIGURE 2.1.7 Calculated absorption (averaged over *x*, *y*, and *z* components) for the lowest-energy tetragonal (I$_0$) and orthorhombic state (P$_0$), as well as for the two most energy-competitive states (I$_1$ and P$_1$) relative to different molecular orientations. The calculated absorption for GaAs is also reported for comparison.

TABLE 2.1.1    Calculated Energy Gap ($E_G$), Radiative Recombination Rate ($R_{rad}$), Intrinsic Carrier Density ($n_i$), B-Factor ($B_{rad}$), and Minority Carrier Radiative Lifetime $\tau_{rad}$ for MAPbI$_3$ in Four Different Structures (see text) and GaAs at $T$ = 298 K

| | $E_G$ (eV) | $R_{rad}$ (s$^{-1}$cm$^{-3}$) | $n_i$ (cm$^{-3}$) | $B_{rad}$ (s$^{-1}$cm$^3$) | $\tau_{rad}$ (ns) |
|---|---|---|---|---|---|
| MAPbI$_3$ P$_0$ | 1.584 | 4.9 | $0.78 \times 10^5$ | $0.80 \times 10^{-9}$ | 64 |
| MAPbI$_3$ P$_1$ | 1.508 | 80.5 | $4.14 \times 10^5$ | $0.47 \times 10^{-9}$ | 106 |
| MAPbI$_3$ I$_0$ | 1.614 | 2.0 | $0.36 \times 10^5$ | $1.57 \times 10^{-9}$ | 31 |
| MAPbI$_3$ I$_1$ | 1.510 | 70.6 | $2.33 \times 10^5$ | $1.30 \times 10^{-9}$ | 38 |
| GaAs | 1.424 | 4446 | $1.87 \times 10^6$ | $1.30 \times 10^{-9}$ | 39 |
| GaAs (exp[a]) | 1.424 | 4500 | $1.80 \times 10^6$ | $1.40 \times 10^{-9}$ | 36 |

Source:    Reprinted with permission from Filippetti, A., P. Delugas, and A. Mattoni. 2014. Radiative recombination and photoconversion of methylammonium lead iodide perovskite by first principles: Properties of an inorganic semiconductor within a hybrid body. *J. Phys. Chem. C* 118; 24843–24853. Copyright © 2014 American Chemical Society.

Note:    The radiative lifetime is calculated for a reference doping $n = 2 \times 10^{16}$ cm$^{-3}$, to be compared with an experimental value for GaAs at the same doping given in the literature (Sell and Casey 1974).

[a]    Sell and Casey (1974).

For energies higher than $E_G$ the calculated absorption $\alpha(E)$ compares well with a recent absorption measurement (PWSIC 2016), which serves as a useful validation of the method. In agreement with the experiment, the calculated $\alpha(E)$ at the onset is ~$10^4$ cm$^{-1}$ and grows linearly (on a logarithmic scale) with the energy. Nicely, a slope change is visible near $E$ = 2.5 eV, reproducing a similar feature of the experimental spectrum. The MAPbI$_3$ and GaAs absorption curves are similar in a wide energy range, except at the absorption onset where the difference in $E_G$ is well visible. For energies lower than $E_G$ the absorption can be modeled with an Urbach exponential tail (Urbach 1953), following the finding of PWSIC (2016), which estimates an Urbach energy (15 meV) similar to that of GaAs (Blakemore 1982; Sturge 1962).

A comparison between calculated optical properties for MAPbI$_3$ and MAPbI$_2$Cl is reported in Figure 2.1.8. For both systems, a large tetragonal anisotropy in both dielectric function ($\varepsilon$) and refractive index ($n$) can be observed. The two systems show different anisotropy because for MAPbI$_3$, the reflectivity and absorption ($\alpha$) are larger along the z direction, while for MAPbI$_2$Cl the planar components prevail, consistent with the picture emerging from the electronic properties. Regarding absorption, the most striking result is the large calculated value (up to 0.05–0.06 nm$^{-1}$) in the 400–800 nm wavelength range for both perovskites. Absorption values for some typical semiconductors (GaAs, Ge, InP) are also reported (Palik 1985). Both perovskites compare very well with these typical optical absorbers, not only in the infrared region (above 700 nm), but even in the visible range from 500 nm and beyond, that is, in the spectral region of maximum sunlight irradiance. This result is favored by the perovskite band structure: the band gap is direct and occurs at the $\Gamma$ point, thus a very robust absorption process can start immediately at the onset of the fundamental interband transition (1.51 eV, corresponding to 820 nm for MAPbI$_2$Cl).

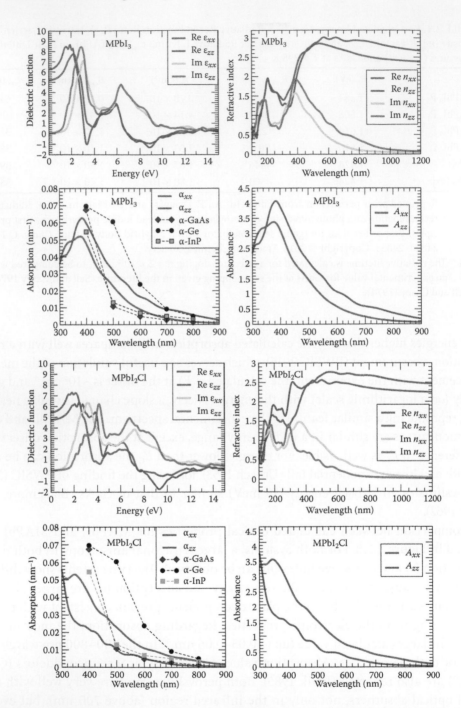

FIGURE 2.1.8 Calculated optical properties for MAPbI$_3$ and MAPbI$_2$Cl: real and imaginary dielectric function ($\varepsilon$), refractive index ($n$), absorption coefficient ($\alpha$), and absorbance ($A$) through a perovskite film of 150 nm (see text). For each quantity, planar ($xx = yy$) and orthogonal ($zz$) components are shown separately. Absorption for typical semiconductors is reported for comparison (Palik 1985). M is methylammonium (MA). (Reprinted with permission from Filippetti, A., and A. Mattoni. 2014. Hybrid perovskites for photovoltaics: Insights from first principles. *Phys. Rev. B* 89; 125203–125208. Copyright 2014 American Physical Society.)

To quantify the performance of the perovskites in actual devices and have a meaningful term of comparison with the experiments, the absorbance $A^*$ is calculated over a perovskite film thickness of $L = 150$ nm, and compared to the one reported in the experiments of Lee et al. (2012). The curves calculated for $A_{zz}$ for MAPbI$_3$ and $A_{xx} = A_{yy}$ for MAPbI$_2$Cl are both in reasonable agreement with the measurement of Lee et al. (2012). For example, they measure $A = 1.8$ at 500 nm, versus the calculated values of $A = 2.2$ and 1.4 for MAPbI$_3$ and MAPbI$_2$Cl, respectively. This is a quite satisfying agreement, considering that calculations refer to intrinsic bulk systems, while actual samples may be affected by structural and chemical inhomogeneity or disorder. Notice that $A = 2.2$ and 1.4 correspond to very large relative absorptions I/I$_0$ = $10^{-A}$ (99.3% and 96%, respectively).

### 2.1.3.2 Radiative Recombination and Photoconversion Limit of MAPbI$_3$

2.1.3.2.1 Recombination Rate and Lifetime  The recombination properties of MAPbI$_3$ calculated according to Van Roosbroek and Shockley at $T = 298$ K are reported in Table 2.1.1 for the four perovskite structures and for GaAs. Filippetti et al. (2014) report the same quantities as a function of $T$; $R_{rad}$ is controlled mainly by $E_G$ and by the absorption coefficient at the onset of the interband transition; for the examined perovskites $R_{rad}$ increases by more than one order of magnitude while going from the ground states ($I_0$, $P_0$) to the excited states ($I_1$, $P_1$), as a consequence of the smaller $E_G$ of the latter ones. Because both $R_{rad}$ and $n_i$ decrease exponentially with $E_G$, these two large and opposite variations mostly compensate in $B_{rad}$ which is thus much less $E_G$-dependent than $R_{rad}$ or $n_i^2$ separately. However, sensible differences in $B_{rad}$ for the examined perovskites still remain (see Table 2.1.1); typically, factor $n_i^2$ overcompensates the difference in $R_{rad}$, so that the larger $B_{rad}$ occurs for the system with larger $E_G$.

By assuming quasi-equilibrium conditions $\left(np \sim n_i^2\right)$, the radiative lifetime (also called minority lifetime) can be determined as $\tau_{rad} = (B_{rad}\, n)^{-1}$. In Table 2.1.1 the lifetimes for the perovskite are compared with that of GaAs at a doping concentration $n = 2 \times 10^{16}$ cm$^{-3}$ for which an experimental reference (Sell and Casey 1974) is available for validation. For this doping level the radiative lifetime of the perovskites is in the range from a few tens up to ~100 nanoseconds, thus similar or larger than in GaAs.

In the work of Stranks et al. (2013) the reported lifetime is 10 ns for pure MAPbI$_3$ and 280 ns for MAPbI$_{3-x}$Cl$_x$. This huge difference may appear surprising because the amount of Cl actually embodied in the perovskite is believed to be only a few percent, thus a fraction that can hardly change the intrinsic properties of the perovskite. To better understand the PL measurements, it is interesting to link the calculated $\tau_{rad}$ as a function of doping at room temperature with the measured lifetimes. The MAPbI$_3$ lifetime intercepts the calculated $\tau_{rad}$ for a doping interval between $n \sim 7 \times 10^{16}$ cm$^{-3}$ and $n \sim 2 \times 10^{17}$ cm$^{-3}$ [see Filippetti et al. (2014)], while the Cl-doped lifetime matches $\tau_{rad}$ in the range between $n \sim 2\times 10^{15}$ cm$^{-3}$ and $n \sim 8 \times 10^{15}$ cm$^{-3}$. Thus, a possible understanding of the PL measurements is to think of them as corresponding to two different doping levels, the lowest of which corresponding to the Cl-doped perovskite. The scenario implied in this hypothesis is that inclusion of Cl in

---

* $A = \alpha L/2.303$, where $L$ is the film thickness, and the factor 2.303 converts from the natural to the common logarithm.

solution somehow helps to clean up the processing, allowing the production of perovskites with a much reduced (one or two orders of magnitude smaller) amount of native defects. Clearly, a limit of the lifetime estimate is that nonradiative contributions are not included. However, there are indications in the literature that lead iodide perovskites are not easily affected by deep electronic traps, as typical defects such as molecule or cation vacancies produce shallow acceptors and donors, respectively (Talapin and Murray 2005) whereas isovalent Cl–I substitutions are ineffective with regard to the introduction of electronic traps (Colella et al. 2013). Molecular fragments are the only reported notable exception that can induce localized levels in the gap. However, they are characterized by relatively high formation energies (Delugas et al. 2015). Thus, the assumption of electron–hole recombination dominated (or primarily determined) by radiative processes should be seen as a sound possibility.

2.1.3.2.2 Current-Voltage Characteristics and Power Conversion Efficiency    The calculation of current-voltage characteristics according to Shockley–Queisser as a function of the thickness ($L$) of the absorber is reported in Figure 2.1.9 for the $P_0$ structure of MAPbI$_3$. The thickness is a crucial parameter, which is reported to vary in the range of a few hundred (100–400) nm, in order not to overcome the diffusion length. Thus, the calculations are performed on a set of different thicknesses in this range.

The calculated $J_{sc}$ substantially matches the values reported in literature, ranging from 15 mA cm$^{-2}$ up to about 20–21 mA cm$^{-2}$ (Liu and Kelly 2014; Liu et al. 2013). As an example, $J_{sc}$ calculated for $L$ = 300 and $L$ = 400 is in satisfactory agreement with the very high 21.5 mA cm$^{-2}$ obtained in Liu and Kelly (2014) for a 330-nm-thick perovskite layer, vapor-deposited within a planar heterojunction solar cell (Liu et al. 2013). According to these results, going above these $J_{sc}$ values would require excessively large thicknesses. As a useful indication, in Figure 2.1.9 the calculation for $L$ = 800 nm is also reported, for which $J_{sc}$ becomes close to the maximum (100% absorption) limit of 25.7 mA cm$^{-2}$. The good agreement with the experimental $J_{sc}$ is an indication that radiative recombination is the dominant mechanism of electron–hole recombination in the perovskite. The second fundamental parameter for the solar cell performance is the open-circuit potential ($V_{oc}$), that is, the value for which short-circuit and bias currents cancel. The obtained value is $V_{oc}$ ~ 1.28 eV, against typical measured values around 1.0 eV [e.g., a value of $V_{oc}$ =1.05 eV was found in solar cells with organic transport layers (Liu et al. 2013, 2014; Sturge 1962) and 1.07 eV in vapor-deposited perovskite films (Malinkiewicz et al. 2014)]. This overestimation is rather obvious, considering the ideality of the calculation, which does not include current leakage due to shunt resistance, series resistances, and most importantly, the voltage loss due to band offsets between perovskite and electron–hole transporters (Schulz et al. 2014). A direct consequence of larger $V_{oc}$ is that the calculated PCEs are larger than the presently measured efficiencies [to the best of our knowledge the record-high value reported in the literature is 22% (NREL 2016)]. Looking at the calculated $P_{out}$ in Figure 2.1.9 (since $P_{in}$ = 100 mW cm$^{-2}$, $P_{out}$ is also the PCE in percent), a PCE = 16% is obtained for $L$ = 100 nm, 21% for $L$ = 200 nm, and more than 23% for $L$ = 300 nm.

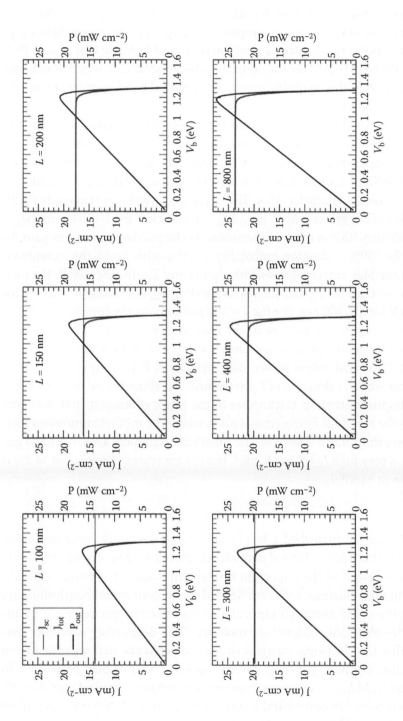

FIGURE 2.1.9 Current-voltage characteristic (Equations 2.1.5 through 2.1.7) of the solar cell with $P_0$ perovskite as absorber. Each panel shows short-circuit ($J_{sc}$) and total ($J_{tot}$) current densities (current per unit surface), as well as output power density ($P_{out}$) (power per unit surface) for a given perovskite layer thickness ($L$). For each calculation the considered power input is $P_{in} = 100$ mW cm$^{-2}$, that is, the power under maximum AM1.5 solar illumination spectrum; because PCE $= P_{out}/P_{in}$, $P_{out}$ also represents the percent efficiency. (Reprinted with permission from Filippetti, A., P. Delugas, and A. Mattoni. 2014. Radiative recombination and photoconversion of methylammonium lead iodide perovskite by first principles: Properties of an inorganic semiconductor within a hybrid body. *J. Phys. Chem. C* 118; 24843–24853. Copyright © 2014 American Chemical Society.)

We remark that these PCE values represent the "intrinsic" limit to the PCE, that is, the efficiency of the material itself and due solely to the fundamental absorption and recombination properties of the absorber. Although probably not reachable in practice (as shunt and series resistances could never be made completely negligible), they nevertheless represent a useful reference point for the device performance, and indicate that a large PCE improvement should still be expected for the perovskite-based solar cell, in terms of better morphological characteristics, device optimization, and optimal band alignment with the carrier transporting materials.

Another fundamental aspect for what concerns the device functionality is the energy gap. To explore its impact on the perovskite efficiencies, the currents and powers for $P_0$ and $P_1$ perovskites can be compared (see Filippetti et al. 2014). The gap of $P_1$ is about 0.08 eV lower than that of $P_0$ (see Table 2.1.1), a difference large enough to produce remarkable differences in the current-voltage characteristics. The current $J_{sc}$ increases remarkably with the gap closure and this effect grows with $L$ (i.e., $\Delta J_{sc} \sim 1$ mA cm$^{-2}$ for $L = 100$ nm, $\Delta J_{sc} \sim 2$ mA cm$^{-2}$ for $L = 400$ nm). This is easily understood, as the maximum current gain for decreasing $E_G$ occurs for 100% absorption probability. On the other hand, the recombination rate, and in turn the bias current, grows with decreasing $E_G$. Indeed, for $P_1$ the total current has a faster decay than for $P_0$, and the corresponding $V_{oc}$ decreases to ~1.25 eV for $L = 100$ nm, and 1.20 eV for $L = 400$ nm. For the $L = 100$ nm sample, the combined increase of $J_{sc}$ and decrease of $V_{oc}$ nearly compensate in the resulting power, which thus remains substantially constant for the two perovskites. Only for $L = 400$ nm the gain in $J_{sc}$ overcomes the decrease of $V_{oc}$, so that output power and efficiency of $P_1$ becomes larger than in $P_0$ by about 1%. In other words, a decrease of $E_G$ in the order of about 0.1 eV does not bring significant efficiency improvements for thicknesses in the range of interest (100–300 nm). This is consistent with the results of Eperon et al. (2014), where the methylammonium molecule is replaced by formamidinium. The molecule substitution induces a smaller band gap (1.48 eV) and, in turn, a very high $J_{sc} = 23$ mA cm$^{-2}$, but the measured efficiency of 14.2% is not sizably higher than in MAPbI$_3$.

## 2.1.4 Conclusions

In this part of the chapter we furnished a brief overview of the results obtained using *ab initio* calculations for the study of hybrid perovskites, giving evidence of the capability of this approach to unveil some of the important fundamental aspects of these materials, especially in relation to their intrinsic behavior. Several major conceptual conclusions are nicely highlighted by the calculations: (1) Optical and transport properties are substantially independent of the electronic states of the molecule. This decoupling is a direct consequence of the fact that the electronic energies of the molecule are well separated from the important region (i.e., the band gap) that governs optical and transport properties. In other words, the molecule, MA, acts as a pure donor whose electronic levels do not directly couple with the bands involved in optical and transport properties. (2) Nevertheless, albeit indirectly, the orientation of MA can significantly influence optics and transport, as it largely affects the amplitude of the octahedral rotations, which in turn governs Pb–I–Pb hopping and ultimately the Pb bandwidth. (3) The $Pb^{2+}$ ion plays a paramount role in the optical properties of these systems because the band gap (direct at $\Gamma$ point) opens between Pb(6s)–I(5p) valence states and a conduction band manifold of exclusively Pb(6p) character. Thus the delocalized character of these orbitals is the basis of the large optical absorption in both the visible and infrared range. Not by chance, this fortunate presence of delocalized Pb s states at the VBT is shared by other plumbates (PbS, PbSe, PbTe), popularly regarded as promising vehicles of photoconversion efficiency (Debnath et al. 2011; Jeong et al. 2012; Piliego et al. 2012; Talapin and Murray 2005). (4) Electron–hole recombination is a key quantity to provide a solid conceptual background to the interpretation of PL lifetime measurements, and an essential factor in the accurate determination of what we call "intrinsic" power conversion efficiency: strong absorption coefficient, small radiative recombination rate, long radiative recombination lifetimes (on the order of $10^3$ ns for $n = 10^{15}$ cm$^{-3}$ doping concentration) and large short-circuit currents all contribute to determine virtually huge power conversion efficiencies, ranging from 16% for $L = 100$ nm perovskite thickness, up to 25% for $L = 400$ nm.

Ultimately, these excellent values may be linked to peculiar fundamental features of these materials: although "hybrids" in their capability of being processed in solutions, they are nevertheless fully "inorganic" for in terms of optical, photoconversion, and transport properties, as the electronic states in the important energy region around the band gap are represented entirely by extended states (bands) derived by Pb and I orbitals. One could visualize the perovskite in this region as a GaAs with inverted band character (mainly s-states in the valence, p-states in conduction) and increased band gap. The increase of band gap largely reduces both the recombination rate and the intrinsic carrier concentration; these two changes mainly compensate in the radiative lifetime, which is thus similar in the perovskite and in GaAs; most importantly, the gap increase is not detrimental for power conversion efficiency, as, for the thicknesses of interest, the decrease of short-circuit current due to the higher band gap and lower illumination is compensated by lower dark current and higher open-circuit voltage.

Finally, it is useful to emphasize the key difference of these perovskites from the much more popular oxide perovskites (e.g., titanates, manganites, or nickelates): in the latter, the band gap opens between fairly localized O(2p) valence and B-site (3d) conduction bands. As a consequence, the charge mobility at the band edges is not particularly satisfying, and the typically small bandwidth does not play favorably for absorption in the useful spectral region.

# REFERENCES

Baikie, T., Y. Fang, J. M. Kadro, M. Schreyer, F. Wei, S. G. Mhaisalkar, M. Grätzel, and T. J. White. 2013. Synthesis and crystal chemistry of the hybrid perovskite $(CH_3NH_3)PbI_3$ for solid-state sensitised solar cell applications. *J. Mater. Chem. A* 1; 5628–5641.

Ball, J. M., M. M. Lee, A. Hey, and H. J. Snaith. 2013. Low-temperature processed meso-superstructured to thin-film perovskite solar cells. *Energy Environ. Sci.* 6; 1739–1743.

Bi, D., L. Yang, G. Boschloo, A. Hagfeldt, and E. M. J. Johansson. 2013. Effect of different hole transport materials on recombination in $CH_3NH_3PbI_3$ perovskite-sensitized mesoscopic solar cells. *J. Phys. Chem. Lett.* 4; 1532–1536.

Blakemore, J. S. 1982. Semiconducting and other major properties of gallium arsenide. *J. Appl. Phys.* 53; R123–R181.

Brivio, F., K. T. Butler, A. Walsh, and M. van Schilfgaarde. 2014. Relativistic quasiparticle self-consistent electronic structure of hybrid halide perovskite photovoltaic absorbers. *Phys. Rev. B* 89; 155204–155206.

Brivio, F., A. B.Walker, and A. Walsh. 2013. Structural and electronic properties of hybrid perovskites for high-efficiency thin-film photovoltaics from first-principles. *APL Mater.* 1; 042111–042115.

Burschka, J., N. Pellet, S.-J. Moon, R. Humphry-Baker, P. Gao, Md. K. Nazeeruddin, and M. Grätzel. 2013. Sequential deposition as a route to high-performance perovskite-sensitized solar cells. *Nature* (London) 499; 316–319.

Cai, B., Y. Xing, Z. Yang, W.-H. Zhang, and J. Qiu. 2013. High performance hybrid solar cells sensitized by organolead halide perovskites. *Energy Environ. Sci.* 6; 1480–1485.

Chang, Y. H., C. H. Park, and K. Matsuishi. 2004. First-principles study of the structural and the electronic properties of the lead-halide-based inorganic-organic perovskites $(CH_3NH_3)PbX_3$ and $CsPbX_3$ (X = Cl, Br, I). *J. Korean Phys. Soc.* 44; 889–893.

Chen, H., X. Pan, W. Liu, M. Cai, D. Kou, Z. Huo, X. Fang, and S. Dai 2013. Efficient panchromatic inorganic–organic heterojunction solar cells with consecutive charge transport tunnels in hole transport material. *Chem. Commun.* 49; 7277–7279.

Colella, S., E. Mosconi, P. Fedeli, A. Listorti, F. Gazza, F. Orlandi, P. Ferro, T. Besagni, A. Rizzo, G. Calestani, G. Gigli et al. 2013. $MAPbI_{3-x}Cl_x$ mixed halide perovskite for hybrid solar cells: The role of chloride as dopant on the transport and structural properties. *Chem. Mater.* 25; 4613–4618.

Debnath, R., O. Bakr, and E. H. Sargent. 2011. Solution-processed colloidal quantum dot photovoltaics: A perspective. *Energy Environ. Sci.* 4; 4870–4881.

Delugas, P., A. Filippetti, and A. Mattoni. 2015. Methylammonium fragmentation in amines as source of localized trap levels and the healing role of Cl in hybrid lead-iodide perovskites. *Phys. Rev. B* 92; 45301–45311.

D'Innocenzo, V., G. Grancini, M. J. P. Alcocer, A. R. S. Kandada, S. D. Stranks, M. M. Lee, G. Lanzani, H. J. Snaith, and A. Petrozza. 2014. Excitons versus free charges in organo-lead tri-halide perovskites. *Nat. Commun.* 5; 3586.

Edri, E., S. Kirmayer, D. Cahen, and G. Hodes. 2013. High open-circuit voltage solar cells based on organic–inorganic lead bromide perovskite. *J. Phys. Chem. Lett.* 4; 897–902.

Eperon, G. E., S. D. Stranks, C. Menelaou, M. B. Johnston, L. M. Herz, and H. J. Snaith. 2014. Formamidinium lead trihalide: A broadly tunable perovskite for efficient planar heterojunction solar cells. *Energy Environ. Sci.* 7; 982–988.

Etgar, L., P. Gao, Z. Xue, Q. Peng, A. K. Chandiran, B. Liu, Md. K. Nazeeruddin, and M. Grätzel. 2012. Mesoscopic $CH_3NH_3PbI_3$/$TiO_2$ heterojunction solar cells. *J. Am. Chem. Soc.* 134; 17396–17399.

Even, J., L. Pedesseau, J.-M. Jancu, and C. Katan. 2013. Importance of spin-orbit coupling in hybrid organic/inorganic perovskites for photovoltaic applications, *J. Phys. Chem. Lett.* 4; 2999–3005.

Filippetti, A., P. Delugas, and A. Mattoni. 2014. Radiative recombination and photoconversion of methylammonium lead iodide perovskite by first principles: Properties of an inorganic semiconductor within a hybrid body. *J. Phys. Chem. C* 118; 24843–24853.

Filippetti, A., P. Delugas, M. I. Saba, and A. Mattoni. 2015. Entropy-suppressed ferroelectricity in hybrid lead-iodide perovskites. *J. Phys. Chem. Lett.* 6; 4909–4915.

Filippetti, A.. and V. Fiorentini. 2009. A practical first-principles band-theory approach to the study of correlated materials. *Eur. Phys. J. B* 71; 139–183.

Filippetti, A., and A. Mattoni. 2014. Hybrid perovskites for photovoltaics: Insights from first principles. *Phys. Rev. B* 89; 125203–125208.

Filippetti, A., A. Mattoni, C. Caddeo, M. I. Saba, and P. Delugas. 2016. Low electron-polar optical phonon scattering as a fundamental aspect of carrier mobility in methylammonium lead halide $CH_3NH_3PbI_3$ perovskites. *Phys. Chem. Chem. Phys.* 18; 15352–15362.

Filippetti, A., C. D. Pemmaraju, S. Sanvito, P. Delugas, D. Puggioni, and V. Fiorentini. 2011. Variational pseudo-self-interaction-corrected density functional approach to the ab initio description of correlated solids and molecules. *Phys. Rev. B* 84; 195127-1–195127-22.

Filippetti, A., and Spaldin, N. A. 2003. Self-interaction-corrected pseudopotential scheme for magnetic and strongly-correlated systems. *Phys. Rev. B* 67; 125109–125115.

Goldschmidt, V. M. 1926. Die Gesetze der Krystallochemie. *Naturwissenschaften* 14; 477–485.

Im, J.-H., C.-R. Lee, J.-W. Lee, S.-W. Park, and N.-G. Park. 2011. 6.5% efficient perovskite quantum-dot-sensitized solar cell. *Nanoscale* 3; 4088–4093.

Jeong, K. S., J Tang, H. Liu, J. Kim, A. W. Schaefer, K. Kemp, L. Levina, X. Wang, S. Hoogland, R. Debnath, L. Brzozowski et al. 2012. Enhanced mobility-lifetime products in PbS colloidal quantum dot photovoltaics. *ACS Nano* 6; 89–99.

Kawamura, Y., H. Mashiyama, and K. Hasebe. 2002. Structural study on cubic–tetragonal transition of $CH_3NH_3PbI_3$. *J. Phys. Soc. Jpn.* 71; 1694–1697.

Kim, H.-S., C.-R. Lee, J.-H. Im, K.-B. Lee, T. Moehl, A. Marchioro, S.-J. Moon, R. Humphry-Baker, J.-H. Yum, J. E. Moser, M. Grätzel, and N.-G. Park. 2012. Lead iodide perovskite sensitized all-solid-state submicron thin film mesoscopic solar cell with efficiency exceeding 9%. *Sci. Rep.* 2; 591–597.

Kojima, A., K. Teshima, Y. Shirai, and T. Miyasaka. 2009. Organometal halide perovskites as visible-light sensitizers for photovoltaic cells. *J. Am. Chem. Soc.* 131; 6050–6051.

Koutselas, I. B., L. Ducasse, and G. C. Papavassiliou. 1996. Electronic properties of three- and low-dimensional semiconducting materials with Pb halide and Sn halide units. *J. Phys.: Condens. Matter* 8; 1217–1227.

Lee, M. M., J. Teuscher, T. Miyasaka, T. N. Murakami, and H. J. Snaith. 2012. Efficient hybrid solar cells based on meso-superstructured organometal halide perovskites. *Science* 338; 643–647.

Liu, M., M. B. Johnston, and H. J. Snaith. 2013. Efficient planar heterojunction perovskite solar cells by vapour deposition. *Nature* 501; 395–398.

Liu, D., and T. L. Kelly. 2014. Perovskite solar cells with a planar heterojunction structure prepared using room-temperature solution processing techniques. *Nat. Photonics* 8; 133–138.

Malinkiewicz, O., A. Yella, Y. H. Lee, G. M. Espallargas, M. Grätzel, Md. K. Nazeeruddin, and H. J. Bolink. 2014. Perovskite solar cells employing organic charge-transport layers. *Nat. Photonics* 8; 128–132.

Mattoni, A., A. Filippetti, M. Saba, C. Caddeo, and P. Delugas. 2016. Temperature evolution of methylammonium trihalide vibrations at the atomic scale. *J. Phys. Chem. Lett.* 7; 529–535.

Mattoni, A., A. Filippetti, M. I. Saba, and P. Delugas. 2015. Methylammonium rotational dynamics in lead halide perovskite by classical molecular dynamics: The role of temperature. *J. Phys. Chem. C* 119; 17421–17428.

Mosconi, E., A. Amat, M. K. Nazeeruddin, M. Grätzel, M., and F. De Angelis. 2013. First-principles modeling of mixed halide organometal perovskites for photovoltaic applications. *J. Phys. Chem. C* 117; 13902–13913.

Norris, D. J., and E. S. Aydil. 2012. Getting more from solar cells. *Science* 338; 625–626.

NREL (National Renewable Energy Laboratory. 2016. NREL Research cell efficiency records. http://www.nrel.gov/ncpv/images/efficiency chart.jpg (Accessed January 17, 2017).

Palik, E. D. 1985. *Handbook of optical constants of solids.* New York: Academic Press.

Piliego, C., M. Manca, R. Kroon, M. Yarema, K. Szendrei, M. R. Andersson, W. Heiss, and M. A. Loi. 2012. Charge separation dynamics in a narrow band gap polymer–PbS nanocrystal blend for efficient hybrid solar cells. *J. Mater. Chem.* 22; 24411–24416.

PWSIC. 2016. The shareware PWSIC code is a modified version of the elder PWSCF code included in the ESPRESSO package; PWSIC is developed and maintained by the CNR-IOM Group in Cagliari.

Qiu, J., Y. Qiu, K. Yan, M. Zhong, C. Mu, H. Yan, and S. Yang. 2013. All-solid-state hybrid solar cells based on a new organometal halide perovskite sensitizer and one-dimensional $TiO_2$ nanowire arrays. *Nanoscale* 5; 3245–3248.

Quarti, C., G. Grancini, E. Mosconi, P. Bruno, J. M. Ball, M. M. Lee, H. J. Snaith, A. Petrozza, and F. De Angelis. 2014. The Raman spectrum of the $CH_3NH_3PbI_3$ hybrid perovskite: Interplay of theory and experiment. *J. Phys. Chem. Lett.* 5; 279–284.

Schulz, P., E. Edri, S. Kirmayer, G. Hodes, D. Cahen, and A. Kahn. 2014. Interface energetics in organo-metal halide perovskite-based photovoltaic cells. *Energy Environ. Sci.* 7; 1377–1381.

Sell, D. D., and H. C. Casey. 1974. Optical absorption and photoluminescence studies of thin GaAs layers in GaAs–$Al_xGa_{1-x}As$ double heterostructures. *J. Appl. Phys.* 45; 800.

Shockley, W., and H. J. Queisser. 1961. Detailed balance limit of efficiency of P-N junction solar cells. *J. Appl. Phys.* 32; 510–519.

Stoumpos, C. C., C. D. Malliakas, and M. G. Kanatzidis. 2013. Semiconducting tin and lead iodide perovskites with organic cations: Phase transitions, high mobilities, and near-infrared photoluminescent properties. *Inorg. Chem.* 52; 9019–9038.

Stranks, S. D., G. E. Eperon, G. Grancini, C. Menelaou, M. J. P. Alcocer, T. Leijtens, L. M. Herz, A. Petrozza, and H. J. Snaith. 2013. Electron-hole diffusion lengths exceeding 1 micrometer in an organometal trihalide perovskite absorber. *Science* 342; 341–344.

Sturge, M. 1962. Optical absorption of gallium arsenide between 0.6 and 2.75 eV. *Phys. Rev.* 127; 768–773.

Talapin, D. V., and C. B. Murray. 2005. PbSe nanocrystal solids for n- and p-channel thin film field-effect transistors. *Science* 310; 86–89.

Umari, P., E. Mosconi, and F. De Angelis. 2014. Relativistic GW calculations on $CH_3NH_3PbI_3$ and $CH_3NH_3SnI_3$ perovskites for solar cell applications. *Sci. Rep.* 4.

Urbach, F. 1953. The long-wavelength edge of photographic sensitivity and of the electronic absorption of solids. *Phys. Rev.* 92; 1324.

Vanderbilt, D. 1990. Soft self-consistent pseudopotentials in a generalized eigenvalue formalism. *Phys. Rev. B* 41; 7892(R).

Van Roosbroeck, W., and W. Shockley. 1954. Photon-radiative recombination of electrons and holes in germanium. *Phys. Rev.* 94; 1558–1560.

Wang, Y., T. Gould, J. F. Dobson, H. Zhang, H. Yang, X. Yao, and H. Zhao. 2014. Density functional theory analysis of structural and electronic properties of orthorhombic perovskite $CH_3NH_3PbI_3$. *Phys. Chem. Chem. Phys.* 16; 1424–1429.

## 2.2 HYBRID ORGANIC–INORGANIC HALIDE PEROVSKITES: ELECTRONIC STRUCTURE, DIELECTRIC PROPERTIES, NATIVE DEFECTS, AND THE ROLE OF $NS^2$ IONS

*Mao-Hua Du*

Materials Science and Technology Division, Oak Ridge
National Laboratory, Oak Ridge, Tennessee, USA

The second part of this chapter discusses density functional calculations of the electronic structure, dielectric properties, defect properties, and their relation to the carrier transport in $CH_3NH_3PbI_3$. The results show that Pb (an $ns^2$ ion) plays a central role in a wide range of material properties, that is, small effective masses, enhanced lattice polarization and large static dielectric constant, high defect concentration, and the suppression of charge localization at defects. We also show generally the effects of the $ns^2$ ions in dielectric, defect, and carrier transport properties of halides containing $ns^2$ ions. The discrepancies in defect calculations of $CH_3NH_3PbI_3$ in the literature are discussed. The hybrid functional calculation including the spin–orbit coupling is shown to provide reliable results on defect level positions. The iodine interstitial and its complexes are found to be the only native point defects that introduce deep levels in the band gap but their carrier capture probabilities may be low, consistent with the observed long carrier diffusion lengths. Furthermore, we reveal the electronic and crystal structural factors behind the shallow nature of the halogen vacancy in $CH_3NH_3PbI_3$, which are important for future search of new halide electronic materials free of deep F centers.

### 2.2.1 Halide Electronic and Optoelectronic Materials

Solar cells based on hybrid organic–inorganic halide perovskites, in particular, methylammonium (MA) lead iodide ($CH_3NH_3PbI_3$), and related mixed halides (e.g., $CH_3NH_3PbI_{3-x}Cl_x$), have recently undergone rapid development (Green et al. 2014). The power conversion efficiency of $CH_3NH_3PbI_3$ solar cells has been reported to exceed 22% (NREL 2016). Halides usually have soft lattices and a high density of defects. As a result, they are usually not good electronic materials. However, $MAPbI_3$ has been shown to have long carrier diffusion lengths (>100 μm in single crystals and >1 μm in polycrystals) (Dong et al. 2015; Shi et al. 2015; Stranks et al. 2013; Xing et al. 2013), which is unusual for a halide. In fact, $MAPbI_3$ possesses an interesting combination of properties, that is, efficient carrier transport (Dong et al. 2015; Stranks et al. 2013; Xing et al. 2013), high density of defects (Walsh et al. 2015) [which are nevertheless benign in terms of carrier trapping (Du 2014, 2015; Shi and Du 2014; Yin et al. 2014)], large static dielectric constant (~60) (Onoda-Yamamuro et al. 1992), and significant ion migration (Aspiroz et al. 2015; Eames et al. 2015; Haruyama et al. 2015; Xiao et al. 2015; Yang et al. 2015).

These properties have important effects on the solar cell performance and are unusual for a photovoltaic (PV) material. However, they are not unique; they have been reported for some other halides. For example, TlBr, a semiconductor radiation detection material, has efficient electron transport (Churilov et al. 2009; Kim et al. 2009), high density of defects (Du 2010), a large dielectric constant (Lowndes 1972; Lowndes and Martin 1969), and significant ionic conductivity (Bishop et al. 2011, 2012; Samara 1981; Secco and Secco 1999). Furthermore, efficient carrier transport has also been reported for many other halides, for example, $CH_3NH_3SnI_3$ (Hao et al. 2014; Noel et al. 2014), $CsSnI_3$ (Chung et al. 2012a, 2012b) (which are being developed as PV materials), TlBr (Churilov et al. 2009), $Tl_6SeI_4$ (Johnsen et al. 2011), $Tl_6SI_4$ (Nguyen et al. 2013), and $CsPbBr_3$ [which are being developed as semiconductor radiation detection materials (Stoumpos et al. 2013)]. The recent surge in the discovery of new halide electronic and optoelectronic materials requires fundamental understanding of physics behind the observed novel material properties in these materials. Such understanding will provide rationales for future material optimization, expedite the discovery of new halide electronic and optoelectronic materials, and unleash the potential of halides in a wide range of applications such as photovoltaics, transistors, radiation detection, and so forth.

### 2.2.2 Halides Containing $ns^2$ Ions

#### 2.2.2.1 Mixed Ionic-Covalent Character

There is one common characteristic in the aforementioned halide electronic and optoelectronic materials, that is, they all contain $ns^2$ ions (whose outer electron configuration is $ns^2$) such as $Pb^{2+}$, $Tl^+$, and $Sn^{2+}$. The presence of the $ns^2$ ions is the key to the understanding of the electronic, dielectric, and defect properties in these halides. In a halide containing $ns^2$ cations, the outermost cation $s$ states are fully occupied below the halogen-$p$-dominated valence band (see Figure 2.2.1). Consequently, the conduction band is mainly made up of the cation $p$ states, rather than the cation $s$ states that are commonly observed in $s$–$p$ semiconductors. The cation $p$ states are spatially more extended than the cation $s$ states, thereby enhancing

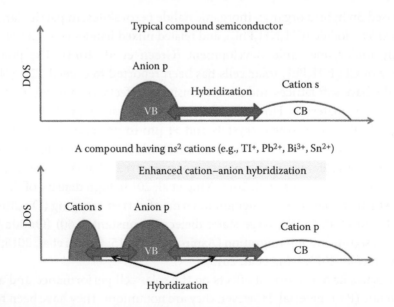

FIGURE 2.2.1 Schematic of density of states for a typical compound semiconductor (not including transition-metal and rare-earth compounds with significant $d$ or $f$ character in the conduction band) and a compound that contains $ns^2$ cations.

the covalency of the material. Furthermore, the $p$ orbital is directional, in contrast to the isotropic $s$ orbital. In the perovskite structure, the cation $p$ and the anion $p$ orbitals point directly to each other, further increasing the orbital overlap and the covalency. Therefore, there is usually significant cross-band-gap hybridization in halides containing the $ns^2$ cations. This is evidenced by the density of states (DOS) of $\beta$-$CH_3NH_3PbI_3$, in which there is significant overlap between the Pb–$6p$ and the I–$5p$ states in both the valence and conduction bands, as shown in Figure 2.2.2. (Here, $\beta$-$CH_3NH_3PbI_3$ is the room temperature phase of $CH_3NH_3PbI_3$ [Baikie et al. 2013; Stoumpos et al. 2013b], which has a tetragonal structure, as shown in Figure 2.2.3. The DOS was calculated using density functional theory (DFT) with Perdew–Burke–Ernzerhof [PBE] functionals [Perdew et al. 1996].) In addition to the cross-band-gap hybridization, there is also significant hybridization between the occupied cation $s$ and the halogen $p$ states (as evidenced by the significant Pb–$6s$ and I–$6p$ overlap in Figure 2.2.2), which increases delocalization of the valence band electrons. The enhanced cation–anion hybridization in halides containing $ns^2$ cations gives these halides mixed ionic-covalent character, as opposed to the strong ionicity observed in many other halides.

The significant covalency in halides containing $ns^2$ ions is reflected by their relatively small band gaps, for example, 1.51 eV [$CH_3NH_3PbI_3$ (Baikie et al. 2013)], 2.25 eV [$CsPbBr_3$ (Stoumpos et al. 2013)], 2.68 eV [TlBr (Owens and Peacock 2004)], 2 eV [InI (Farrell et al. 1997)], 1.86 eV [$Tl_6SeI_4$ (Johnsen et al. 2011)], and 2.1 eV [$Tl_6SI_4$ (Nguyen et al. 2013)], and by their dispersive conduction and valence bands, which enable efficient electron and hole

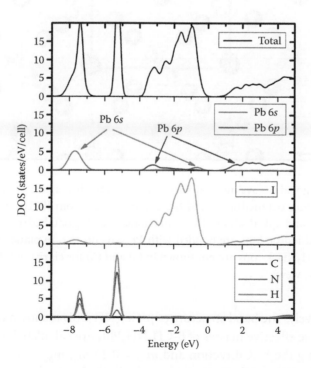

FIGURE 2.2.2 Density of states (DOS) of $\beta$-$CH_3NH_3PbI_3$ calculated using PBE functionals. The total DOS is projected into Pb ($6s$ and $6p$), I, C, N, and H atoms. The energy of valence band maximum is set to zero. Note that the band gap is underestimated owing to the use of the PBE functional.

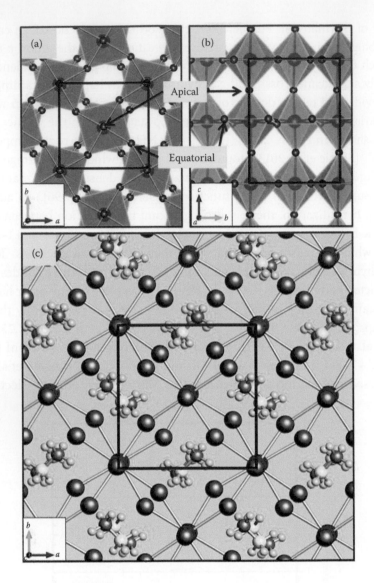

FIGURE 2.2.3 Tetragonal perovskite structure of β-CH$_3$NH$_3$PbI$_3$ (space group *I4cm*) with out-of-phase tilting of the PbI$_6$ octahedra around the c-axis viewed from [001] (a) and [100] (b) directions. The equatorial and apical sites of the iodine ions in a PbI$_6$ octahedron are also shown. (The equatorial and apical iodine ions are bonded with Pb ions on the *ab* plane and along the *c*-axis, respectively.) The (CH$_3$NH$_3$)$^+$ ions are not shown in (a) and (b) for clarity. (c) Ball-and-stick plot of β-CH$_3$NH$_3$PbI$_3$ ([001] view).

transport, respectively. The band structure of β-CH$_3$NH$_3$PbI$_3$ is shown in Figure 2.2.4. The electron and the hole effective masses of β-CH$_3$NH$_3$PbI$_3$ are calculated to be $m_{e,x} = 0.32\ m_0$, $m_{h,x} = 0.37\ m_0$ along the Γ–X direction and $m_{e,z} = 0.20\ m_0$, $m_{h,z} = 0.27\ m_0$ along the Γ–Z direction (Du 2014). The small effective masses for both the electron and the hole are consistent with the observed long diffusion length for both electrons and holes (Stranks et al. 2013; Xing et al. 2013).

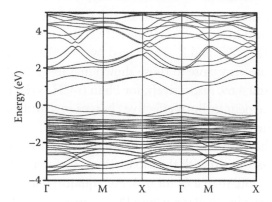

FIGURE 2.2.4   Band structure of β-$CH_3NH_3PbI_3$ calculated using PBE functionals. The energy of valence band maximum is set to zero. Note that the band gap is underestimated owing to the use of the PBE functional.

### 2.2.2.2 Enhanced Born Effective Charge and Strong Lattice Polarization

The mixed ionic-covalent character is known to enhance Born effective charges and lattice polarization (Cohen 1992; Du and Singh 2010a,b; Ghosez et al. 1998). This is illustrated schematically in Figure 2.2.5 using a binary compound as an example. In a largely ionic binary compound in which charges are mostly localized around ions, a displacement of the cation sublattice causes the polarization of the lattice. If there is electron sharing between cations and anions to some extent, the cation displacement would cause electron flow in the opposite direction, thereby enhancing the polarization (Ghosez et al. 1998). This is why the mixed ionic-covalent character can enhance the lattice polarization. The Born effective charge measures how lattice polarization develops with atomic displacements. Indeed, enhanced Born effective charges are found for $CH_3NH_3PbI_3$ (Du 2014). β-$CH_3NH_3PbI_3$ has body-centered tetragonal structure, which implies that $(CH_3NH_3)^+$ ions must freely rotate and take random orientations. In the ideal body-centered tetragonal structure (space group *I4cm*) of β-$CH_3NH_3PbI_3$, $Z^*_{xz}$, $Z^*_{zx}$, $Z^*_{yz}$, and $Z^*_{zy}$ of the Born effective charge tensor should be zero. However, in calculations, $(CH_3NH_3)^+$ ions are relaxed into fixed configurations, which lower the symmetry. As a result, the calculated Born charges in β-$CH_3NH_3PbI_3$ show differences on lattice sites that should be equivalent in *I4cm* structure and $Z^*_{zy}$, $Z^*_{xz}$, $Z^*_{zx}$, and $Z^*_{yz}$ are nonzero. For simplicity, $(CH_3NH_3)^+$ is replaced by a uniform positive charge

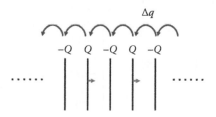

FIGURE 2.2.5   Schematic of electron transfer responding to a displacement of the cation sublattice in a binary compound.

background with total charge equal to that of the $(CH_3NH_3)^+$ ions. The resulting Born effective charges are shown in Table 2.2.1. If the $(CH_3NH_3)^+$ ions are explicitly included, the Born effective charges ($Z_{xx}^*, Z_{yy}^*$, and $Z_{zz}^*$) averaged over the lattice sites that are equivalent under $I4cm$ symmetry are very close to those shown in Table 2.2.1 (e.g., averaged $Z_{xx}^*$, $Z_{yy}^*$, and $Z_{zz}^*$ are 4.35, 4.35, and 4.69, respectively, for Pb in $\beta$-$CH_3NH_3PbI_3$). This is not surprising because the enhancement in Born effective charges is mainly due to the mixed ionic-covalent character of the Pb–I bonds.

The Born effective charges for Pb are more than doubled from its nominal ionic charge of +2. The Pb–$I_{4a}$ bond is aligned with the $c$ axis. Thus, $Z_{zz}^*$ for apical iodine ions ($I_{4a}$) is significantly enhanced by more than a factor of 3. For equatorial iodine ions, the Born charge enhancement is found on the $ab$ plane where Pb–$I_{8c}$ bonds lie. (The equatorial and apical iodine ions are bonded with Pb ions on the $ab$ plane and along the $c$-axis, respectively, as shown in Figure 2.2.3.)

The enhanced Born effective charges in $CH_3NH_3PbI_3$ lead to a large static dielectric constant in which the lattice contribution is much larger than the electronic contribution. The calculated static dielectric tensor with $(CH_3NH_3)^+$ ions explicitly included in the calculation is $\varepsilon_0^{xx} = 25.40$, $\varepsilon_0^{yy} = 21.88$, $\varepsilon_0^{zz} = 17.83$, $\varepsilon_\infty^{xx} = 5.30$, $\varepsilon_\infty^{yy} = 5.27$, and $\varepsilon_\infty^{zz} = 4.62$ where $\varepsilon_0$ and $\varepsilon_\infty$ are static and optical dielectric constants, respectively (Du 2014). $\varepsilon_0^{xx}$ and $\varepsilon_\infty^{xx}$ differ from $\varepsilon_0^{yy}$ and $\varepsilon_\infty^{yy}$, respectively, due to fixing the orientations of $(CH_3NH_3)^+$ ions, which reduces the symmetry. It is clear that the static dielectric constant is large (~20) compared to other compound semiconductors with similar band gaps and the dielectric screening is dominated by lattice (rather than electron) polarization. Note that these calculations do not consider the contribution by the dipoles associated with the MA molecules, which can freely rotate at room temperature. Experimentally, when the excitation frequency is sufficiently low such that the molecular dipoles can respond to the electric field, the static

TABLE 2.2.1　Born Effective Charge ($Z^*$) Tensors (Cartesian Coordinates) for $\beta$-$CH_3NH_3PbI_3$ with $(CH_3NH_3)^+$ Ions Replaced by a Uniform Charge Background (Space Group $I4cm$).

| Atom Type | WP | Born Effective Charge Tensor | | |
|---|---|---|---|---|
| Pb | 4a | 4.38 | −0.84 | 0.0 |
| | | 0.84 | 4.38 | 0.0 |
| | | 0.0 | 0.0 | 4.86 |
| I (apical) | 4a | −0.53 | 0.0 | 0.0 |
| | | 0.0 | −0.53 | 0.0 |
| | | 0.0 | 0.0 | −3.45 |
| I (equatorial) | 8c | −2.43 | 1.79 | 0.0 |
| | | 1.79 | −2.43 | 0.0 |
| | | 0.0 | 0.0 | −0.61 |

*Note:* The Wyckoff positions (WP) are shown and only the charges for the inequivalent atoms are listed.

TABLE 2.2.2    Born Effective Charges ($Z^*$) and Dielectric Constants for a Number of Halides and Chalcohalides That Contain $ns^2$ Ions

|  | Born Charge | Nominal Ionic Charge | $\varepsilon_0$ (exp.) | $\varepsilon_\infty$ (exp.) |
|---|---|---|---|---|
| TlCl | 2.02[c] | 1 | 32.7[a] | 4.76[a] |
| TlBr | 2.1[c] | 1 | 30.6[a,b] | 5.34[a,b] |
| TlI | 2.21[c] | 1 |  |  |
| InCl | 1.8 (*a*); 2.3 (*b*); 2.2 (*c*)[d] | 1 |  |  |
| InBr | 2.1 (*a*); 2.4 (*b*); 2.4 (*c*)[d] | 1 | 34.7[e] | 6.5[e] |
| InI | 2.4 (*a*); 2.5 (*b*); 2.5 (*c*) | 1 | 25.7[e] | 7.7[e] |
| PbI$_2$ | 4.0 (*a*, *b*); 1.92 (*c*)[d] | 2 | 26.4 (*a*, *b*)[f]; 6.5 (c)[g] |  |
| BiI$_3$ | 5.2 (*a*, *b*); 2.8 (*c*)[d] | 3 | 54 (*a*, *b*); 8.6 (*c*)[h] |  |
| Tl$_6$SeI$_4$ Tl$_6$SI$_4$ | ~2[i] | 1 |  |  |

*Note:* For simplicity, only the Born charges for cations are shown.
[a] Lowndes (1972).
[b] Lowndes and Martin (1969).
[c] Du and Singh (2010a).
[d] Du and Singh (2010b).
[e] Clayman et al. (1982).
[f] Lucovsky et al. (1976).
[g] Dugan and Henisch (1967).
[h] Vits (1973).
[i] Biswas et al. (2012), Shi and Du (2015).

dielectric constant was measured to be greater than 60 (Onoda-Yamamuro et al. 1992). As the excitation frequency increases, the molecular dipoles no longer respond and consequently the static dielectric constant is reduced to 29.7 (Poglitsch and Weber 1987), in reasonable agreement with the calculated value.

In addition to $CH_3NH_3PbI_3$, significantly enhanced Born effective charges have been found in a large number of halides and chalcohalides that contain $ns^2$ ions, as shown in Table 2.2.2. The enhanced Born effective charge leads to a large static dielectric constant, which is mainly due to large lattice (rather than electronic) polarization as evidenced by the relatively small high-frequency dielectric constant (see Table 2.2.2). Some materials in Table 2.2.2 such as $PbI_2$ and $BiI_3$ have layered structures. In these materials, enhanced Born charges and large static dielectric constants are found within the layer where there is mixed ionic-covalent bonding network. The results in Table 2.2.2 show that it is common that the presence of $ns^2$ ions is accompanied by strong dielectric screening, which has important effects on defect properties and carrier transport.

### 2.2.2.3 Effects of a Large Static Dielectric Constant on Carrier Transport

How the presence of $ns^2$ ions affects the dielectric and defect properties and carrier transport in MAPbI$_3$ is illustrated in Figure 2.2.6. The strong enhancement in Born effective charges and the soft lattice of the halide leads to strong lattice polarization in halides containing $ns^2$ ions. In MAPbI$_3$, the rotational freedom of the dipolar MA molecules further increases the static dielectric constant. This is one of the most important roles of the MA molecule in MAPbI$_3$. A large static dielectric constant provides strong screening

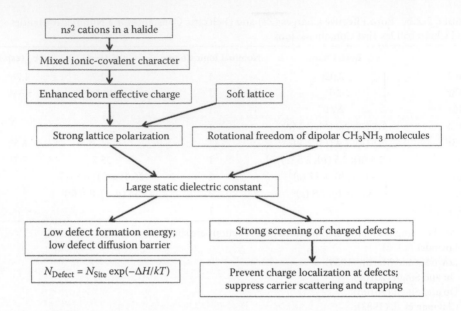

FIGURE 2.2.6　A flow chart that illustrates the effects of the $ns^2$ ion in $MAPbI_3$ on the performance of the $MAPbI_3$ solar cell.

of the charged defects and impurities, therefore reducing carrier scattering and trapping. The strong screening also prevents charge localization at defects. Deep trapping of charges at a defect may be due to either long-range electrostatic potential or short-range potential or a combination of both. The former is more important in ionic materials while the latter is more dominant in covalent materials. The suppressed long-range electrostatic potential in $MAPbI_3$ due to the large static dielectric constant of greater than 60 (Onoda-Yamamuro et al. 1992) should be partially responsible for the lack of deep defects, as shown by first-principles calculations (Du 2014, 2015; Shi and Du 2014; Yin et al. 2014).

Although a large static dielectric constant suppresses carrier scattering and trapping, it also promotes the formation of charged defects. The fact that creating a charged defect costs energy is partly due to the electrostatic potential applied on the crystal lattice by the defect. The strong screening effectively reduces the formation energy of charged defects due to the suppression of the defect-induced electrostatic potential. The low formation energies of vacancies in $MAPbI_3$ has been confirmed by density functional calculations (Walsh et al. 2015). Similarly, the low formation energies of vacancies have also been found in TlBr, which has a large static dielectric constant of 30.6 (Lowndes 1972; Lowndes and Martin 1969).

Under thermal equilibrium, the defect concentration is given by $N_D = N_0 \exp(-\Delta H/kT)$, where $N_D$ is the defect concentration, $N_0$ is the available defect sites in the crystal, $\Delta H$ is the defect formation enthalpy, $k$ is the Boltzmann constant, and $T$ is the temperature. Apparently, reduced defect formation energy can potentially increase the defect concentration. On the other hand, a low temperature can reduce the defect concentration. Therefore, the combination of a large static dielectric constant and a low growth temperature would help reduce defect-induced carrier scattering and trapping. $MAPbI_3$ solar cells are typically made by solution based low-temperature synthesis. The large static dielectric constant and the low growth temperature should be related to the efficient carrier transport in $MAPbI_3$.

The low formation energy of a defect implies that the energy cost of cutting energy bonds is low. Therefore, the low formation energy of a defect is often accompanied by the low defect diffusion barrier. There are several published theoretical works on the calculations of diffusion barriers of the iodine and MA vacancies ($V_I$ and $V_{MA}$) (Aspiroz et al. 2015; Eames et al. 2015; Haruyama et al. 2015).

The published results do not agree with each other. For example, the reported diffusion barriers of $V_I$ vary from ~0.1 eV to ~0.6 eV (Aspiroz et al. 2015; Eames et al. 2015; Haruyama et al. 2015). This discrepancy should be resolved by future calculations. However, these reported diffusion barriers are generally low, consistent with the experimentally observed ionic migration in $MAPbI_3$ at room temperature (Xiao et al. 2015; Yang et al. 2015). It is shown that the ionic conductivity is much higher than the electronic conductivity in dark conditions (Yang et al. 2015). The temperature-dependent bulk conductivity measurement shows an activation energy of 0.4–0.43 eV (Knop et al. 1990; Yang et al. 2015). For ionic conduction induced by defect diffusion, the activation energy should be equal to the sum of the defect formation energy and the defect diffusion barrier. Therefore, the defect diffusion barrier should be lower than ~0.4 eV. Significant ionic conductivity at room temperature has also been observed in TlBr (Bishop et al. 2011, 2012; Samara 1981; Secco and Secco 1999), which, similar to $MAPbI_3$, also contains an $ns^2$ ion and has a large static dielectric constant (Lowndes 1972; Lowndes and Martin 1969), low vacancy formation energies, and low vacancy diffusion barriers (Du 2010, 2013). $MAPbI_3$ and TlBr have many common characteristics, such as efficient carrier transport, large density of defects (which are nevertheless mostly shallow), large static dielectric constant, and significant ionic conductivity. The detailed calculations of electronic structure, dielectric properties, and defect properties suggest that these observed phenomena are related to the presence of $ns^2$ ions in these halides, as illustrated in Figure 2.2.6. The presence of a dipolar molecular ion in $MAPbI_3$ is also highly important, as the molecular dipoles can rotate to respond to static charges such as charged defects and impurities but cannot respond fast enough to diffusing electrons and holes. The presence of an $ns^2$ ion ($Pb^{2+}$) and a dipolar molecular ion ($MA^+$) is the key to efficient carrier transport in $MAPbI_3$.

### 2.2.3 Native Defects in $CH_3NH_3PbI_3$

*2.2.3.1 Defect Levels and Effects of Spin–Orbit Coupling and Self-Interaction Error*

Defects scatter and trap charge carriers and lower the carrier transport efficiency. Density functional calculations show that the formation energies of many defects in $MAPbI_3$ are low (Agiorgousis et al. 2014; Buin et al. 2015; Du 2014, 2015; Shi and Du 2014; Walsh et al. 2015; Yin et al. 2014), which suggest that the defect concentration in $MAPbI_3$ should be high. However, efficient carrier transport in $MAPbI_3$ has been demonstrated by many experimental studies as evidenced by long carrier diffusion lengths in single- and polycrystalline $MAPbI_3$ (Dong et al. 2015; Shi et al. 2015; Stranks et al. 2013; Xing et al. 2013). The concentration of deep defects, which are most detrimental to carrier transport, is found to be low (Shi et al. 2015), consistent with the observed long carrier diffusion length.

There are many theoretical studies of defect levels in $MAPbI_3$ based on DFT (Agiorgousis et al. 2014; Buin et al. 2015; Du 2014, 2015; Shi and Du 2014; Yin et al. 2014). Some of these studies used functionals based on generalized gradient approximation (GGA) and neglected the spin–orbit coupling (SOC) (Agiorgousis et al. 2014; Buin et al. 2015; Kim et al. 2014; Yin et al. 2014). Some others used the Heyd–Scuseria–Ernzerhof (HSE) hybrid functional (Heyd et al. 2003; Krukau et al. 2006) and included the SOC in the calculations (Du 2014, 2015). Both approaches produce band gaps close to that of the experimental value of 1.51 eV (Baikie et al. 2013). However, the valence and conduction band edges are completely different in the two approaches, leading to different defect level positions relative to the band edges. The GGA calculation is known to underestimate band gaps of semiconductors and insulators due partially to the self-interaction error, which raises the valence band maximum (VBM). The GGA band gap of $\beta$-$CH_3NH_3PbI_3$ is 0.60 eV, which is calculated including the SOC. The SOC effect is important in the electronic structure of $CH_3NH_3PbI_3$ due to the heavy Pb and I elements (Brivio et al. 2014; Even et al. 2013; Menéndez-Proupin et al. 2014; Umari et al. 2014). Neglecting the SOC raises the CBM by 0.81 eV and lowers the VBM by 0.17 eV, thereby increasing the band gap to 1.59 eV, which is in good agreement with the experimental value. Hence, many previous GGA calculations neglected the SOC for obtaining the correct band gap as well as for reducing computational time (Agiorgousis et al. 2014; Buin et al. 2015; Kim et al. 2014; Yin et al. 2014). However, this is a correction of the band gap based on error cancelation (Mosconi et al. 2013). Both the CBM and the VBM are in fact artificially high due to the absence of the SOC and the self-interaction error of GGA, respectively, although their difference, which is the band gap, is in good agreement with the experimental value.

On the other hand, a hybrid density functional calculation incorporating a fraction ($\alpha$) of Fock exchange (Heyd et al. 2003; Krukau et al. 2006) partially removes the self-interaction error, thereby lowering the VBM substantially relative to that of a GGA calculation. Hybrid functional calculations have recently been applied to many semiconductors and have generally shown improvement in calculated structural, electronic, dielectric, and defect properties (Alkauskas et al. 2008a, 2008b; Biswas and Du 2011a; Du and Zhang 2009; Janotti et al. 2010; Komsa et al. 2010; Muñoz Ramo et al. 2007; Paier et al. 2006; Ramo et al. 2007). A band gap of 1.50 eV is obtained for $\beta$-$CH_3NH_3PbI_3$ by using HSE hybrid functionals (Heyd et al. 2003; Krukau et al. 2006) ($\alpha = 0.43$) including SOC (Du 2014, 2015). The band gap correction in a hybrid functional calculation is attributed mostly to the lowering of the VBM, which is important for predicting defect properties. For example, the existence of AX centers in several $p$-doped II–VI semiconductors is correctly predicted by hybrid functional calculations but is shown to be unstable by LDA (local density approximation) and GGA calculations due to the artificially high VBM in LDA and GGA calculations (Biswas and Du 2011a).

Although the GGA–non-SOC and the HSE–SOC calculations both give good band gap of $CH_3NH_3PbI_3$, the difference in the calculated band edges lead to drastically different predictions of defect level positions relative to the band edges. Figure 2.2.7 shows a schematic that illustrates two cases in which qualitatively different results in defect levels are

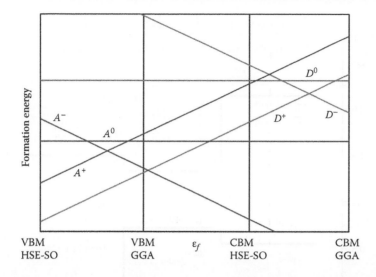

FIGURE 2.2.7 A schematic of formation energies of a single-electron donor ($D$) and a single-electron acceptor ($A$) as a function of Fermi level ($\varepsilon_f$) in $CH_3NH_3PbI_3$. The VBM and the CBM calculated using GGA–non-SOC and HSE–SOC calculations are shown. The defect formation energy is a function of the Fermi level ($\Delta H_f = A + q\varepsilon f$, where $q$ is the charge state of the defect and $A$ is arbitrary for the schematic figure shown here) (Van de Walle and Neugebauer 2004). The slope of a formation energy line indicates the charge state of the defect. The Fermi level at which the formation energy lines for two different charge states of a defect intersect is the charge transition level.

obtained by GGA–non-SOC and HSE-SOC calculations. In Figure 2.2.7, the electron trapping level, the (+/0) level, of donor defect, $D$, is deep in GGA–non-SOC calculations but shallow in HSE–SOC calculations mainly because the former neglects the SOC whereas the latter includes it. On the other hand, the hole trapping level, the (0/–) level, of acceptor defect, $A$, is shallow in GGA calculations but deep in HSE calculations due to the self-interaction error in GGA, which raises VBM and consequently lowers the energy of a free hole relative to a trapped hole. The failure of GGA calculations to predict the deep (+/–) transition for the defect A in Figure 2.2.7 is analogous to the failure to predict the stability of AX centers in II–VI semiconductors (Biswas and Du 2011a).

Figure 2.2.8 shows the charge transition levels for several native defects in $MAPbI_3$. These levels relative to the band edges are drastically different when calculated using three different methods, that is, GGA–non-SOC, GGA–SOC, and HSE–SOC. The charge transition level $\varepsilon(q/q')$ for a defect is determined by the Fermi level ($\varepsilon_f$) at which the formation energies of the defect with charge states $q$ and $q'$ are equal to each other (Van de Walle and Neugebauer 2004). $\varepsilon(q/q')$ can be calculated using

$$\varepsilon(q/q') = \frac{E_{D,q'} - E_{D,q}}{q - q'} \qquad (2.2.1)$$

where $E_{D,q}(E_{D,q'})$ is the total energy of the supercell that contains the relaxed structure of a defect at charge state $q$ ($q'$).

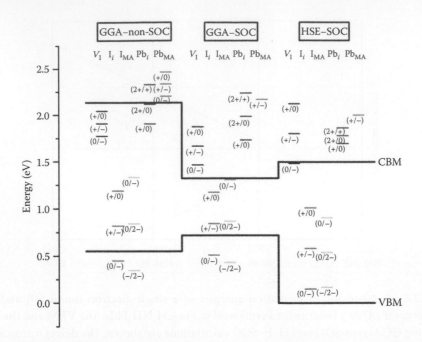

FIGURE 2.2.8  Charge transition levels of $V_I$, $I_i$, $I_{MA}$, $Pb_i$, and $Pb_{MA}$ in β-$CH_3NH_3PbI_3$ calculated using GGA–non-SOC, GGA–SOC, and HSE–SOC (α = 0.43) calculations. For $Pb_{MA}$, only the (+/–) level is shown when the SOC is included because a localized $Pb^0_{MA}$ cannot be stabilized when the SOC is included.

The general trends in Figure 2.2.8 are that (1) some defect levels are below the CBM in GGA–non-SOC calculations (e.g., those of $V_I$), but are above the CBM after including the SOC because the SOC lowers the CBM substantially; (2) some acceptor levels are below the VBM in GGA calculations [e.g., the (0/–) level of $I_i$], but are inside the band gap in HSE calculations due to the reduction of the self-interaction error that lowers the VBM substantially. It can be seen that neglecting the SOC in GGA calculations is an *ad hoc* correction of the band gap, which results in incorrect VBM and CBM positions and consequently incorrect defect level positions relative to the band edges.

The published GGA–non-SOC calculations do not agree with each other in some defect levels (e.g., the important iodine vacancy defect, $V_I$) despite having used the same method. The (+/0) and the (0/–) levels are shown to be inside the band gap by Agiorgousis et al. (2014) but above the CBM by several others (Kim et al. 2014; Yin et al. 2014). This is likely because the former predicted the correct defect structures for neutral and negatively charged $V_I$, that is, $V_I^0$ and $V_I^-$ (in which the Pb–Pb distances are short) while the latter used the incorrect defect structures (with long Pb–Pb distances). Nevertheless, the incorrect defect structure leads to high defect levels, even above the already artificially high CBM in the GGA–non-SOC calculations, thereby, accidentally reaching the correct conclusion that $V_I$ is a shallow donor, in agreement with the HSE–SOC calculations.

Note that we focus on charge trapping at localized defect states with substantial local structural relaxation because such local trapping may lead to deep charge trapping. The charge transition levels shown in Figure 2.2.8 all correspond to the local charge trapping (within the length scale of lattice constant). Shallow hydrogenic levels, which are close to the CBM or the VBM and correspond to charge trapping by long-range Coulomb attraction with negligible structural relaxation, are not calculated.

Among all the native point defects considered, including vacancies ($V_{MA}$, $V_{Pb}$, $V_I$), interstitials ($MA_i$, $Pb_i$, $I_i$), and antisites ($Pb_{MA}$, $MA_{Pb}$, $Pb_I$, $MA_I$, $I_{Pb}$, $I_{MA}$), remarkably, only $I_i$ and $I_{MA}$ introduce deep levels in the band gap. Note that $I_{MA}$ is in fact a $V_{MA}$–$I_i$ complex, whose charge transition levels are close to those of $I_i$. All the vacancies ($V_{MA}$, $V_{Pb}$, $V_I$), cation interstitials ($MA_i$, $Pb_i$), and some antisite defects ($MA_I$, $Pb_{MA}$, $MA_{Pb}$) create only shallow levels. Specifically, $V_I$, $MA_i$, $Pb_i$, $MA_I$, and $Pb_{MA}$ are shallow donors while $V_{MA}$, $V_{Pb}$, and $MA_{Pb}$ are shallow acceptors. $I_{Pb}$, $I_{MA}$, and $Pb_I$ antisite defects are $V_{Pb}$–$I_i$, $V_{MA}$–$I_i$, and $V_I$–$Pb_i$ complexes, respectively.

$I_i$ and $I_{MA}$ can each assume two different stable charge states. $I_i^+$ and $I_{MA}^0$ are stable when the Fermi level is low, while $I_i^-$ and $I_{MA}^{2-}$ are stable when the Fermi level is high. The structures of $I_i^+$ and $I_i^-$ are shown in Figure 2.2.9. The HSE–SOC calculations show that the (+/−) and the (0/2−) transition levels for $I_i$ and $I_{MA}$ are 0.57 eV and 0.54 eV above the VBM, respectively, as shown in Figure 2.2.8. The Fermi levels in $CH_3NH_3PbI_3$ were shown to be either near midgap in planar thin film device architectures or near the CBM in meso-superstructured device architectures (Leijtens et al. 2014). Under these conditions, $I_i^-$ and $I_{MA}^{2-}$ are stable and their acceptor levels are deep, i.e., the (0/−) level for $I_i$ and the (−/2−) level for $I_{MA}$ are 0.15 eV and 0.17 eV above the VBM, respectively. These calculated deep acceptor level positions of $I_i$ and $I_{MA}$ are close to an observed defect level ∼ 0.16 eV above the VBM measured using admittance spectroscopy (Duan et al. 2015).

HSE–SOC calculations further show that neither $I_i^-$ nor $I_{MA}^{2-}$ has single-particle levels inside the band gap although their thermodynamic hole trapping levels are inside the band gap. Therefore, $I_i^-$ and $I_{MA}^{2-}$ cannot be excited by sub-band-gap excitation and cannot be easily seen in optical absorption experiments (De Wolf et al. 2014). More importantly, with single-particle levels below the VBM, the hole trapping at $I_i^-$ and $I_{MA}^{2-}$ likely involves kinetic barriers, thereby significantly reducing the hole trapping cross sections.

The aforementioned defect calculations show that the native point defects are mostly shallow; the ones that do introduce deep levels (i.e., $I_i$ and $I_{MA}$) may have low carrier

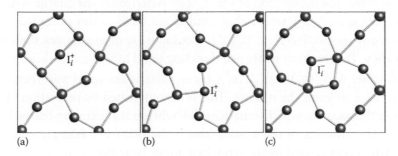

FIGURE 2.2.9 Two $I_i^+$ structures (a) and (b) and the structure of $I_i^-$ (c). The energies of the two $I_i^+$ structures differ by less than 0.01 eV at the GGA–SO level. Only one Pb–I layer (parallel to the *ab* plane) is shown for clarity. Red and blue balls represent Pb and I atoms, respectively.

trapping probabilities. These defect properties are consistent with the observed long carrier diffusion lengths and the high solar cell efficiencies.

### 2.2.3.2 Why the Halogen Vacancy Is Shallow and the Implication for the Search for New Halide Electronic Materials

The halogen vacancy is one of the most important native defects in halides. In large-gap ionic halides, such as alkali metal halides, the halogen vacancy is usually a deep center, known as the F center. The deep centers are detrimental to carrier transport due to the carrier trapping and nonradiative recombination. In halides with $ns^2$ ions, which are more covalent, the band gaps are relatively small. The halogen vacancy is shallow in some materials but deep in some others. For example, in InI, $Tl_6SeI_4$, $Tl_6SI_4$ [with band gaps of 2 eV for InI (Farrell et al. 1997), 1.86 eV for $Tl_6SeI_4$ (Johnsen et al. 2011), 2.1 eV for $Tl_6SI_4$ (Nguyen et al. 2013)], $V_H^-$ (with two electrons trapped at a deep level of the halogen vacancy) has been found to be stable when the Fermi level is high in the band gap (Biswas and Du 2011b; Biswas et al. 2012; Shi and Du 2015). On the other hand, the halogen vacancies in $CH_3NH_3PbI_3$ and TlBr [with band gaps of 1.51–1.52 eV (Baikie et al. 2013; Stoumpos et al. 2013b) and 2.68 eV (Owens and Peacock 2004), respectively] are shallow donors. The band gap alone cannot explain why the halogen vacancy is shallow in $CH_3NH_3PbI_3$ and TlBr because the halogen vacancy in TlBr (with a band gap of 2.68 eV) is shallow while those in smaller-gap halides; that is, InI, $Tl_6SeI_4$, $Tl_6SI_4$, are deep. In light of the outstanding carrier transport properties of $CH_3NH_3PbI_3$ (Du 2014, 2015; Shi and Du 2014; Yin et al. 2014) and TlBr (Churilov et al. 2009, 2010) compared to most of other halides, it is important to understand the material structure and chemistry that determine the shallow or deep nature of $V_H$ in halides. Such understanding is important to the discovery and development of new halide-based electronic and optoelectronic materials.

As discussed in Section 2.2.2.3, a charge can be localized at a defect by the long-range electrostatic potential, the short-range potential due to orbital overlapping and hybridization, and a combination of both. The electrostatic potential in $CH_3NH_3PbI_3$ is significantly reduced by screening as a result of the large static dielectric constant of ~60 (Onoda-Yamamuro et al. 1992). The defect states at $V_H$ can also be formed by hybridization of dangling bond orbitals from the nearby cations. The lowest defect state at $V_H$ is typically an $a_1$ state. The occupation of the $a_1$ state determines the charge state of $V_H$, that is, $V_H^+(a_1^0), V_H^0(a_1^1),$ and $V_H^-(a_1^2)$, where $a_1^0, a_1^1,$ and $a_1^2$ are the $a_1$ states occupied by zero, one, and two electrons, respectively. At $V_H^+(a_1^0)$, Coulomb repulsion moves the cations away from the vacancy. The resulting weakened dangling bond hybridization leads to a relatively high-lying $a_1^0$ state. At $V_H^0(a_1^1)$ and $V_H^-(a_1^2)$, the $a_1$ state is occupied and therefore there is energetic incentive to lower the $a_1$ state. This is accomplished by relaxation of the nearby cations toward the vacancy, which enhances the hybridization among the cation dangling bond orbitals. The lowering of the $a_1$ state lowers the electronic energy, which offsets the increased strain energy caused by the stretched bonds near the vacancy.

In $MAPbI_3$, the perovskite structure determines that each iodine ion has only two Pb neighbors. Therefore, there are only two Pb $6p$ orbitals hybridizing with each other at $V_I$. Furthermore, the MA molecular ion occupying the A site of the $ABC_3$ perovskite structure has a large size and expands the lattice, leading to long Pb–Pb distances. These two factors may be the causes to the weak Pb–Pb interaction at $V_I$, which makes the deep electron trapping at $V_I$ unfavorable. In contrast, the Pb–Pb hybridization at the halogen vacancy should be stronger in $CsPbBr_3$ due to the smaller halogen ion (Br) and the smaller A-site ion (Cs). Indeed, calculations show that $V_{Br}$ in $CsPbBr_3$ can trap two electrons locally forming $V_{Br}^-$ (Shi and Du 2014).

In halides with $ns^2$ cations, the conduction band states are mostly made up of the cation $p$ orbitals, which are directional, unlike the isotropic $s$ orbitals that make up the conduction bands of many $s–p$ semiconductors. Therefore, the crystal structure can affect the hybridization of directional cation $p$ orbitals at the halogen vacancy. One notable example is the comparison between simple cubic (SC) and rocksalt (RS) TlBr. In SC TlBr, the eight Tl $6p$ orbitals near a $V_{Br}$ point to nearby Tl ions rather than the vacancy. In contrast, in RS TlBr, the six Tl $6p$ orbitals around a $V_{Br}$ point directly to the vacancy. As a result, $V_{Br}$ is a shallow donor in SC TlBr but a deep donor and recombination center in RS TlBr (Shi and Du 2014).

The preceding results suggest that shallow halogen vacancies can exist in halides to give rise to good carrier transport properties. The key is to find the right crystal structures and compounds, such as those with large cation–cation distances and low anion coordination numbers, and those with crystal symmetry that prevents strong hybridization among cation orbitals in the halogen vacancy.

## 2.2.4 Conclusions

Halides are usually not good electronic materials. However, many halides with $ns^2$ ions have dispersive conduction and valence bands, which enable fast carrier diffusion. The large static dielectric constant resulting from the presence of the $ns^2$ ions and the molecular dipoles is related to a range of interesting material properties in $CH_3NH_3PbI_3$, for example, the high defect concentration, defect tolerance, and the defect diffusion. The defect tolerance is due to the strong screening and the lack of deep defect levels. The defect calculations show that the hybrid functional calculations including the spin–orbit coupling provide reliable results on defect level positions relative to band edges. Most native point defects are found to be shallow except the iodine interstitial and its complexes, which introduce deep levels in the band gap but nevertheless may have low carrier capture probabilities, in consistence with the observed long carrier diffusion lengths in $CH_3NH_3PbI_3$. The important donor defect, halogen vacancy, in halides with $ns^2$ ions can be shallow if given suitable crystal structures that have long cation–cation distance near the halogen vacancy and that have crystal symmetries preventing strong hybridization between the directional cation $p$ orbitals around the halogen vacancy. These insights are useful for searching for new halide-based electronic and optoelectronic materials.

## 2.2.5 Acknowledgment

Mao-Hua Du was supported by the U.S. Department of Energy, Office of Science, Basic Energy Sciences, Materials Sciences and Engineering Division.

# REFERENCES

Agiorgousis, M. L., Y.-Y. Sun, H. Zeng, and S. Zhang. 2014. Strong covalency-induced recombination centers in perovskite solar cell material $CH_3NH_3PbI_3$. *J. Am. Chem. Soc.* 136; 14570–14575.

Alkauskas, A., P. Broqvist, F. Devynck, and A. Pasquarello. 2008a. Band offsets at semiconductor-oxide interfaces from hybrid density-functional calculations. *Phys. Rev. Lett.* 101; 106802–106804.

Alkauskas, A., P. Broqvist, and A. Pasquarello. 2008b. Defect energy levels in density functional calculations: Alignment and band gap problem. *Phys. Rev. Lett.* 101; 046405.

Azpiroz, J. M., E. Mosconi, E., J. Bisquert, and F. De Angelis. 2015. Defect migration in methylammonium lead iodide and its role in perovskite solar cell operation. *Energy Environ. Sci.* 8; 2118–2127.

Baikie, T., Y. Fang, J. M. Kadro, M. Schreyer, F. Wei, S. G. Mhaisalkar, M. Grätzel, and T. J. White. 2013. Synthesis and crystal chemistry of the hybrid perovskite $(CH_3NH_3)PbI_3$ for solid-state sensitised solar cell applications. *J. Mater. Chem. A* 1; 5628–5641.

Bishop, S. R., W. Higgins, G. Ciampi, A. Churilov, K. S. Shah, and H. L. Tuller. 2011. The defect and transport properties of donor doped single crystal TlBr. *J. Electrochem. Soc.* 158; J47–J51.

Bishop, S. R., H. L. Tuller, G. Ciampi, W. Higgins, J. Engel, A. Churilov, and K. S. Shah. 2012. The defect and transport properties of acceptor doped TlBr: Role of dopant exsolution and association. *Phys. Chem. Chem. Phys.* 14; 10160–10167.

Biswas, K., and M.-H. Du. 2011a AX centers in II-VI semiconductors: Hybrid functional calculations. 2011. *Appl. Phys. Lett.* 98; 181913.

Biswas, K., and M.-H. Du. 2011b. First principles study of native defects in InI. 2011. *J. Appl. Phys.* 109; 113518.

Biswas, K., M. H. Du, and D. J. Singh. 2012. Electronic structure and defect properties of $Tl_6SeI_4$: Density functional calculations. *Phys. Rev. B* 86; 144108.

Brivio, F., K. T. Butler, A. Walsh, and M. van Schilfgaarde. 2014. Relativistic quasiparticle self-consistent electronic structure of hybrid halide perovskite photovoltaic absorbers. *Phys. Rev. B* 89; 155204–155206.

Buin, A., R. Comin, J. X. Xu, A. H. Ip, and E. H. Sargent. 2015. Halide-dependent electronic structure of organolead perovskite materials. *Chem. Mater.* 27; 4405–4412.

Chung, I., B. Lee, J. He, R. P. H. Chang, and M. G. Kanatzidis. 2012a. All-solid-state dye-sensitized solar cells with high efficiency. *Nature* 485; 486–490.

Chung, I., J.-H Song, J. Im, J. Androulakis, C. D. Malliakas, H. Li, A. J. Freeman, J. T. Kenney, and M. G. Kanatzidis. 2012b. $CsSnI_3$: Semiconductor or metal? High electrical conductivity and strong near-infrared photoluminescence from a single material. High hole mobility and phase-transitions. *J. Am. Chem. Soc.* 134; 8579–8587.

Churilov, A. V., G. Ciampi, H. Kim, L. J. Cirignano, W. M. Higgins, F. Olschner, and K. S. Shah. 2009. Thallium bromide nuclear radiation detector development. *IEEE Trans. Nucl. Sci.* 56; 1875–1881.

Churilov, A. V., G. Ciampi, H. Kim, W. M. Higgins, L. J. Cirignano, F. Olschner, V. Biteman, M. Minchello, and K. S. Shah. 2010. TlBr and $TlBr_xI_{1-x}$ crystals for gamma-ray detectors. *J. Cryst. Growth* 312; 1221–1227.

Clayman, B. P., R. J. Nemanich, J. C. Mikkelsen, and G. Lucovsky. 1982. Lattice dynamics of the layered compounds InI and InBr. *Phys. Rev. B* 26; 2011–2015.

Cohen, R. E. 1992. Origin of ferroelectricity in perovskite oxides. *Nature* 358; 136–138.

De Wolf, S., J. Holovsky, S.-J. Moon, P. Loeper, B. Niesen, M. Ledinsky, F.-J. Haug, J.-H. Yum, and C. Ballif. 2014. Organometallic halide perovskites: Sharp optical absorption edge and its relation to photovoltaic performance. *J. Phys. Chem Lett.* 5; 1035–1039.

Dong, Q., Y. Fang, Y. Shao, P. Mulligan, J. Qiu, L. Cao, and J. Huang. 2015. Electron-hole diffusion lengths >175 µm in solution-grown $CH_3NH_3PbI_3$ single crystals. *Science* 347; 967–970.

Du, M.-H. 2010. First-principles study of native defects in TlBr: Carrier trapping, compensation, and polarization phenomenon. *J. Appl. Phys.* 108; 053506.

Du, M.-H. 2013. Effects of impurity doping on ionic conductivity and polarization phenomenon in TlBr. *Appl. Phys. Lett.* 102; 082102–082104.

Du, M. H. 2014. Efficient carrier transport in halide perovskites: Theoretical perspectives. *J. Mater. Chem. A* 2; 9091–9098.

Du, M. H. 2015. Density functional calculations of native defects in $CH_3NH_3PbI_3$: Effects of spin-orbit coupling and self-interaction error. *J. Phys. Chem. Lett.* 6; 1461–1466.

Du, M.-H., and D. J. Singh. 2010a. Enhanced Born charge and proximity to ferroelectricity in thallium halides. *Phys. Rev. B* 81; 144114–144115.

Du, M.-H., and D. J. Singh. 2010b. Enhanced Born charges in III–VII, IV–VII$_2$, and V–VII$_3$ compounds. *Phys. Rev. B* 82; 045203–045205.

Du, M.-H., and S. B. Zhang. 2009. Impurity-bound small polarons in ZnO: Hybrid density functional calculations. *Phys. Rev. B* 80; 115217.

Duan, H.-S., H. Zhou, Q. Chen, P. Sun, S. Luo, T.-B. Song, B. Bob, and Y. Yang. 2015. The identification and characterization of defect states in hybrid organic-inorganic perovskite photovoltaics. *Phys. Chem. Chem. Phys.* 17; 112–116.

Dugan, A. E., and H. K. Henisch. 1967. Dielectric properties and index of refraction of lead iodide single crystals. *J. Phys. Chem. Solids* 28; 971–976.

Eames, C., J. M. Frost, P. R. F. Barnes, B. C. O'Regan, A. Walsh, and M. S. Islam. 2015. Ionic transport in hybrid lead iodide perovskite solar cells. *Nat. Commun.* 6; 7497.

Even, J., L. Pedesseau, J.-M. Jancu, and C. Katan. 2013. Importance of spin-orbit coupling in hybrid organic/inorganic perovskites for photovoltaic applications. *J. Phys. Chem. Lett.* 4; 2999–3005.

Farrell, R., F. Olschner, K. Shah, and M. R. Squillante. 1997. Advances in semiconductor photodetectors for scintillators. *Nucl. Instrum. Methods Phys. Res. A* 387; 194–198.

Ghosez, P., J. P. Michenaud, and X. Gonze. 1998. Dynamical atomic charges: The case of ABO$_3$ compounds. *Phys. Rev. B* 58; 6224–6240.

Green, M. A., A. Ho-Baillie, and H. J. Snaith. 2014. The emergence of perovskite solar cells. *Nat. Photonics* 2014, 8; 506–514.

Hao, F., C. C. Stoumpos, D. H. Cao, R. P. H. Chang, and M. G. Kanatzidis. 2014. Lead-free solid-state organic-inorganic halide perovskite solar cells. *Nat. Photonics* 8; 489–494.

Haruyama, J., K. Sodeyama, L. Han, and Y. Tateyama. 2015. First-principles study of ion diffusion in perovskite solar cell sensitizers. *J. Am. Chem. Soc.* 137; 10048–10051.

Heyd, J., G. E. Scuseria, and M. Ernzerhof. 2003. Hybrid functionals based on a screened Coulomb potential. *J. Chem. Phys.* 118; 8207–8215.

Janotti, A., J. B. Varley, P. Rinke, N. Umezawa, G. Kresse, and C. G. Van de Walle. 2010. Hybrid functional studies of the oxygen vacancy in TiO$_2$. *Phys. Rev. B* 81; 085212–085217.

Johnsen, S., Z. Liu, J. A. Peters, J.-H Song, S. L. Nguyen, C. D. Malliakas, H. Jin, A. J. Freeman, B. W. Wessels, and M. G. Kanatzidis. 2011. Thallium chalcohalides for x-ray and γ-ray detection. *J. Am. Chem. Soc.* 133; 10030–10033.

Kim, H., L. Cirignano, A. Churilov, G. Ciampi, W. Higgins, F. Olschner, and K. Shah. 2009. Developing larger TlBr detectors-detector performance. *IEEE Trans. Nucl. Sci.* 56; 819–823.

Kim, J., S. H. Lee, J. H. Lee, and K. H. Hong. 2014. The role of intrinsic defects in methylammonium lead iodide perovskite. *J. Phys. Chem. Lett.* 5; 1312–1317.

Knop, O., R. E. Wasylishen, M. A. White, T. S. Cameron, and M. J. Vanoort. 1990. Alkylammonium lead halides 0.2. $CH_3NH_3PbCl_3$, $CH_3NH_3PbBr_3$, $CH_3NH_3PbI_3$ Perovskites: Cuboctahedral halide cages with isotropic cation reorientation. *Can. J. Chem.* 68; 412–422.

Komsa, H.-P., P. Broqvist, and A. Pasquarello. 2010. Alignment of defect levels and band edges through hybrid functionals: Effect of screening in the exchange term. *Phys. Rev. B* 81; 205118.

Krukau, A. V., O. A. Vydrov, A. F. Izmaylov, and G. E. Scuseria. 2006. Influence of the exchange screening parameter on the performance of screened hybrid functionals. *J. Chem. Phys.* 125; 224106.

Leijtens, T., S. D. Stranks, G. E. Eperon, R. Lindblad, E. M. J. Johansson, I. J. McPherson, H. Rensmo, J. M. Ball, M. M. Lee, and H. J. Snaith. 2014. Electronic properties of meso-superstructured and planar organometal halide perovskite films: Charge trapping, photodoping, and carrier mobility. *ACS Nano* 8; 7147–7155.

Lowndes, R. P. 1972. Anharmonicity in the silver and thallium halides: Low-frequency dielectric response. *Phys. Rev. B* 6; 4667–4674.

Lowndes, R. P., and D. H. Martin. 1969. Dielectric dispersion and the structures of ionic lattices. *Proc. Roy. Soc. A* 308; 473–496.

Lucovsky, G., R. M. White, W. Y. Liang, R. Zallen, and P. Schmid. 1976. Lattice polarizability of $PbI_2$. 1976. *Solid State Commun.* 18; 811–814.

Menéndez-Proupin, E., P. Palacios, P. Wahnón, and J. C. Conesa. 2014. Self-consistent relativistic band structure of the $CH_3NH_3PbI_3$ perovskite. *Phys. Rev. B* 90; 045207.

Mosconi, E., A. Amat, M. K. Nazeeruddin, M. Grätzel, M., and F. De Angelis. 2013. First-principles modeling of mixed halide organometal perovskites for photovoltaic applications. *J. Phys. Chem. C* 117; 13902–13913.

Muñoz Ramo, D., A. L. Shluger, J. L. Gavartin, and G. Bersuker. 2007. Theoretical prediction of intrinsic self-trapping of electrons and holes in monoclinic $HfO_2$. *Phys. Rev. Lett.* 99; 155504.

Nguyen, S. L., C. D. Malliakas, J. A. Peters, Z. Liu, J. Im, L.-D. Zhao, M. Sebastian, H. Jin, H. Li, S. Johnsen, B. W. Wessels et al. 2013. Photoconductivity in $Tl_6SI_4$: A novel semiconductor for hard radiation detection. *Chem. Mater.* 25; 2868–2877.

Noel, N. K., S. D. Stranks, A. Abate, C. Wehrenfennig, S. Guarnera, A. A. Haghighirad, A. Sadhanala, G. E. Eperon, S. K. Pathak, M. B. Johnston, A. Petrozza et al. 2014. Lead-free organic-inorganic tin halide perovskites for photovoltaic applications. *Energy Environ. Sci.* 7; 3061–3068.

NREL (National Renewable Energy Laboratory). 2016. Research cell efficiency records. http://www.nrel.gov/ncpv/images/efficiency chart.jpg (Accessed January 17, 2017).

Onoda-Yamamuro, N., T. Matsuo, and H. Suga. 1992. Dielectric study of $CH_3NH_3PbX_3$ (X = Cl, Br, I). *J. Phys. Chem. Solids* 53; 935–939.

Owens, A., and A. Peacock. 2004. Compound semiconductor radiation detectors. *Nucl. Instrum. Methods Phys. Res. A* 531; 18–37.

Paier, J., M. Marsman, K. Hummer, G. Kresse, I. C. Gerber, and J. G. Angyan. 2006. Screened hybrid density functionals applied to solids. *J. Chem. Phys.* 124; 154709–154713.

Perdew, J. P., K. Burke, and M. Ernzerhof. 1996. Generalized gradient approximation made simple. *Phys. Rev. Lett.* 77; 3865–3868.

Poglitsch, A., and D. Weber. 1987. Dynamic disorder in methylammoniumtrihalogenoplumbates(II) observed by millimeter-wave spectroscopy. *J. Chem. Phys.* 87; 6373–6378.

Ramo, D. M., J. L. Gavartin, A. L. Shluger, and G. Bersuker. 2007. Spectroscopic properties of oxygen vacancies in monoclinic $HfO_2$ calculated with periodic and embedded cluster density functional theory. *Phys. Rev. B* 75; 205336.

Samara, G. A. 1981. Pressure and temperature dependences of the ionic conductivities of the thallous halides TlCl, TlBr, and TlI. *Phys. Rev. B* 23; 575–586.

Secco, E. A., and R. A. Secco. 1999. Cation conductivity in mixed thallium halides. *Solid State Ionics* 118; 37–42.

Shi, D., V. Adinolfi, R. Comin, M. J. Yuan, E. Alarousu, A. Buin, Y. Chen, S. Hoogland, A. Rothenberger, K. Katsiev, Y. Losovyj et al. 2015. Low trap-state density and long carrier diffusion in organolead trihalide perovskite single crystals. *Science* 347; 519–522.

Shi, H. L., and M. H. Du. 2014. Shallow halogen vacancies in halide optoelectronic materials. *Phys. Rev. B* 90; 174103–174106.

Shi, H. L., and M. H. Du. 2015. Native defects in $Tl_6SI_4$: Density functional calculations. *J. Appl. Phys.* 117; 175701–175705.

Stoumpos, C. C., C. D. Malliakas, and M. G. Kanatzidis. 2013. Semiconducting tin and lead iodide perovskites with organic cations: Phase transitions, high mobilities, and near-infrared photoluminescent properties. *Inorg. Chem.* 52; 9019–9038.

Stoumpos, C. C., C. D. Malliakas, J. A. Peters, Z. Liu, M. Sebastian, J. Im, T. C. Chasapis, A. C. Wibowo, D. Y. Chung, A. J. Freeman, B. W. Wessels, and M. G. Kanatzidis. 2013. Crystal growth of the perovskite semiconductor CsPbBr$_3$: A new material for high-energy radiation detection. *Cryst. Growth Des.* 13; 2722–2727.

Stranks, S. D., G. E. Eperon, G. Grancini, C. Menelaou, M. J. P. Alcocer, T. Leijtens, L. M. Herz, A. Petrozza, and H. J. Snaith. 2013. Electron-hole diffusion lengths exceeding 1 micrometer in an organometal trihalide perovskite absorber. *Science* 342; 341–344.

Umari, P., E. Mosconi, and F. De Angelis. 2014. Relativistic GW calculations on CH$_3$NH$_3$PbI$_3$ and CH$_3$NH$_3$SnI$_3$ perovskites for solar cell applications. *Sci. Rep.* 4; 4467, 7 pp.

Van de Walle, C. G., and J. Neugebauer. 2004. First-principles calculations for defects and impurities: Applications to III-nitrides. *J. Appl. Phys.* 95; 3851–3879.

Vits, P. 1973. Diplomarbeit, RWTH Aachen.

Walsh, A., D. O. Scanlon, S. Y. Chen, X. G. Gong, and S.-H. Wei. 2015. Self-regulation mechanism for charged point defects in hybrid halide perovskites. *Angew. Chem. Int. Ed.* 54; 1791–1794.

Xiao, Z., Y. Yuan, Y. Shao, Q. Wang, Q. Dong, C. Bi, P. Sharma, A. Gruverman, and J. Huang. 2015. Giant switchable photovoltaic effect in organometal trihalide perovskite devices. *Nat. Mater.* 14; 193–198.

Xing, G., N. Mathews, S. Sun, S. S. Lim, Y. M. Lam, M. Grätzel, S. Mhaisalkar, and T. C. Sum. 2013. Long-range balanced electron- and hole-transport lengths in organic-inorganic CH$_3$NH$_3$PbI$_3$. *Science* 342; 344–347.

Yang, T. Y., G. Gregori, N. Pellet, M. Grätzel, and J. Maier. 2015. The significance of ion conduction in a hybrid organic-inorganic lead-iodide-based perovskite photosensitizer. *Angew. Chem. Int. Ed.* 54; 7905–7910.

Yin, W.-J., T. Shi, and Y. Yan. 2014. Unusual defect physics in CH$_3$NH$_3$PbI$_3$ perovskite solar cell absorber. *Appl. Phys. Lett.* 104; 063903.

# CHAPTER 3

# Electric Properties of Organic–Inorganic Halide Perovskites and Their Role in the Working Principles of Perovskite-Based Solar Devices

Claudio Quarti, Domenico Di Sante, Liang Z. Tan, Edoardo Mosconi,
Giulia Grancini, Alessandro Stroppa, Paolo Barone, Filippo De Angelis,
Silvia Picozzi, and Andrew M. Rappe

## CONTENTS

## 3.1 EXTREMELY SLOW CHANGES IN PHOTOLUMINESCENCE AND RAMAN IN ORGANOHALIDE HYBRID PEROVSKITES: A FIRST PRINCIPLES INVESTIGATION

*Claudio Quarti,[1] Edoardo Mosconi,[2] Giulia Grancini,[3] and Filippo De Angelis[2,4]*

[1]Laboratory for Chemistry of Novel Materials, Universite de Mons, Belgium

[2]Computational Laboratory for Hybrid/Organic Photovoltaics, CNR-ISTM, Perugia, Italy

[3]Group for Molecular Engineering of Functional Materials, Institute of Chemical Sciences and Engineering, École Polytechnique Fédérale de Lausanne, Sion, Switzerland

[4]CompuNet, Istituto Italiano di Tecnologia, Genova, Italy

### 3.1.1 Introduction

The recent swift surge of hybrid metal–halide perovskites for photovoltaic applications is revolutionizing the field of the renewable energies thanks to an unprecedented increase of the performance of perovskite-based photovoltaic devices, from 3.8% in 2009 (Kojima et al. 2009) up to the current approximately 20% (Burschka et al. 2013; Im et al. 2011; Jeon et al. 2015; Lee et al. 2012; Liu et al. 2013; Zhou et al. 2014). However, if this class of materials seems to hold for the attainment of clean, renewable energy from the sun, it also presents many open issues. The positive optical (De Wolf et al. 2014; Kitazawa et al. 2002) and physical properties (Stranks et al. 2013; Xing et al. 2013; Wehrenfennig et al. 2014a,b) of hybrid metal–halide perovskites in fact are paralleled by very unusual phenomena, which do not find similar analogues in the traditional inorganic semiconductors. In 2014, Snaith and co-workers reported on an unusual hysteretic effect in the measured current–voltage ($J$–$V$) curve of the $CH_3NH_3PbI_{3-x}Cl_x$ perovskite (Snaith et al. 2014). These authors observed a different photocurrent with respect to the direction of the potential scan, from zero to open-circuit voltage or vice versa, together with a dependence of the photocurrent on the scanning rate and the device architecture (Snaith et al. 2014). Later studies reported on many similar effects, where the response of the material showed a non-steady-state, time-dependent response, generally depending on the illumination conditions and external parameters, such as the presence of moisture in the atmosphere and/or the presence of electric fields.

It is worth mentioning that the amazing achievements in the area of hybrid perovskites within a short time of five years were not paralleled by a comparable improvement in the basic understanding of this class of materials and most of their inherent electronic-optical-chemical properties, and an understanding of the connection with the material structure is still required. In this regard, vibrational spectroscopy techniques, Raman and infrared (IR), are unique tools, complementary to x-ray diffraction (XRD) to investigate the fundamental structure of the materials at the atomic scale (Weber and Merlin 2000), providing information on the crystalline disorder (Noda et al. 1999) and the electronic properties (Castiglioni et al. 2006). Raman vibrational spectroscopy was recently used to extract useful information, not only in terms of basic material characterization (Brivio et al. 2015; Quarti et al. 2014a) but also to gain insight into the material structure in working devices and under practical working conditions, unveiling the structure (Quarti et al. 2014a), the morphology (Grancini et al. 2014), and the presence of chlorine doping (Park et al. 2015), together with the structural-electronic characterization. The aim of the present chapter is to provide a wide perspective on the Raman investigations conducted so far on this hybrid perovskite, with a special focus on the structural-electronic properties of perovskite in working devices and, more specifically, on their response under illumination, external field, and ambient conditions.

The chapter is organized as follows. We first present a complete description of the Raman spectrum of the methylammonium lead iodide perovskite (MAPbI$_3$), taken as the prototypical hybrid perovskite for photovoltaic applications, together with a systematic assignment of the spectroscopic marker of the perovskite, on the basis of the theoretical simulations reported in the literature. Then, we illustrate some recent achievements on the structural and electronic properties of hybrid perovskites under practical operative conditions (externally applied electric field and presence of moisture). Finally, we discuss the slow changes in the photoresponse and photoluminescence of hybrid perovskites, their origin, and their effects on the basic photovoltaic working mechanism.

### 3.1.2 Raman Spectrum of MAPbI$_3$

In the following, we discuss the experimental spectrum of the MAPbI$_3$ perovskite, considered as prototypical of this class of highly efficient hybrid perovskites for photovoltaic applications, followed by a thorough assignment based on the Density Functional Theory (DFT) calculations reported in the literature.

Before presenting the experimental Raman data, a few comments are necessary. As their fully inorganic counterpart, hybrid perovskites are characterized by the presence of various phase transitions. In particular, MAPbI$_3$ presents two phase transitions, at 162 K and at 327.4 K (Kawamura et al. 2002; Poglitsch and Weber 1987; Stoumpos et al. 2013). At room temperature, MAPbI$_3$ is stable under a tetragonal crystalline structure, with proposed *I4mcm* (Kawamura et al. 2002; Poglitsch and Weber 1987) and *I4cm* space groups (Stoumpos et al. 2013). Noteworthy, nuclear magnetic resonance (Wasylishen et al. 1985) and millimeter wave spectroscopic measurements (Poglitsch and Weber 1987) demonstrated that the MA cations are not fixed in this crystalline phase but are free to rotate. This is consistent also with symmetry arguments. The $C_{3v}$ space group, characteristic of the isolated MA cation, in fact, is incompatible with both the *I4cm* and *I4mcm* space groups proposed for the tetragonal phase of $MAPbI_3$, and thus, these space groups result from the rotation of the MA cations from one orientation to an equivalent one with respect to the tetragonal symmetry. As a result, $MAPbI_3$ not only has one minimum structure but also has several structures that lie close in energies (Quarti et al. 2014b). In addition, DFT demonstrated that the inorganic framework is less rigid than the molecule and that specific orientations of the molecules impose different local distortions on the inorganic framework (Borriello et al. 2008). Thus, in light of the inherent soft nature of this material, $MAPbI_3$ and hybrid lead halide perovskites in general should be considered as materials with a large degree of local disorder, from a nanometer to a mesoscopic scale, as polymeric materials (Quarti et al. 2013) and biological systems (Siebert and Hildebrandt 2008), even if they retain a large degree of long-range structural coherence, as shown by the XRD measurements (Stoumpos et al. 2013).

### 3.1.2.1 Raman Spectrum of MAPbI$_3$: An Experimental Overview

In Figure 3.1.1, we compare different Raman spectra, as reported in the literature, measured on the $MAPbI_3$ perovskite at room temperature (Gottesmann et al. 2015; Grancini et al. 2014; Ledinský et al. 2015; Park et al. 2015; Quarti et al. 2014a; Wei et al. 2014). All the spectra are measured in the low-frequency range, below 300 $cm^{-1}$, where the marker bands of the inorganic framework are expected. In fact, several investigations on chemically analogous components, including $PbI_2$ (Baibarac et al. 2004; Preda et al. 2006), the $CsPbCl_3$ perovskite (Calistru et al. 1997), the hybrid $MAPbCl_3$ perovskite (Maalej et al. 1997), and many 2D hybrid lead–halide perovskites (Dammak et al. 2007, 2009; Elleuch et al. 2008), locate the vibrational marker of the inorganic framework in the low-frequency region below 100 $cm^{-1}$. The inner modes of the MA cations instead fall at the usual frequencies probed in structural chemical analyses (from a few hundred to ~3500 $cm^{-1}$). The low-lying inner vibration of the MA cation is the torsion around the CN axis, that is, the relative rotation of the $NH_3$ and $CH_3$ groups around the CN chemical bond. The exact position of this peak is discussed in the following text.

Noteworthy, all the Raman spectra reported in Figure 3.1.1 are measured under resonance conditions whereas nonresonance measurements have not been reported yet, at least to the best of the authors' knowledge. As it emerges, the various spectra show some small differences in the position and relative intensity of the various peaks, but they all agree on a semiquantitative level. All the spectra present two main Raman bands: a low-frequency band, between 52 $cm^{-1}$ and 71 $cm^{-1}$, and a high-frequency band, between 108 $cm^{-1}$ and 119 $cm^{-1}$. In addition, some Raman spectra show also two less pronounced signals, around 94 $cm^{-1}$ and 154 $cm^{-1}$. Finally, we note the presence in the spectrum of Figure 3.1.1a of a broad and unresolved band at 250 $cm^{-1}$.

We focus first on the two most intense bands. The vibrational frequency of the lowest band falls within a wide range of values, between 59 $cm^{-1}$ and 71 $cm^{-1}$, but this discrepancy is not completely unexpected. Noteworthy, this band floats above the dominating Raleigh signal, introducing a strong background signal under which the Raman bands float. Thus precise assignment of the band at approximately 60 cm$^{-1}$ is inherently complicated and depends on the procedure of subtraction of this signal. The high-frequency band instead is found at 109 ± 1 cm$^{-1}$ on all the spectra measured on thin film samples, while it is located

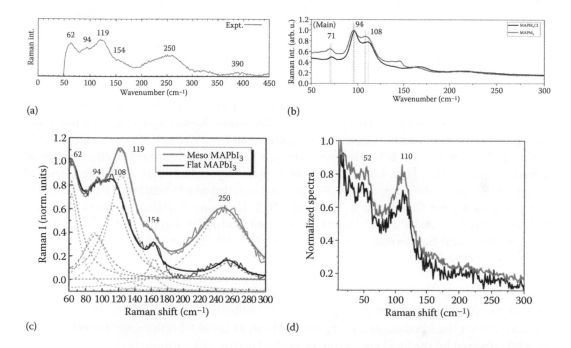

FIGURE 3.1.1 Summary of the Raman spectra of MAPbI$_3$ at room temperature (tetragonal phase) reported in the literature. (a) Resonance Raman spectrum (532-nm laser), recorded on a meso-structured MAPbI$_3$ sample, deposited over a porous Al$_2$O$_3$ substrate, obtained with a 532-nm Raman line. (Reprinted with permission from Quarti, C., G. Grancini, E. Mosconi, P. Bruno, J. M. Ball, M. M. Lee, H. J. Snaith, A. Petrozza, and F. De Angelis. 2014a. The Raman spectrum of the CH$_3$NH$_3$PbI$_3$ hybrid perovskite: Interplay of theory and experiment. *J. Phys. Chem. Lett.* 5; 279–284. Copyright © 2014 American Chemical Society.) (b) Resonance Raman spectrum (532-nm laser), recorded on a thin-film MAPbI$_3$ sample. (Reprinted with permission from Park, B.-W., S. M. Jain, X. Zhang, A. Hagfeldt, G. Boschloo, and T. Edvinsson. 2015. Resonance Raman and excitation energy dependent charge transfer mechanism in halide-substituted hybrid perovskite solar cells. *ACS Nano* 9; 2088–2101. Copyright © 2015 American Chemical Society.) (c) Resonance Raman spectrum (532-nm laser) obtained for a mesoporous ("meso") and thin-film ("flat") sample of MAPbI$_3$. (Reprinted with permission from Grancini, G., S. Marras, M. Prato, C. Giannini, C. Quarti, F. De Angelis, M. De Bastiani, G. E. Eperon, H. J. Snaith, L. Manna, and A. Petrozza. 2014. The impact of the crystallization processes on the structural and optical properties of hybrid perovskite films for photovoltaics. *J. Phys. Chem. Lett.* 5; 3836–3842. Copyright © 2014 American Chemical Society.) (d) Resonance Raman spectra (514-nm laser) recorded on a MAPbI$_3$ thin-film sample. (Reprinted with permission from Ledinský, M., P. Löper, B. Niesen, J. Holovský, S. J. Moon, J. H. Yum, S. De Wolf, A. Fejfar, and C. Ballif. 2015. Raman spectroscopy of organic-inorganic halide perovskites. *J. Phys. Chem. Lett.* 6; 401–406. Copyright © 2015 American Chemical Society.) (*Continued*)

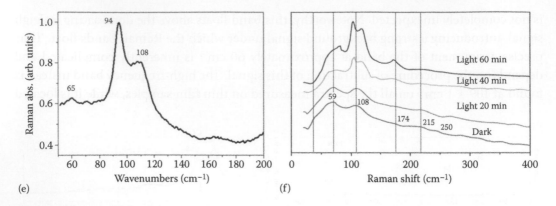

(e)     (f)

FIGURE 3.1.1 (CONTINUED)   Summary of the Raman spectra of MAPbI$_3$ at room temperature (tetragonal phase) reported in the literature. (e) Resonance Raman spectrum (514-nm laser) recorded on a MAPbI$_3$ thin film. (Reprinted with permission from Wei, J., Y. Zhao, H. Li, G. Li, P. J., D. Xu, Q. Zhao, and D. Yu. 2014. Hysteresis analysis based on the ferroelectric effect in hybrid perovskite solar cells. *J. Phys. Chem. Lett.* 5; 3937–3945. Copyright © 2014 American Chemical Society.) (f) Resonance Raman spectrum (532-nm laser) recorded on a MAPbI3 thin film. (Reprinted with permission from Gottesman, R., L. Gouda, B. S. Kalanoor, E. Haltzi, S. Tirosh, E. Rosh-Hodesh, Y. Tischler, A. Zaban, C. Quarti, E. Mosconi, and F. De Angelis. 2015. Photoinduced reversible structural transformation in free-standing CH$_3$NH$_3$PbI$_3$ perovskite thin films. *J. Phys. Chem. Lett.* 6; 2332–2338. Copyright © 2015 American Chemical Society.)

at 119 cm$^{-1}$ for a mesostructured sample deposited on a mesoporous Al$_2$O$_3$ scaffold. This dependence on the sample morphology inherently suggests that the 108 cm$^{-1}$ or 119 cm$^{-1}$ is associated with the libration of the organic cations, as these vibrations are expected to be strongly affected by the local environment probed by the MA cation.

Dealing with the minor features at 94 and 154 cm$^{-1}$, we notice that the first is present not only in Figure 3.1.1a, but also in Figure 3.1.1b, c, and e. On the other hand, a strong signal of the PbI$_2$ is known to take place at the same frequency and Ledinský et al. (2015), who do not find the present Raman signal in their spectra (Figure 3.1.1a), associate the band at 94 cm$^{-1}$ with the PbI$_2$ forming from the degradation of the sample. Whether or not a certain contribution to the intensity of the 94 cm$^{-1}$ could result from the PbI$_2$, several measurements agree on the presence of this peak and theoretical calculations confirm the presence of vibrational markers of MAPbI$_3$ around 94 cm$^{-1}$ (see later), which brings a non-negligible Raman intensity. In light of this, it is reasonable to consider the band at 94 cm$^{-1}$ as informative of the perovskite. The band at 154 cm$^{-1}$ instead, as observed by Quarti et al. (2014a) and confirmed by Grancini et al. (2014), is present also in the spectrum of Park et al. (2015). Also Gottesman et al. (2015) find a weak, unresolved band around 176 cm$^{-1}$ that could be related to the band at 154 cm$^{-1}$. Noteworthy, the morphology of the sample affects the shape and resolution of this band, which is better resolved in more ordered thin film samples of Figure 3.1.1b, c, and e. For this reason, the band at 154 cm$^{-1}$ was proposed as a genuine signal from the perovskite and as a marker of the crystalline order.

Finally, a broad, unresolved band was originally observed by Quarti et al. (2014a) in the region between 200 and 300 cm$^{-1}$, with a maximum at 250 cm$^{-1}$ (see Figure 3.1.1a).

The assignment of this band is a bit tricky, as it is quite unresolved and is present only in some measurements (see Figure 3.1.1a, c). On the one hand, the width of the present band suggested some contribution from the disordered phase, as widely happens in the case of polymers (Quarti et al. 2013). Again, note that for the "meso" sample in Figure 3.1.1c, where this band is broader and more intense, pointing to an increased local disorder due to the degree of freedom of the organic cation. On the other hand, theoretical DFT calculations on the crystalline phase of the MAPbI$_3$ do not predict any vibrational feature in this frequency region, neither Raman nor IR, finding only the torsional vibration of the MA cation at approximately 300 cm$^{-1}$. A first, tentative assignment by Quarti et al. (2014a) ascribed the 250 cm$^{-1}$ unresolved band to the torsional mode of the MA cation, which these authors demonstrated to down-shift and increase in intensity when the MA cation senses a more asymmetric environment. In this perspective then, the assignment of the band at 250 cm$^{-1}$ as another marker of the order of the perovskite is confirmed by DFT. The more recent investigation by Ledinský et al. (2015) proposed instead that the present peak is due mainly to the formation of PbI$_2$, as result of the degradation of the perovskite. A previous investigation by Baibarac et al. provides an interesting indication in this regard. These authors in fact observed that under nonresonance conditions, the Raman spectrum of PbI$_2$ does not show signals between 200 and 300 cm$^{-1}$. At the opposite, under a resonance condition, using a Raman line at 514 nm, a strong band at 212 cm$^{-1}$ appears (Baibarac et al. 2004). The appearance of this peculiar signal in the Raman spectrum of PbI$_2$ only under resonant conditions strongly supports the assignment by Ledinský et al. (2015).

### 3.1.2.2 Assignment of the Raman Spectrum from DFT Simulations

We now proceed to a detailed assignment of the experimental Raman features, on the basis of the DFT simulations reported in the work of Quarti et al. (2014a). In Figure 3.1.2, we report the comparison from Quarti et al. between the experimental Raman spectrum of Figure 3.1.1a (sample deposited on a mesoporous Al$_2$O$_3$ scaffold, under resonance conditions) and the spectrum obtained from DFT simulations.

Quarti et al. (2014a) performed the calculation of the vibrational spectrum using the harmonic approximation (Baroni et al. 2001). The dynamical matrix and the Raman intensities are calculated within periodic boundary conditions, using the crystalline structure proposed by Kawamura et al. (2002), properly relaxed. DFT calculations are performed resorting to the plane-wave/pseudopotential approach, as implemented in the Quantum-Espresso suite (Giannozzi et al. 2009), with the Local Density Approximation (LDA) for the exchange-correlation potential. Unfortunately, the calculation of the Raman intensities with more involved levels of theory, as the Generalized Gradient Approximations (GGA) or hybrid functionals, is not implemented in Quantum-Espresso, yet (Lazzeri and Mauri 2003). For more computational details, we refer the reader to the original paper (Quarti et al. 2014a).

The simulation of the Raman spectrum of the MAPbI$_3$ in the tetragonal phase from DFT simulations in this case presents at least three complications. On a "conceptual side," (1) the experimental Raman spectrum is in resonance condition, while only nonresonance calculations are currently available. Resonance effects are expected to seriously affect the intensity of the band, and thus the theoretical spectrum is not expected to perfectly reproduce

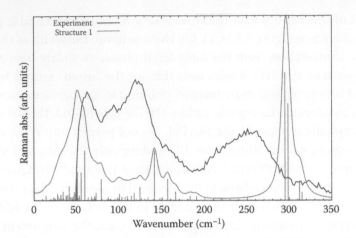

FIGURE 3.1.2 Comparison between the experimental spectrum of the MAPbI$_3$ perovskite, as in Figure 3.1.1a, and the DFT simulated spectrum. The theoretical spectrum is calculated at the DFT level, within the LDA approximation, using periodic boundary conditions, within a planewave/pseudopotential approach. For more details, see (Quarti et al. 2014a). (Reprinted with permission from Quarti, C., G. Grancini, E. Mosconi, P. Bruno, J. M. Ball, M. M. Lee, H. J. Snaith, A. Petrozza, and F. De Angelis. 2014a. The Raman spectrum of the CH$_3$NH$_3$PbI$_3$ hybrid perovskite: Interplay of theory and experiment. *J. Phys. Chem. Lett.* 5; 279–284. Copyright © 2014 American Chemical Society.)

the experiment, but it provides some useful guidelines for the assignment. (2) Normal modes, or phonons, in the low-frequency range are affected by serious anharmonic effects. The small energy of the vibrational quanta allows for the population of higher lying vibrational levels for all the phonons. In addition, the presence of several vibrational modes in this frequency region is such that band overtones can contribute significantly to the intensity of the Raman spectrum, especially under resonance conditions. On a "simulation side," (3) the MA cations are not fixed but they can rotate among different orientations, as discussed in the preceding text (Poglitsch and Weber 1987; Wasylishen et al. 1985). This implies the absence of a univocal minimum structure but several structures at similar energies and the actual Raman spectrum is associated with a thermal average of spectra of all the possible local minima that differ only in the orientation of the cations. Quarti et al. (2014a) simulated the Raman spectra for two structures, characterized by a different orientation of the MA cations. In Figure 3.1.2, we report only the structure with the best agreement with the experiment. In spite of all these complications, the theoretical spectrum in Figure 3.1.2 is in reasonable qualitative agreement with the experiment, reproducing the most important experimental features. Noteworthy, the DFT simulated spectra (Quarti et al. 2014a) reported in the Figure 3.1.2, together with that in Brivio et al. (2015) are the only two simulations that use periodic models for MAPbI$_3$ perovskite, which is the correct approach for the simulation of the Raman spectrum of an extended crystalline structure.

Comparison between experiment and simulation in Figure 3.1.2 allows the following assignments: the low-frequency marker at 62 cm$^{-1}$ found in the experiments is safely associated with the band at 52 cm$^{-1}$, while the high-frequency marker at 119 cm$^{-1}$ (or 109 cm$^{-1}$

in the other experiments) could be associated with the theoretical band at 141 cm⁻¹. As shown in Figure 3.1.2, the marker at 62 cm⁻¹ is associated with several normal modes of vibration, in the language of chemistry, or with several phonons in the center of the first Brillouin zone, in the language of physics. Analysis of the eigenvectors of the dynamical matrix, that is, of the direction of the atomic vibrations, demonstrates that these normal modes, or phonons, are due mainly to the vibration of the inorganic $PbI_6$ framework, with the organic cations that readapt as consequence of the variation of the surrounding potential, that is, by distortions of the $PbI_6$ cage. For the high-frequency 119 cm⁻¹ signal, instead, analysis of the vibrational eigenvectors demonstrates that this is essentially due the libration of the MA cations, that is, the vibration of the MA cation in terms of the position of its center of mass within the PbI6 cavity, plus the rotation of the CN axis. This is also consistent with the strong red-shift of this signal, from 119 cm⁻¹ for a mesostructured sample to 108 cm⁻¹ for a thin film sample, as discussed previously. The peak at 94 cm⁻¹ can be assigned to the quite strong band at 89 cm⁻¹ and it is associated mainly with the libration of the MA cation but with a significant readaptation of the inorganic framework. Thus, the present band is still informative of the inorganic framework of the perovskite. Finally, the mode at 154 cm⁻¹ is clearly a libration mode of the MA cation.

The assignment of the 62 cm⁻¹ and of the 94 cm⁻¹ bands as marker bands of the inorganic framework is also confirmed by other studies. The simulated DFT in Park et al. (2015), computed with a cluster approach, also found that the vibrational modes of the inorganic framework fall at a low frequency (~90 cm⁻¹) that is the high-frequency band associated with the inorganic framework. Also, previous work by Abid and co-workers on very similar 2D hybrid lead–halide perovskites confirms the present assignment. In their work, both Elleuch and Dammak (Dammak et al. 2009; Elleuch et al. 2008) assigned modes at 104 cm⁻¹ and at 80 cm⁻¹ to the inorganic framework, with the support of DFT cluster calculations. In particular, they assigned the bending modes of the inorganic components to bands around 50 cm⁻¹, and the stretching modes to bands predicted between 90 and 120 cm⁻¹, the former for the Pb–I bonds and the latter for the Pb–Br bonds. More recent theoretical calculations carried out on the orthorhombic (Pérez-Osorio et al. 2015), and both on the orthorhombic and tetragonal phases of the $MAPbI_3$ perovskite (Brivio et al. 2015) confirm this separation between vibrations of the inorganic cage (below 100 cm⁻¹) and inner vibrations of the MA cations, at larger frequencies. Both these works provide a nice assignment of the vibrations of the organic and of the inorganic component of the perovskite, on the basis of a quantitative analysis of the eigenvectors of the dynamical matrix obtained from DFT calculations (see Figure 3.1.3). These analyses confirm that the motion of the inorganic framework is still limited below the 100 cm⁻¹ region, thus supporting the assignment of the 94 cm⁻¹ to the vibration of the inorganic framework and the band at 108–119 cm⁻¹ and at 154 cm⁻¹ to the libration of the MA cations.

The lowest frequency inner vibration of the MA cations is the torsional normal mode and its assignment still represents an open issue. Waldron proposed the first assignment of this mode for the methylammonium chloride salt, MACl (Waldron 1953). Measuring the infrared (IR) spectrum of the α, β, and γ phases of this compound, he observed a sharp and clear band in the β phase at 487 cm⁻¹, that was found also in the γ phase at 478 cm⁻¹ but

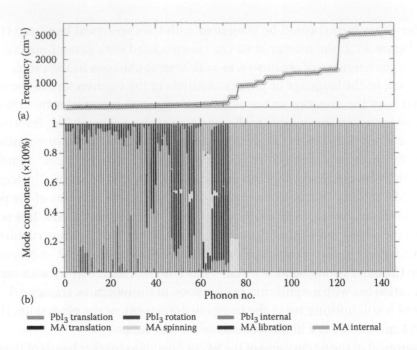

FIGURE 3.1.3 (a) DFT (GGA) vibrational frequencies calculated for MAPbI$_3$ in the orthorhombic phase. (b) Decomposition of the eigenvectors of the dynamical matrix (the atomic displacements) in contributions from different kinds of motions, from the organic and the inorganic components: the PbI$_3$ translation (gray), PbI$_3$ rotation (brown), PbI$_3$ internal (blue), MA translation (black), MA spinning (yellow), MA librations (red), MA internal (green). (Reprinted with permission from Pérez-Osorio, M. A., R. L. Milot, M. R. Filip, J. B. Patel, L. M. Herz, M. B. Johnston, and F. Giustino. 2015. Vibrational properties of the organic-inorganic halide perovskite CH$_3$NH$_3$PbI$_3$ from theory and experiment: Factor group analysis, first-principles calculations, and low-temperature infrared spectra. *J. Phys. Chem. C* 119; 25703–25718. Copyright © 2015 American Chemical Society.)

with lower intensity, while it completely disappears in the α phase. This was in part supported by the observation of many overtone bands in the IR spectrum that were explained on the basis of the combinations of various normal modes with a mode at about 480 cm$^{-1}$. On the contrary, Waldron also pointed out the large difference in the vibrational frequency of this mode, compared to the torsional frequency of similar compounds (for ethane in solution, this falls at 275 cm$^{-1}$). A later work by Theoret and Sandorfy (1967) confirmed the assignment by Waldron. They observe in fact that in the IR spectra of the δ phase of methylammonium bromide and iodide a band was found respectively around 312 and 281 cm$^{-1}$. These bands are close to the torsional frequency of ethane in solution, thus making their assignment to the torsional mode of the cation more plausible. In addition, through isotopic substitution studies, these authors excluded the possible assignment of these bands to lattice vibrations. In more recent works, this mode was associated to the band at 427 cm$^{-1}$ for the CH$_3$NH$_3$HgCl$_3$ and to the band at 487 cm$^{-1}$ for the CH$_3$NH$_3$PbCl$_3$ perovskite in the low temperature orthorhombic phase (Jiang et al. 1995). To the best of the authors' knowledge, no experimental data are available for the torsional frequency of

TABLE 3.1.1    Summary of the Raman Marker Bands for the MAPbI₃ and Their Assignment

| Raman Band (cm⁻¹) | Marker | Detailed Assignment |
|---|---|---|
| 59–71 | Inorganic framework | Pb–I bending |
| 94 | Mixed organic–inorganic | Pb–I stretching |
| 108–119 | Organic | MA librations |
| 154 | Organic component | MA librations |
| 250 | Debated | Mainly PbI₂ Possible contributions from the torsional mode of the MA cations |

the methylammonium cation in solution or in the gas phase, but a few theoretical calculations in literature locate this mode at 267 and 302 cm⁻¹ [calculations at the Hartree Fock level of theory (DeFrees and McLead 1985)], and at 316 and 324 cm⁻¹ [MP2 calculation (Zeroka and Jensen 1998)]. In Figure 3.1.2, the strong signal in the theoretical spectrum at approximately 300 cm⁻¹ is assigned to the torsional vibration of the MA cations. The intensity predicted from theoretical simulation is largely overestimated, likely because of the inaccurate charge flux associated to the formation of the hydrogen bond between the ammonium group and the iodide atoms, which is well known for DFT for the case of the X–H (with X=C, N, O) stretching vibrations (Galimberti et al. 2013a,b). Basic symmetry arguments in fact predicts that this vibrational mode is both IR and Raman silent, for the isolated molecule, since it has $A_2$ irreducible representation; similarly, it is supposed to have small Raman activity also in the case of MAPbI₃. Quarti et al. (2014a) tentatively assigned the torsional mode to the broad band at 250 cm⁻¹, and they also demonstrated by means of DFT calculations that this band red-shifts and gains intensity when the MA cation probes a less symmetric environment, thus confirming this band as a marker of the crystalline order. In contrast, Ledinský et al. (2015) assigned the band at 250 cm⁻¹ to the formation of PbI₂, following the material degradation. On the other hand, these authors do not propose any assignment for the torsional mode of the MA cation. In light of the work of Ledinský et al. and of the peculiar Raman response of the PbI₂ under resonance conditions reported by Baibarac et al., it is quite likely that the broad, unresolved band at 250 cm⁻¹ is actually due to PbI₂ formation (Baibarac et al. 2004). The assignments of the spectroscopic Raman markers of the MAPbI₃ perovskite are summarized in Table 3.1.1.

### 3.1.2.3 Structure/Electronic Properties and Degradation of Hybrid Perovskites in Working Photovoltaic Devices

In the previous sections, we presented an exhaustive overview of the Raman spectroscopic data for the MAPbI₃ perovskite, the most studied hybrid metal–halide perovskite for photovoltaic applications. In particular, we aimed to underline the fact that the present material presents a large degree of local and dynamic structural disorder. Although it has been reported that the organic cations do not participate as frontier orbitals in the conduction and valence band, their degrees of freedom have been proven to strongly impact the optoelectronic properties and energetic landscape of the semiconductor (De Bastiani et al. 2014). The optical-absorption edge, that is, the band gap, and the exciton binding energy are a few of the fundamental parameters that are influenced by the local disorder induced by the organic cation (Grancini et al. 2015a). Together, this class of materials also shows poor stability, especially under illumination. In the following section, we summarize some interesting investigations on the optoelectronic properties and the photovoltaic working mechanism of hybrid lead–halide perovskites, mainly $MAPbI_3$ and $MAPbI_{3-x}Cl_x$, carried out with Raman spectroscopy. First, we present the impact of the local order and how the motion of the MA cation may affect the optoelectronic properties of the $MAPbI_3$ perovskite. We then deal with the photovoltaic properties of this class of materials, considering the effect of the environment, in particular, humidity, and of an externally applied electric field. In both cases, different electronic/photovoltaic responses are connected with structural information using Raman spectroscopy. In this perspective, the assignments reported in Table 3.1.1 are fundamental to interpreting the following results.

### 3.1.2.4 How Structure Impacts the Optical and Photophysical Properties

The role of disorder in hybrid perovskites, intrinsic in the polycrystalline nature of the material, has been often associated to the degrees of freedom of the organic cations within the inorganic cage. Their orientation order, indeed, affects the structural deformation of the crystal lattice. In Figure 3.1.4a, the Raman spectrum of $MAPbI_3$ is compared for two different sample morphologies, a mesostructured morphology (hereafter named "meso"),

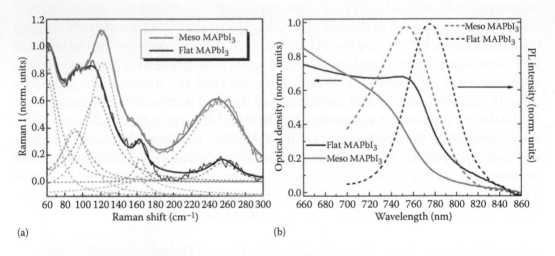

(a)                                              (b)

FIGURE 3.1.4   (a) Resonant Raman spectra of two $MAPbI_3$ samples with different morphologies: A sample grown on a mesoporous $Al_2O_3$ scaffold ("meso") and a thin-film sample, grown on a glass substrate ("flat"). Dashed lines represent the Gaussian model used to fit the Raman peaks; the thick lines are the results of the fit, as the sum of the individual Gaussian functions. (b) Normalized optical absorption (at 700 nm) for the "meso" $MAPbI_3$ and for the "flat" $MAPbI_3$ (solid lines) and corresponding CW photoluminescence spectra (dashed lines). (Reprinted with permission from Grancini, G., S. Marras, M. Prato, C. Giannini, C. Quarti, F. De Angelis, M. De Bastiani, G. E. Eperon, H. J. Snaith, L. Manna, and A. Petrozza. 2014. The impact of the crystallization processes on the structural and optical properties of hybrid perovskite films for photovoltaics. *J. Phys. Chem. Lett.* 5; 3836–3842. Copyright © 2014 American Chemical Society.)

using an $Al_2O_3$ mesoporous scaffold as substrate, and a thin-film one (hereafter named "flat"), using a flat glass substrate. On one side, the crystallization process affects the opto-electronic properties through the modulation of the lattice strain (Grancini et al. 2014). Raman analysis of $MAPbI_3$ crystals constrained to growth within the nanometer-scale mesoporous oxide scaffold presents a more distorted structural arrangement with respect to micrometer-size grains grown in a flat substrate that, in contrast, exhibit a larger degree of long-range order. As a result, the organic cations in small crystals confined by the $Al_2O_3$ scaffold might be more randomly oriented within the inorganic cage.

Raman spectroscopy is a useful tool to investigate the material structure. In particular, the different degrees of order can be retrieved considering the Raman mode at 154 cm$^{-1}$ and 250 cm$^{-1}$, as discussed previously (see Table 3.1.1). The latter spectroscopic signal especially, being sensitive to the specific organic–inorganic interactions (occurring mainly through Coulomb and hydrogen bond interactions between the $NH_3$ groups of MA and the electronegative iodine atoms), represents an important marker of the orientation disorder of the material. In particular, in the more disordered "meso" morphology of $MAPbI_3$ (whose extreme is represented by the cubic phase for $T > 327$ K) the MA cations are arranged in a random orientation within the inorganic cage owing to their large orientation mobility. This leads to a broadening and a red-shifting of the band at around 250 cm$^{-1}$. On the contrary, in the more ordered "head to tail" arrangement of the MA cations, this band loses intensity (by symmetry selection rules) and considerably blue-shifts. The broad shoulder peaking at <160 cm$^{-1}$ is assigned to libration of the MA cations, calculated at 156 cm$^{-1}$. From the simulation it is additionally expected to become sharper and more intense, in the more ordered "head-to-tail" arrangement. Analyzing the evolution of the Raman spectrum from the "meso" to the "flat" $MAPbI_3$ sample a clear trend is visible at first glance: (1) the band at around 250 cm$^{-1}$ blue-shifts and its intensity strongly decreases; (2) the peak at 119 cm$^{-1}$ decreases in intensity and red shifts to 115 cm$^{-1}$; (3) the band at around 160 cm$^{-1}$ becomes sharper and more resolved; and (4) the peak at around 94 cm$^{-1}$ gains strength.

According to the preceding rationalization, these observations highlight the different organic–inorganic interactions affecting the orientation order of the organic cation in the unit cell in relation to the different morphologies of the sample. The experimental findings point to a clear trend, that is, an ordered ("head-to-tail" like) arrangement of the MA cations in "flat" samples versus a more disordered cation arrangement in the "meso" sample. A similar effect was found by Mosconi et al. (2014) investigating the evolution of the IR spectrum of the $MAPbI_{3-x}Cl_x$ with temperature. These authors observed a clear blue-shift of a broad, unresolved band at 220 cm$^{-1}$ and a decrease of its IR intensity with decreasing temperature, thus corresponding with an expected increase of the orientation order of the MA cations. It is important to note that, as a consequence of increasing structural order, the peak at 94 cm$^{-1}$ increases in intensity and becomes sharper along with the reduction of the close peak at 119 cm$^{-1}$. This appears in the spectrum as an inversion of their relative amplitudes. A similar effect was already observed in the literature when comparing pure ordered $PbI_2$ crystals [for which the mode at 96 cm$^{-1}$ represents the main peak (Baibarac et al. 2004)] with ammonia-intercalated $PbI_2$ compounds, which present distorted Pb–I bonds. Overall, our analysis suggests an ordered arrangement of the organic cation in

MAPbI$_3$ crystallites grown on "flat" substrates, whereas it results in fully random orientation in the crystallites grown in the "meso" scaffold.

Although the MA cations do not directly participate to the frontier orbitals of the semiconductor, their interaction with the inorganic cage and their displacement affect the electronic properties of the compound. In Figure 3.1.1b, we report for both the "meso" and "flat" samples of the MAPbI$_3$ perovskite the absorption and the photoluminescence measurements, informative of the optoelectronic properties of the material. Indeed, a clear difference at the onset of the UV–VIS spectra is observed for the two different morphologies: the "flat" sample, characterized by large crystals, exhibits a red shifted UV–VIS and photoluminescence spectra indicative of a shrinking of the band edge. On the contrary, the "meso" sample, characterized by small crystals, presents an expanded band gap, mainly due to disorder. In addition, note that a clear excitonic transition is observed only for the "flat" sample, already evident just below room temperature. No excitonic transition instead is observed, even at 4K, for the "meso" film. Thus, the degree of material order imposed by the MA cations and their collective motions impact on the exciton binding energy affecting the exciton screening as discussed in the papers of Grancini et al. and D'Innocenzo et al. (D'Innocenzo et al. 2014; Grancini et al. 2015a). This univocally proves that the degree of freedom of the organic cation is the dynamical trigger able to induce, through hydrogen bonding interaction, the local distortion of the lattice, and to tune the optoelectronic properties of the material modifying the local dielectric properties.

In the "meso" sample, where the small grains are grown within the scaffold, the local disorder induced by the degree of freedom of the organic cations prevents the formation of a stable exciton population (Grancini et al. 2015a). In the other case, the preferential order of the organic cation in the large crystalline film may possibly slow down the rotational motion of the MA cations, which in turn affects the screening of the excitonic transition at the onset of the absorption spectrum. In particular, the rotation of the organic cation, because it possesses a permanent dipole moment, can respond in a dielectric manner modulating the electrostatic potential within the crystal, as recently modeled (Filippetti et al. 2014; Frost et al. 2014; Grancini et al. 2015a). In particular, as for the case of the small-grain morphology, this would result in a higher dielectric constant, responsible for screening the electron–hole interaction and thus reducing the exciton binding energy. If one considers the Mott–Wannier exciton binding energy as $E_b = \dfrac{m^* e^4}{\hbar^2 \varepsilon^2}$ the dielectric response of the organic cation will increase $\varepsilon$, thus reducing the binding energy (Poglitsch and Weber 1987). This would result in the exciton sampling a higher dielectric constant, thus decreasing its binding energy from around 40 meV [as calculated for the large crystal in D'Innocenzo et al. (2014)] to 2 meV and increasing the size to 19 nm, and thus resulting in the immediate dissociation of the exciton into free charges. This dynamical behavior can be monitored by transient absorption spectroscopy that follows the photoexcited state dynamics. As summarized in Figure 3.1.5c, the transient absorption spectra look completely different as functions of the sample morphology. On 1 ps the positive band is associated with band edge photo bleaching on carrier thermalization. No exciton feature is present. On the contrary, the sample made

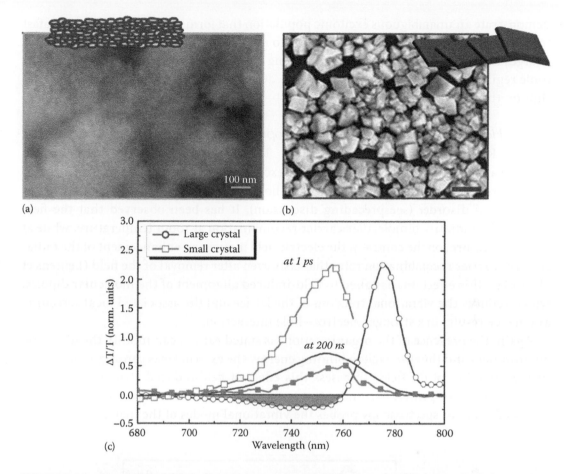

FIGURE 3.1.5    Top: SEM images of MAPbI$_3$ sample with different crystalline sizes (scale bar: 2 μm). (a) "Small crystals" composed of small grains of less than 150 nm. (b) "Large crystals" composed of large micrometer size cuboids are grown. (c) Comparison on transient absorption spectra at 1 ps and 200 ns on photoexcitation for the two different morphologies as indicated in the legend. (Reprinted by permission from Macmillan Publishers Ltd., Nature Photonics, Grancini, G., A. R. Srimath Kandada, J. M. Frost, A. J. Barker, M. De Bastiani, M. Gandini, S. Marras, G. Lanzani, A. Walsh, and A. Petrozza. 2015a. Role of microstructure in the electron-hole interaction of hybrid lead-halide perovskites. 9; 695–701, Copyright 2015.)

of approximately 1 μm large crystals exhibits different spectral features. The transient-absorption spectrum formed in 1 ps on photoexcitation above the band gap resembles the one obtained in quantum confined two-dimensional perovskite (Zhu et al. 2014), where a clear stable excitonic state is also present at the band edge. Importantly, at longer time delays (200 ns) this red-shifted feature disappears and the free carrier bleach is observed. This dynamic reflects the decay of the excitonic population that appears to be shorter lived with respect to the free carrier population. The electron–hole separation due to electrostatic disorder can be significant in small crystals (but weaker in large crystals, allowing Wannier exciton formation). Thus, the local disorder found for smaller crystals suppresses exciton formation, while larger crystals of the same composition demonstrate an unambiguous excitonic population that forms in less than 1 ps. Ultrafast transient absorption spectra have been used to monitor the dynamics of exciton and free charges in these different structures, providing evidence that both free carrier and excitonic regimes are possible, depending on the degree of order of the organic cations and thus on the crystalline structure.

### 3.1.2.5 How Structure Reflects the Device Properties

3.1.2.5.1 Raman Under an Electric Field Only    The presence of an external electric field can also affect the photophysics of the photoexcited species of $MAPbI_3$ thin films, manifested through the variation of the electron–hole interaction, with a similar role played by the local disorder (see preceding discussion). It has been observed that the field reduces the radiative bimolecular carrier recombination at room temperature, while at low temperature, on the contrary, the electric field induces an enhancement of the radiative free carrier recombination rates that lasts even after removal of the field (Leijtens et al. 2015a). This effect was assigned to field-induced alignment of the molecular dipoles, which reduces the vibrational freedom of the lattice and the associated local screening and hence results in a stronger electron–hole interaction.

Again, the presence of the organic cation, as stated earlier, can modify the dielectric environment and thus the exciton binding energy. The external bias can align the dipoles in the direction of the field. To assess this effect of field-induced structural changes, Raman spectra of perovskites under the applied bias were measured (see Figure 3.1.6). As noted, Raman spectroscopy probes the vibrational modes of the material and hence

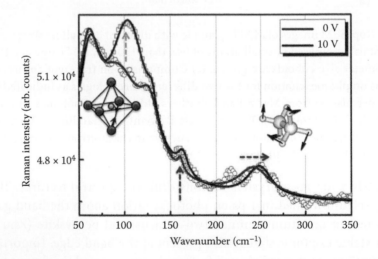

FIGURE 3.1.6    Raman spectra collected at 0 V (blue open circles) and at 10 V (black open squares) bias applied across the perovskite film. The solid lines represent the fit. (Reprinted with permission from Leijtens, T., A. R. Srimath Kandada, G. E. Eperon, G. Grancini, V. D'innocenzo, J. M. Ball, S. D. Stranks, H. J. Snaith, and A. Petrozza. 2015a. Modulating the electron-hole interaction in hybrid lead halide perovskite with an electric field. *J. Am. Chem. Soc.* 137; 15451–15459. Copyright © 2015 American Chemical Society.)

the interaction between the organic and inorganic moieties. As discussed in Section 3.1.2.2, the low-frequency region of the Raman spectrum can be divided in two main regions: below and above approximately 90 cm$^{-1}$, which are assigned respectively to the vibration framework and to the libration of the MA cations within the cubo-octahedral cavities. Both spectra in Figure 3.1.6 show the features in agreement with the assignment discussed in the previous section. However, one could observe that in the presence of an electric field a shift and narrowing of the transitions related to the organic modes occurs. This behavior is typical when the degrees of freedom of the cation are reduced. These results demonstrate that the field reduces the orientation mobility of the organic cation, thus limiting the degrees of freedom of the vibrational modes in the perovskite lattice. This correlates with an enhanced electron–hole Coulomb interaction. Alignment of dipoles induces dipole twinning and restricts their rotations. This will reduce the effective dielectric permittivity of the material at frequencies (GHz–THz) that are relevant for carrier interaction. Similarly to what was discussed in the previous paragraph, the electron–hole interaction is tuned albeit the control is achieved by the applied electric field.

3.1.2.5.2 Raman Under Air/Humidity    Considering the relation between structure and properties of the hybrid perovskite device under operative condition one could not neglect the potentially enormous effect induced by the presence of water/humidity/moisture in the atmosphere. In this regard, specific Raman measurements of a single MAPbI$_3$ crystal in an inert N$_2$ atmosphere and under high-humidity conditions have been performed. As one would expect, the results of humidity exposure particularly affect the border of the crystal (Grancini et al. 2015b), leading to a shrinking of the band edge and a local distortion of the crystal surface (Grancini et al. 2015b). As shown in Figure 3.1.7, focusing on the structural changes with humidity, it has been demonstrated that the Raman spectra clearly show marked differences. In particular, the peak at 109 cm$^{-1}$ shifts in energy and increases in intensity, demonstrating evidence of local distortion. Additionally, the mode of the MA cation at 250 cm$^{-1}$ red shifts. This behavior can be due to the effect of hydrogen bonding interactions between water

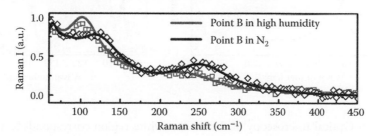

FIGURE 3.1.7    Comparison of the Raman spectra at the crystal edge, under dry N$_2$ versus in a high-humidity environment. (Reprinted from Grancini, G., V. D'Innocenzo, E. R. Dohner, N. Martino, A. R. Srimath Kandada, E. Mosconi, F. De Angelis, H. I. Karunadasa, E. T. Hoke, and A. Petrozza. 2015b. CH$_3$NH$_3$PbI$_3$ perovskite single crystals: Surface photophysics and their interaction with the environment. *Chem. Sci.* 6; 7305–7310. Published by The Royal Society of Chemistry. This article is licensed under a Creative Commons Attribution 3.0 Unported Licence.)

molecules and the organic cations particularly relevant at the crystal edge. They act as an open gate for spontaneous water incorporation, ultimately leading to a widening of the band gap along with structural distortion occurring at the crystal edges. Importantly, it was demonstrated that these effects are reversible, distinct from the irreversible perovskite degradation through its conversion into its hydrated phases. The observed phenomenon can likely be an initial step of structural reorganization toward conversion to a hydrated phase.

3.1.2.5.3 Raman Under an Electric Field and Humidity    Reversible and irreversible effects on application of an electric field under different environmental conditions have been the subject of intense study. It was demonstrated that the application of an electric field under inert conditions leads only to a reversible poling on a time scale of minutes. It was also found that

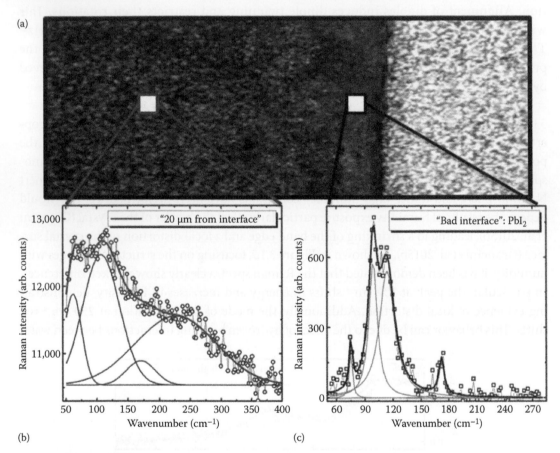

FIGURE 3.1.8    (a) Optical microscopy of the device (white region corresponds to the electrode), (b) Raman spectra taken in the central part of the device, not degraded, and (c) on degraded regions of material close to the left electrode. The thick lines represent the fits of the spectra using four Gaussian peaks. (Reprinted with permission from Leijtens, T., E. T. Hoke, G. Grancini, D. J. Slotcavage, G. E. Eperon, J. M. Ball, M. De Bastiani, A. R. Bowring, N. Martino, K. Wojciechowski, M. D. McGhee, H. J. Snaith, and A. Petrozza. 2015b. Mapping electric field-induced switchable poling and structural degradation in hybrid lead halide perovskite thin films. *Adv. Energy Mater.* 5; 1500962. Copyright © 2015 Wiley-VCH Verlag GmbH & Co. KGaA.)

the presence of moisture results in an irreversible degradation in the presence of an electric field (Christians et al. 2015; Habisreutinger et al. 2014; Tress et al. 2015; Xiao et al. 2015; Yang et al. 2015). The combined presence of humidity and an electric field was demonstrated to lead to irreversible degradation of the semiconductor through the decomposition to the hydrated phase, ultimately leading to decomposition into $PbI_2$. This has direct relevance to perovskite solar cells; hysteretic behavior in current–voltage curves is aggravated by the presence of moisture while devices aged under load accelerate degradation (Leijtens et al. 2015b). It was recently predicted that when an electric field is applied while the device is kept in ambient air, a rapid irreversible degradation occurs. Interestingly, this phenomenon has a spatial dependence: it starts from the region near the positively biased electrode. As shown in the optical microscope image in Figure 3.1.8a, the degraded material (yellow in the original paper) corresponds to PbI2, as confirmed by micro-Raman, and in Figure 3.1.8c that closely resembles the $PbI_2$ spectrum, as discussed at the beginning. It was proposed that in the presence of humidity ion migration occurs more readily in response to an electric field. This is different from the reversible motion of intrinsic defects described for poling under inert conditions (D'Innocenzo et al. 2014), and drastically accelerates the degradation of the perovskite to $PbI_2$. This suggests that the density of mobile ions in a perovskite film will depend on the environmental conditions. Thus hysteresis in the $J$–$V$ curves of perovskite solar cells is strongly affected by the presence of moisture, while stressing a device in the presence of even a small amount of moisture at a working bias severely accelerates device degradation.

### 3.1.3 Extremely Slow Photocurrent and Photoluminescence Response in Hybrid Perovskites

In the last section of this chapter, we describe the slow photocurrent and photoluminescence response of the $MAPbI_3$ perovskite material under bias and illumination, which are peculiar phenomena for this class of material and that can inherently limit the applicability of hybrid perovskite in real technology. Similarly to the well reported hysteretic behavior of the $J$–$V$ curve (Snaith et al. 2014), these effects involve a response of the perovskite on a time scale that is much longer with respect to the typical electronic phenomena taking place in photovoltaic devices, on the order of seconds and minutes. It is also quite likely that this slow photoresponse is only another side of the coin concerning the physics behind the hysteresis of the $J$–$V$ curve.

#### 3.1.3.1 Extremely Slow Photoconductivity Response

In their paper Gottesman et al. (2014) used the device in Figure 3.1.9a to measure the time evolution of the photocurrent of the $MAPbI_3$. The $MAPbI_3$ perovskite is deposited between two gold electrodes spaced approximately 2000 nm apart. A thin layer of $Al_2O_3$ is deposited on top of the gold electrodes, to prevent electrical interactions between the deposited $MAPbI_3$ directly on the top surface of the gold. The whole circuit is deposited on glass and illuminated from below. The novelty of the present device is the use of symmetric electrodes that avoid a response biased by the presence of different contacts. For more details on the experimental setup, we send readers to the original paper (Gottesman et al. 2014).

In the present measurements, the sample is subjected to an external bias, while a white LED is switched on several times for a period of 5 minutes each, followed by a dark period

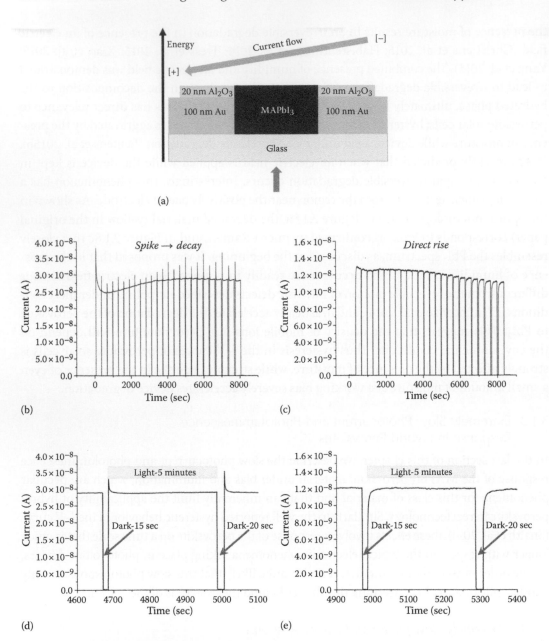

FIGURE 3.1.9 (a) Structure of the device used to measure the photocurrent in the $MAPbI_3$ perovskite. (b, c) Full i–t profiles for the spike → decay and direct rise cases showing the full range of cycles. The dark times are visible by the drop in the current when the light was turned off (each time for a different period) and then turned on for 5 minutes. (d, e) 500-second sections taken from panels a and c, respectively showing a dark time of 15 seconds before illuminating for 5 minutes, then the light is turned off again for 20 seconds, and so on. (Reprinted with permission from Gottesman, R., E. Haltzi, L. Gouda, S. Tirosh, Y. Bouhadna, A. Zaban, E. Mosconi, and F. De Angelis. 2014. Extremely slow photoconductivity response of $CH_3NH_3PbI_3$ perovskites suggesting structural changes under working conditions. *J. Phys. Chem. Lett.* 5; 2662–2669. Copyright © 2014 American Chemical Society.)

of variable duration, from 1 s to 60 s. In these measurements, what is generally observed is a spike in the photoconductivity, followed by a decay. In the present measurements, the authors found this response (Figure 3.1.9b, d) for some samples, but they also found a gradual rise to the steady-state response, on other samples (Figure 3.1.9c, d).

In Figure 3.1.10, the dynamics of the response is reported for both the spike → decay cases (Figure 3.1.10a, b) and for the direct rise cases (Figure 3.1.10d, e), both under 0.1 V (Figure 3.1.10a, d) and 0.3 V (Figure 3.1.10b, e). For both the spike → decay and the direct rise, cases the amount of time required to reach the steady-state current is on the order of many seconds. These values are enormous compared with the times of the typical processes taking place in other well-known semiconductors employed in photovoltaic devices. Moreover, it is surprising that such behavior is characteristic of the isolated MAPbI$_3$ film, as the device in Figure 3.1.9a does not contain selective contacts. Noteworthy, this photocurrent time-response depends (1) on the duration of the dark period before the reillumination, with a slower response for longer dark periods, and (2) on the external potential applied, with a slower response for larger potential.

On the basis of the extremely long time scale of the photocurrent evolution, the authors suggested three possible mechanisms for the observed photoconductivity response: (1) ion migration through the film under illumination and electric field, (2) photoinduced traps for charge carriers in the MAPbI$_3$ (within the bulk or near the surface/interface), and (3) the alignment of the MA cation in the material, induced by a combination of illumination and applied bias [Stoumpos et al. (2013) reported in fact ferroelectric properties for the MAPbI$_3$ perovskite in the tetragonal phase]. In the paper, the first two mechanisms were considered unlikely, in light of the symmetry of the device and of the opposite response of the photocurrent, namely, spike → decay and direct rise. Instead, DFT calculations were reported to support the hypothesis of a change in the degree of alignment of the MA cations. A model structure for the tetragonal MAPbI$_3$ was relaxed both in the ground and in the first triplet state, the latter assumed to mimic the structure resulting from the transition to the excited state following the absorption of light. This assumption is reasonable, as the transition from the ground to the first triplet state is similar to the transition to the excited state, consisting in a transfer of electron density from the iodine to the lead atoms. Calculations showed that the binding energy of the MA ions to the inorganic cage decreases of 0.07 eV going from the ground to the triple state. This reduction in the binding energy likely allows the MA cations to rotate freely on light excitation and justifies the formation of a phase with a different amount of alignment of the MA cations, thus affecting the mechanism of exciton screening by collective motion of the MA cations, as suggested by Even et al. (2014) and affecting the final measured photocurrent.

Foremost, these data point out a non–steady-state response of the MAPbI$_3$ perovskite to an external field, which is likely connected to the hysteretic behavior of the $J$–$V$ curve of this material. However, a unified comprehension at the origin of this effect is still to be achieved. The proposed mechanism of the alignment of the MA cations in fact still would greatly benefit from further simulation activity and from structural characterization studies, aimed to demonstrate the presence and the nature of a structural reorganization. Some recent results in this sense will be shown in the text that follows. On the other hand, the

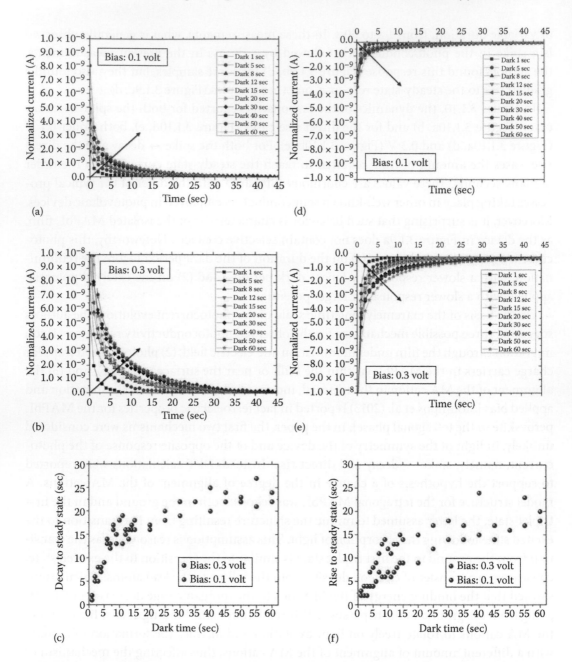

FIGURE 3.1.10 (a, b) Spike → decay and (d, e) direct rise normalized current decay plots to a designated values of 0 ampere (steady-state current) of different dark times. (a, d) Applied bias of 0.1 V. (b, e) 0.3 V (the black arrows point from short to long dark durations). (c, f) Plots showing the time (in seconds) the current takes to decay to a value of 5% from the average peak value at each bias versus the dark time before each illumination period. (Reprinted with permission from Gottesman, R., E. Haltzi, L. Gouda, S. Tirosh, Y. Bouhadna, A. Zaban, E. Mosconi, and F. De Angelis. 2014. Extremely slow photoconductivity response of $CH_3NH_3PbI_3$ perovskites suggesting structural changes under working conditions. *J. Phys. Chem. Lett.* 5; 2662–2669. Copyright © 2014 American Chemical Society.)

mechanism of ion migration has recently received substantial attention (Haruyama et al. 2015; Tress et al. 2015), as it can explain some recent chronoamperometry measurements under various conditions of bias/illuminations (Eames et al. 2015). In the present case, however, the mechanism of ion migration is still not sufficient to explain the presence of two different kinds of response (spike → decay and direct rise), especially for a device with symmetric contacts as in Figure 3.1.9a. The current understanding of the mechanism of ion migration implies in fact that, in correspondence of a switch in the sign of the external bias, we should observe the opposite photoconductivity response, that is, from spike → decay to direct rise and vice versa. Instead, Gottesman et al. clearly state that "...changing the magnitude or the sign of the bias did not shift between the two cases" (Gottesman et al. 2014, p. 2663).

### 3.1.3.2 Evolution of the Raman and Photoluminescence Response of the MAPbI₃ under Illumination

To conclude, we now describe the time-dependent Raman and photoluminescence response of hybrid, lead–halide perovskites, with respect to the most fundamental external parameter for a photovoltaic material: the light. In their work Gottesman et al. (2015) performed the following measurement. Depositing a thin-film sample of MAPbI₃ perovskite, previously deposited on glass, and properly sealing it, these authors measured the Raman spectrum under different cycles of dark–light. The thin film was illuminated with a white LED, with an intensity <0.1 mW cm⁻² (0.1 sun). The authors clearly demonstrate through structural analysis that the illumination does not imply any degradation in the material. The Raman spectrum was measured in dark, then after 20, 40, and 60 minutes of illumination, as shown in Figure 3.1.11. The authors verified that after some minutes under dark conditions, the Raman spectrum completely recovered the original form, thus demonstrating that the sample does not degrade.

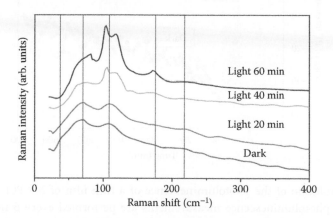

FIGURE 3.1.11   Raman spectrum of the MAPbI₃ perovskite under dark, after 20 minutes, 40 minutes, and 60 minutes of illumination. (Reprinted with permission from Gottesman, R., L. Gouda, B. S. Kalanoor, E. Haltzi, S. Tirosh, E. Rosh-Hodesh, Y. Tischler, A. Zaban, C. Quarti, E. Mosconi, and F. De Angelis. 2015. Photoinduced reversible structural transformation in free-standing CH₃NH₃PbI₃ perovskite thin films. *J. Phys. Chem. Lett.* 6; 2332–2338. Copyright © 2015 American Chemical Society.)

The Raman spectrum clearly shows that the peak at 108 cm$^{-1}$, assigned to the libration of the MA cations (Table 3.1.1), gains intensity and splits in a doublet at approximately 100–110 cm$^{-1}$ under illumination. The high-frequency doublet component is associated with the peak at 108 cm$^{-1}$ (see Table 3.1.1), assigned to the libration of the MA cations. The low-frequency doublet component could be reasonably assigned to the band at 94 cm$^{-1}$, informative of both the inorganic and the organic component of the perovskite. This variation of the Raman spectrum could be the marker of a transition to a new structure, which becomes metastable under illumination, consistent with the mechanism of an alignment of the MA cations, proposed in the aforementioned work of Gottesman et al. (2014). The reorganization of the MA cations in more aligned domains would result in fact in the consequent relaxation of the inorganic framework, which is expected to take place over a long time scale (seconds to minutes) over macroscopic MAPbI$_3$ samples. Unfortunately, the low resolution of the Raman spectrum, and the inherent complications associated with a reliable simulation from first principles, do not allow drawing definitive conclusions on the details of this structural evolution at the microscopic scale. XRD measurements instead provide interesting indications, showing a clear evolution under illumination of the intensity ratio of the 24.55° and the 23.54° signals in the XRD spectrum, which is associated with specific features of the inorganic structure (Quarti et al. 2014b). This ratio increases under illumination and returns to the original values under dark. Noteworthy, this variation of the XRD pattern is reversible, as is the evolution of the Raman spectrum.

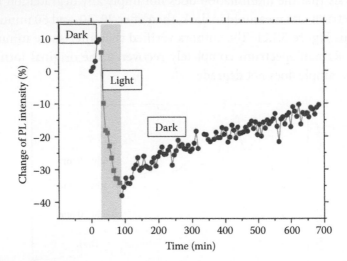

FIGURE 3.1.12 Evolution of the photoluminescence of a thin film of MAPbI$_3$ perovskite under illumination. The photoluminescence measurements are performed every 6 minutes, switching off the white LED, to avoid photon emission due to the LED. (Reprinted with permission from Gottesman, R., L. Gouda, B. S. Kalanoor, E. Haltzi, S. Tirosh, E. Rosh-Hodesh, Y. Tischler, A. Zaban, C. Quarti, E. Mosconi, and F. De Angelis. 2015. Photoinduced reversible structural transformation in free-standing CH$_3$NH$_3$PbI$_3$ perovskite thin films. *J. Phys. Chem. Lett.* 6; 2332–2338. Copyright © 2015 American Chemical Society.)

Together with this variation of the Raman and XRD pattern under illumination, which strongly point out a structural evolution of the material, Gottesman et al. also found impressive evolution of the electronic properties, that is, the photoluminescence. The authors excited the $MAPbI_3$ thin film perovskite under dark, using an excitation at 390 nm and then they illuminated the sample with the LED for long periods measuring the photoluminescence every 6 minutes. During the measurement phase, the authors switched off the LED temporarily, to avoid unwanted emission due to the LED illumination. A typical photoluminescence measurement is reported in Figure 3.1.12. A starting dark period of 25 minutes was followed by 1 hour of LED illumination and then again by a dark period. Figure 3.1.12 shows a strong, accumulative effect of white light illumination on the photoluminescence intensity. Within 1 hour of illumination (<0.1 sun) the photoluminescence gradually decreased by 40%. On turning off the LED, a very slow recovery of the photoluminescence started, reaching 15% less than the original value after 10 hours. The reversibility of the photoluminescence evolution under illumination strongly indicates that there is no sample degradation associated with the LED illumination. This fact is also verified by XRD measurements, which demonstrate the stability of perovskite crystalline structure.

### 3.1.4 Conclusions

This chapter summarized the state of the art of the Raman spectroscopic theoretical investigations on hybrid lead–halide perovskites, both from the material characterization and the application under real operative conditions in a photovoltaic device standpoint. It also reports some interesting results, where the measured slow electronic response of the material to external perturbations is connected with structural information and Raman and XRD measurements. This new electronic-structural perspective provided useful indications about structural modifications of the material that take place under bias and/or illumination. This induced structural modification may contribute to the slow-photocurrent and photoluminescence response of the material and also to the hysteretic behavior of the current–voltage curve, but there still remain many open issues concerning this topic. First, the detailed nature of the aforementioned structural modifications, at a microscopic level, still must be clarified. Second, the relation between the new structure obtained under bias and/or illumination and the time variation of the electronic response should be addressed. Finally, because the ion migration is still considered as the physical reason at the basis of the $J–V$ curve hysteresis, the connection between these two mechanisms must be found—if they are exclusive or if they are interconnected in giving rise to the unusual time-dependent response of hybrid perovskites. We believe that further experimental investigations aimed to characterize the material better under various conditions, together with a comprehensive simulation activity over a wide length scale, from the atomistic up to a mesoscopic level could definitely unveil the physical reasons at the basis of the peculiar properties of hybrid metal–halide perovskite.

# REFERENCES

Baibarac, M., N. Preda, L. Mihut, I. Baltog, S. Lefrant, and J. Y. Mevellec. 2004. On the optical properties of micro- and nanometric size $PbI_2$ particles. *J. Phys. Condens. Matter* 16; 2345–2356.

Baroni, S., S. de Gironcoli, A. Dal Corso, and P. Giannozzi. 2001. Phonons and related crystal properties from density-functional perturbation theory. *Rev. Mod. Phys.* 73; 515–562.

Borriello, I., G. Cantele, and D. Ninno. 2008. Ab initio investigation of hybrid organic-inorganic perovskites based on tin halides. *Phys. Rev. B* 77; 1–9.

Brivio, F., J. M. Frost, J. M. Skelton, A. J. Jackson, O. J. Weber, M. T. Weller, A. R. Goñi, A. M. A. Leguy, P. R. F. Barnes, and A. Walsh. 2015. Lattice dynamics and vibrational spectra of the orthorhombic, tetragonal, and cubic phases of methylammonium lead iodide. *Phys. Rev. B* 92; 144308-8.

Burschka, J., N. Pellet, S.-J. Moon, R. Humphry-Baker, P. Gao, Md. K. Nazeeruddin, and M. Grätzel. 2013. Sequential deposition as a route to high-performance perovskite-sensitized solar cells. *Nature* (London) 499; 316–319.

Calistru, D. M., L. Mihut, S. Lefrant, and I. Baltog. 1997. Identification of the symmetry of phonon modes in $CsPbCl_3$ in phase IV by Raman and resonance-Raman scattering. *J. Appl. Phys.* 82; 5391–5395.

Castiglioni, C., F. Negri, M. Tommasini, E. Di Donato, and G. Zerbi. 2006. Raman spectra and structure of $sp^2$ carbon-based materials: Electron-phonon coupling, vibrational dynamics and Raman activity. In G. Messina and S. Santangelo (eds.), *Carbon* (pp. 381–403). Berlin and Heidelberg: Springer.

Christians, J. A., P. A. Miranda Herrera, and P. V. Kamat. 2015. Transformation of the excited state and photovoltaic efficiency of $CH_3NH_3PbI_3$ perovskite upon controlled exposured to humidified air. *J. Am. Chem. Soc.* 137; 1530–1538.

Dammak, T., N. Fourati, H. Bougzhala, A. Mlayah, and Y. Abid. 2007. X-ray diffraction, vibrational and photoluminescence studies of the self-organized quantum well crystal $H_3N(CH_2)_6NH_3PbBr_4$. *J. Lumin.* 127; 404–408.

Dammak, T., S. Elleuch, H. Bougzhala, A. Mlayah, R. Chtourou, and Y. Abid. 2009. Synthesis, vibrational and optical properties of a new three-layered organic-inorganic perovskite $(C_4H_9NH_3)_4Pb_3I_4Br_6$. *J. Lumin.* 129; 893–897.

De Bastiani, M., V. D'Innocenzo, S. D. Stranks, H. J. Snaith, and A. Petrozza. 2014. Role of the crystallization substrate on the photoluminescence properties of organo-lead mixed halide perovskites. *APL Mater.* 2, 081509-6.

DeFrees, D. J., and A. D. McLead. 1985. Molecular orbital predictions of the vibrational frequencies of some molecular ions. *J. Chem. Phys.* 82; 333–341.

De Wolf, S., J. Holovsky, S.-J. Moon, P. Loeper, B. Niesen, M. Ledinský, F.-J. Haug, J.-H. Yum, and C. Ballif. 2014. Organometallic halide perovskites: Sharp optical absorption edge and its relation to photovoltaic performance. *J. Phys. Chem Lett.* 5; 1035–1039.

D'Innocenzo, V., G. Grancini, M. J. P. Alcocer, A. R. S. Kandada, S. D. Stranks, M. M. Lee, G. Lanzani, H. J. Snaith, and A. Petrozza. 2014. Excitons versus free charges in organo-lead trihalide perovskites. *Nature Commun.* 5; (3586) 1–6.

Eames, C., J. M. Frost, P. R. F. Barnes, B. C. O'Regan, A. Walsh, and M. S. Islam. 2015. Ionic transport in hybrid lead iodide perovskite solar cells. *Nat. Commun.* 6; (7497); 1–7.

Elleuch, S., Y. Abid, A. Mlayah, and H. Bougzhala. 2008. Vibrational and optical properties of a one-dimensional organic-inorganic crystal $[C_6H_{14}N]PbI_3$. *J. Raman Spectrosc.* 39; 786–792.

Even, J., L. Pedesseau, and C. Katan. 2014. Analysis of multivalley and multibandgap absorption and enhancement of free carriers related to exciton screening in hybrid perovskites. *J. Phys. Chem. C* 118; 11566–11572.

Filippetti, A., P. Delugas, and A. Mattoni. 2014. Radiative recombination and photoconversion of methylammonium lead iodide perovskite by first principles: Properties of an inorganic semiconductor within a hybrid body. *J. Phys. Chem. C* 118; 24843–24853.

Frost, J. M., K. T. Butler, and A. Walsh. 2014. Molecular ferroelectric contributions to anomalous hysteresis in hybrid perovskite solar cells. *APL Mater.* 2; 081506–081510.

Galimberti, D. R., A. Milani, and C. Castiglioni. 2013a. Charge mobility in molecules: Charge fluxes from second derivatives of the molecular dipole. *J. Chem. Phys.* 138; 164115–164118.

Galimberti, D. R., A. Milani, and C. Castiglioni. 2013b. Infrared intensities and charge mobility in hydrogen bonded complexes. *J. Chem. Phys.* 139, 074304–074311.

Giannozzi, P., S. Baroni, N. Bonini, M. Calandra, R. Car, C. Cavazzoni, D. Ceresoli, G. L. Chiarotti, M. Coccioni, I. Dabo, A. Dal Corso et al. 2009. QUANTUM ESPRESSO: A modular and open-source software project for quantum simulations of materials. *J. Phys.: Condens. Matt.* 21, 395502.

Gottesman, R., L. Gouda, B. S. Kalanoor, E. Haltzi, S. Tirosh, E. Rosh-Hodesh, Y. Tischler, A. Zaban, C. Quarti, E. Mosconi, and F. De Angelis. 2015. Photoinduced reversible structural transformation in free-standing $CH_3NH_3PbI_3$ perovskite thin films. *J. Phys. Chem. Lett.* 6; 2332–2338.

Gottesman, R., E. Haltzi, L. Gouda, S. Tirosh, Y. Bouhadna, A. Zaban, E. Mosconi, and F. De Angelis. 2014. Extremely slow photoconductivity response of $CH_3NH_3PbI_3$ perovskites suggesting structural changes under working conditions. *J. Phys. Chem. Lett.* 5; 2662–2669.

Grancini, G., A. R. Srimath Kandada, J. M. Frost, A. J. Barker, M. De Bastiani, M. Gandini, S. Marras, G. Lanzani, A. Walsh, and A. Petrozza. 2015a. Role of microstructure in the electron-hole interaction of hybrid lead-halide perovskites. *Nature Photonics* 9; 695–701.

Grancini, G., V. D'Innocenzo, E. R. Dohner, N. Martino, A. R. Srimath Kandada, E. Mosconi, F. De Angelis, H. I. Karunadasa, E. T. Hoke, and A. Petrozza. 2015b. $CH_3NH_3PbI_3$ perovskite single crystals: Surface photophysics and their interaction with the environment. *Chem. Sci.* 6; 7305–7310.

Grancini, G., S. Marras, M. Prato, C. Giannini, C. Quarti, F. De Angelis, M. De Bastiani, G. E. Eperon, H. J. Snaith, L. Manna, and A. Petrozza. 2014. The impact of the crystallization processes on the structural and optical properties of hybrid perovskite films for photovoltaics. *J. Phys. Chem. Lett.* 5; 3836–3842.

Habisreutinger, S. N., T. Leijtens, G. E. Eperon, S. D. Stranks, R. J. Nicholas, and H. J. Snaith. 2014. Carbon nanotube/polymer composites as a highly stable hole collection layer in perovskite solar cells. *Nano Lett.* 14; 5561–5568.

Haruyama, J., K. Sodeyama, L. Han, and Y. Tateyama. 2015. First-principles study of ion diffusion in perovskite solar cell sensitizers. *J. Am. Chem. Soc.* 137; 10048–10051.

Im, J.-H., C.-R. Lee, J.-W. Lee, S.-W. Park, and N.-G. Park. 2011. 6.5% efficient perovskite quantum-dot-sensitized solar cell. *Nanoscale* 3; 4088–4093.

Jeon, N. J., J. H. Noh, W. S. Yang, Y. C. Kim, S. Ryu, J. Seo, and S. I. Seok. 2015. Compositional engineering of perovskite materials for high-performance solar cells. *Nature* 517; 476–480.

Jiang, Z. T., B. D. James, J. Leiesegang, K. L. Tan, R. Gopalakrishnan, and I. Novak. 1995. Investigation of the ferroelectric phase transition in $(CH_3NH_3)HgCl_3$. *J. Phys. Chem. Solids* 56; 277–283.

Kawamura, Y., H. Mashiyama, and K. Hasebe. 2002. Structural study on cubic–tetragonal transition of $CH_3NH_3PbI_3$. *J. Phys. Soc. Jpn.* 71; 1694–1697.

Kitazawa, N., Y. Watanabe, and Y. Nakamura. 2002. Optical properties of $CH_3NH_3PbX_3$ (X=halogen) and their mixed-halide crystals. *J. Mater. Sci.* 37; 3585–3587.

Kojima, A., K. Teshima, Y. Shirai, and T. Miyasaka. 2009. Organometal halide perovskites as visible-light sensitizers for photovoltaic cells. *J. Am. Chem. Soc.* 131; 6050–6051.

Lazzeri, M., and F. Mauri. 2003. First-principle calculation of vibrational Raman spectra in large systems: Signature of small rings in crystalline SiO$_2$. *Phys Rev Lett*. 90; 036401–036404.

Ledinský, M., P. Löper, B. Niesen, J. Holovský, S. J. Moon, J. H. Yum, S. De Wolf, A. Fejfar, and C. Ballif. 2015. Raman spectroscopy of organic-inorganic halide perovskites. *J. Phys. Chem. Lett*. 6; 401–406.

Lee, M. M., J. Teuscher, T. Miyasaka, T. N. Murakami, and H. J. Snaith. 2012. Efficient hybrid solar cells based on meso-superstructured organometal halide perovskites. *Science* 338; 643–647.

Leijtens, T., A. R. Srimath Kandada, G. E. Eperon, G. Grancini, V. D'innocenzo, J. M. Ball, S. D. Stranks, H. J. Snaith, and A. Petrozza. 2015a. Modulating the electron-hole interaction in hybrid lead halide perovskite with an electric field. *J. Am. Chem. Soc*. 137; 15451–15459.

Leijtens, T., E. T. Hoke, G. Grancini, D. J. Slotcavage, G. E. Eperon, J. M. Ball, M. De Bastiani, A. R. Bowring, N. Martino, K. Wojciechowski, M. D. McGhee, H. J. Snaith, and A. Petrozza. 2015b. Mapping electric field-induced switchable poling and structural degradation in hybrid lead halide perovskite thin films. *Adv. Energy Mater*. 5; 1500962.

Liu, M., M. B. Johnston, and H. J. Snaith. 2013. Efficient planar heterojunction perovskite solar cells by vapour deposition. *Nature* 501; 395–398.

Maalej, A., Y. Abid, A. Kallel, A. Daoud, A. Lautié, and F. Romain, Phase transitions and crystal dynamics in the cubic perovskite CH$_3$NH$_3$PbCl$_3$. *Solid State Commun*. 103; 279–284.

Mosconi, E., C. Quarti, T. Ivanovska, G. Ruani, and F. De Angelis. 2014. Structure and electronic properties of organo-halide Lead Perovskites: A combined IR-spectroscopy and ab initio molecular dynamics investigation. *Phys. Chem. Chem. Phys*. 16; 16137–16144.

Noda, I., A. E. Dowrey, C. Marcott, H. W. Siesler, I. Zebger, Ch. Kulinna, S. Okretic, S. Shilov, U. Hoffmann, G. Zerbi, M. Del Zoppo et al. 1999. Vibrational Spectra as a Probe of Structural Order/Disorder in Chain Molecules and Polymers. In G. Zerbi (ed.), *Modern polymer spectroscopy*. Weinheim: Wiley-VCH 87–206.

Park, B.-W., S. M. Jain, X. Zhang, A. Hagfeldt, G. Boschloo, and T. Edvinsson. 2015. Resonance Raman and excitation energy dependent charge transfer mechanism in halide-substituted hybrid perovskite solar cells. *ACS Nano* 9; 2088–2101.

Pérez-Osorio, M. A., R. L. Milot, M. R. Filip, J. B. Patel, L. M. Herz, M. B. Johnston, and F. Giustino. 2015. Vibrational properties of the organic-inorganic halide perovskite CH$_3$NH$_3$PbI$_3$ from theory and experiment: Factor group analysis, first-principles calculations, and low-temperature infrared spectra. *J. Phys. Chem. C* 119; 25703–25718.

Poglitsch, A., and D. Weber. 1987. Dynamic disorder in methylammoniumtrihalogenoplumbates (II) observed by millimeter-wave spectroscopy. *J. Chem. Phys*. 87; 6373–6378.

Preda, N., L. Mihut, I. Baltog, V. Velula, and V. Teodorescu. 2006. Optical properties of low-dimensional PbI$_2$ particles embedded in polyacrylamide matrix. *J. Optoelectron. Adv. M* 8; 909–913.

Quarti, C., G. Grancini, E. Mosconi, P. Bruno, J. M. Ball, M. M. Lee, H. J. Snaith, A. Petrozza, and F. De Angelis. 2014a. The Raman spectrum of the CH$_3$NH$_3$PbI$_3$ hybrid perovskite: Interplay of theory and experiment. *J. Phys. Chem. Lett*. 5; 279–284.

Quarti, C., A. Milani, and C. Castiglioni. 2013. Ab initio calculation of the IR spectrum of PTFE: Helical symmetry and defects. *J. Phys Chem. B* 117; 706–718.

Quarti, C., E. Mosconi, and F. De Angelis. 2014. Interplay of orientational order and electronic structure in methylammonium lead iodide: Implications for solar cells operation. *Chem. Mater*. 26; 6557–6569.

Siebert, F., and P. Hildebrandt. 2008. *Vibrational spectroscopy in life science*. Berlin: Wiley-VCH Verlag.

Snaith, H. J., A. Abate, J. M. Ball, G. E. Eperon, T. Leijtens, N. K. Noel, S. D. Stranks, J. T.-S. Wang, K. Wojciechowski, and W. Zhang. 2014. Anomalous hysteresis in perovskite solar cells. *J. Phys. Chem. Lett*. 5; 1511–1515.

Stoumpos, C. C., C. D. Malliakas, and M. G. Kanatzidis. 2013. Semiconducting tin and lead iodide perovskites with organic cations: Phase transitions, high mobilities, and near-infrared photoluminescent properties. *Inorg. Chem.* 52; 9019–9038.

Stranks, S. D., G. E. Eperon, G. Grancini, C. Menelaou, M. J. P. Alcocer, T. Leijtens, L. M. Herz, A. Petrozza, and H. J. Snaith. 2013. Electron-hole diffusion lengths exceeding 1 micrometer in an organometal trihalide perovskite absorber. *Science* 342; 341–344.

Théoret, A., and C. Sandorfy. 1967. The infrared spectra of solid methylammonium halides-II. *Spectrochim. Acta A* 23; 519–542.

Tress, W., N. Marinova, T. Moehl, S. M. Zakeeruddin, M. K. Nazeeruddin, and M. Grätzel. 2015. Understanding the rate-dependent J-V hysteresis, slow time component, and aging in $CH_3NH_3PbI_3$ perovskite solar cells: The role of a compensated electric field. *Energy Environ Sci.* 8; 995–1004.

Waldron, R. D. 1953. The infrared spectra of three solid phases of methylammonium chloride. *J. Chem. Phys.* 21; 734–741.

Wasylishen, R., O. Knop, and J. Macdonald. 1985. Cation rotation in methylammonium lead halides. *Solid State Commun.* 56; 581–582.

Weber, W. H., and R. Merlin (eds.) 2000. *Raman scattering in material science*. Berlin and Heidelberg: Springer.

Wehrenfennig, C., G. E. Eperon, M. B. Johnston, H. J. Snaith, and L. M. Herz. 2014a. High charge carrier mobilities and lifetimes in organolead trihalide perovskites. *Adv. Mater.* 26; 1584–1589.

Wehrenfennig, C., M. Liu, H. J. Snaith, M. B. Johnston, and L. M. Herz. 2014. Charge-carrier dynamics in vapour-deposited films of the organolead halide perovskite $CH_3NH_3PbI_{3-x}Cl_x$. *Energy Environ. Sci.* 7; 2269–2275.

Wei, J., Y. Zhao, H. Li, G. Li, P. J., D. Xu, Q. Zhao, and D. Yu. 2014. Hysteresis analysis based on the ferroelectric effect in hybrid perovskite solar cells. *J. Phys. Chem. Lett.* 5; 3937–3945.

Xiao, Z., Y. Yuan, Y. Shao, Q. Wang, Q. Dong, C. Bi, P. Sharma, A. Gruverman, and J. Huang. 2015. Giant switchable photovoltaic effect in organometal trihalide perovskite devices. *Nature Mater.* 14; 193–198.

Xing, G., N. Mathews, S. Sun, S. S. Lim, Y. M. Lam, M. Grätzel, S. Mhaisalkar, and T. C. Sum. 2013. Long-range balanced electron- and hole-transport lengths in organic-inorganic $CH_3NH_3PbI_3$. *Science* 342; 344–347.

Yang, J., B. E. Siempelkamp, D. Liu, and T. L. Kelly. 2015. Investigation of $CH_3NH_3PbI_3$ degradation rates and mechanisms in controlled humidity environment using in situ techniques. *ACS Nano* 9; 1955–1963.

Zeroka, D., and J. O. Jensen. 1998. Infrared spectra of some isotopomers of methylamine and the methylammium ion: A theoretical study. *J. Mol. Struct.* 425; 181–192.

Zhou, H., Q. Chen, G. Li, S. Luo, T.-B. Song, H.-S. Duan, Z. Hong, J. You, Y. Liu, and Y. Yang. 2014. Interface engineering of highly efficient perovskite solar cells. *Science* 345; 542–546.

Zhu, Z., J. Ma, Z. Wang, C. Mu, Z. Fan, L. Du, Y. Bai, L. Fan, H. Yan, D. L. Phillips, and S. Yang. 2014. Efficiency enhancement of perovskite solar cells through fast electron extraction: The role of graphene quantum dots. *J. Am. Chem. Soc.* 136; 3760–3763.

## 3.2 FERROELECTRICITY AND SPIN-ORBIT COUPLING IN ORGANIC–INORGANIC PEROVSKITE HALIDES

*Domenico Di Sante,*[1,2] *Alessandro Stroppa,*[2,3] *Liang Z. Tan,*[4]

*Paolo Barone,*[2,5] *Andrew M. Rappe,*[4] *and Silvia Picozzi*[2]

[1]Institut für Theoretische Physik und Astrophysik, Universität Würzburg, Am Hubland Campus Süd, Würzburg, Germany

[2]Consiglio Nazionale delle Ricerche (CNR-SPIN), L'Aquila, Italy

[3]International Centre for Quantum and Molecular Structures and Physics Department, Shanghai University, Shanghai, China

[4]Makineni Theoretical Laboratories, Department of Chemistry, University of Pennsylvania, Philadelphia, Pennsylvania, USA

[5]Graphene Labs, Istituto Italiano di Tecnologia, Genova, Italy

### 3.2.1 Introduction

The possibility to control electron spins by means of external handles, such as electric fields, represents one of the grand challenges in modern electronics. The Rashba effect, relying on the combination of the lack of inversion symmetry and spin-orbit coupling, is currently seen as a promising way to achieve spin manipulation. Of particular relevance is the recent proposal of coupling the Rashba effect with ferroelectricity: the control over Rashba spin splitting in the electronic structure of ferroelectric materials might indeed represent a breakthrough in spintronics. So far, the search for materials hosting the coexistence between ferroelectricity and Rashba effects has focused mostly on inorganic systems. The presence of relatively heavy elements in tin- and lead-based halide perovskites (including hybrid organic–inorganic materials) suggests that large spin-orbit coupling might play a fundamental role. If spin-orbit coupling is linked with the lack of inversion symmetry inherent to ferroelectricity, this class of materials might exhibit exotic and technologically appealing spin splitting phenomena. Furthermore, the possibility to permanently tune and switch the direction of polarization via an electric field—possible in principle in ferroelectric perovskite halides—would imply a corresponding nonvolatile control over the spin texture. Interestingly, dipole and spin degrees of freedom, as well as their interplay, have been proposed to perform important functions in understanding the high power generation efficiency of perovskite halides, and as such they might open new avenues in halide-based photovoltaics.

In this chapter, after a brief introduction on spin-orbit coupling and spin-splitting effects in nonmagnetic and noncentrosymmetric systems, we review the current literature on ferroelectric properties of organic–inorganic perovskite halides, both from experimental and theoretical points of view. We focus our attention mainly on the most promising tin- and lead-based compounds, that is, $MA(Pb,Sn)I_3$ and $FASnI_3$, where MA and FA refer to methylammonium $(CH_3NH_3^+)$ and formamidinium $((NH_2)_2CH^+)$ cations, respectively. Furthermore, we will show that such systems show an intriguing interplay between ferroelectricity and spin-orbit coupling, leading to the coexistence of Dresselhaus and Rashba spin splittings and the possibility of their electrical tuning. Eventually, the possible role played by ferroelectricity and spin-orbit coupling effects in determining the photovoltaic properties of perovskite halides is summarized. The enhancement of photocurrents has already been proposed for ferroelectric materials, due to the so-called bulk photovoltaic effect (Spanier et al. 2016), to the domain structure inherent to ferroelectrics, and to the space charge regions at interfaces. On the other hand, the Rashba spin-orbit interaction has been proposed to enhance the carrier's lifetime and reduce the electron–hole recombination rate, an effect that has been related to the shift of the spin-split bands and to modified spin selection rules determining the recombination probability, and that puts forward organometal halide perovskites as promising solar-cell materials for next-generation photovoltaic applications.

### 3.2.2 Spin-Orbit Coupling in Nonmagnetic Systems Lacking Inversion Symmetry

#### 3.2.2.1 Rashba and Dresselhaus Spin-Orbit Interactions

To explain the peculiarities of electron spin resonance experiments in two-dimensional semiconductors, Bychkov and Rashba proposed in 1984 a simple form of spin-orbit coupling (SOC) (Bychkov and Rasbha 1984; Rashba 1960). That seminal work marked the beginning of an intense research activity focused on the so-called Rashba SOC, which led to the formulation of concepts and predictions far beyond the semiconductor world. In particular, the possibility of controlling the electron's motion by acting on its spin degree of freedom is of great technological interest, paving the way to the realization of novel spintronic devices.

Remarkably, the lack of inversion symmetry is the key ingredient at the heart of these innovative concepts. In nonmagnetic crystals without an inversion center, electronic energy bands are generally spin-split by SOC. This effect can be naïvely understood by realizing that, even in the absence of an external magnetic field (but taking into account relativistic corrections), electrons experience an effective magnetic field in their frame of motion, arising from the nonvanishing gradient of the crystalline electric potential in a noncentrosymmetric environment (Winkler 2003). Such a field, usually referred to as *spin-orbit field*, couples to the electron's magnetic moment, therefore lifting the spin degeneracy. The study and investigation of the Rashba physics are giving rise to a new area of spintronics, dubbed *spin-orbitronics*, aimed at the control of transport properties by using SOC (Hoffmann and Bader 2015).

Let us now briefly review the essential features of the Rashba spin-orbit interaction. In the case of a nonmagnetic and centrosymmetric electronic system, up and down spin channels of a given energy band are degenerate, that is, $\varepsilon(\mathbf{k}, \uparrow) = \varepsilon(\mathbf{k}, \downarrow)$ (Kramers' degeneracy). This is an important consequence of time-reversal and inversion symmetries, and the Kramers pair can be lifted if either the former or the latter symmetry is broken. In nonmagnetic systems, the easiest and most common way to achieve spin splittings is to break inversion symmetry. For a two-dimensional (2D) free-electron gas, such as that occurring at a metallic surface or in low-dimensional semiconductor heterostructures, inversion symmetry is broken perpendicularly to the 2D plane, creating an interfacial electric field $\mathbf{E} = E_z\mathbf{z}$. When SOC is properly taken into account, the essential physics of the resulting spin-split electronic states is captured by the Rashba Hamiltonian:

$$H_R = \frac{\hbar^2}{2m^*}\left(k_x^2 + k_y^2\right)\sigma_0 + \alpha_R\left(\sigma \times \mathbf{k}\right) \cdot \hat{\mathbf{z}} \qquad (3.2.1)$$

with $\alpha_R$ known as the Rashba parameter, and $\sigma_0$ and $\sigma$ the identity and spin Pauli matrices respectively. As Equation 3.2.1 is strictly correct only for Bloch states, which are appropriate for describing the 2D free-electron gas at surfaces or interfaces in semiconducting heterostructures where the inversion symmetry is structurally broken, the term *structure inversion asymmetry* (SIA) is commonly used to identify a Rashba spin-orbit interaction (Winkler 2003). However, the exact form of the spin-momentum coupling in bulk materials is determined by the symmetry properties of the wave functions in reciprocal space. As recently pointed out, some materials possessing bulk inversion asymmetry (BIA) may also display signatures of Rashba spin-orbit interaction, the most remarkable examples being the polar semiconductors BiTeI and GeTe (Di Sante et al. 2013b; Ishizaka et al. 2011). The magnitude of the phenomenological parameter $\alpha_R$ has been estimated for a wide range of materials, showing either structural or bulk inversion symmetry breaking. Values around $0.5 \times 10^{-11}$ eV m have been predicted for InAlAs/InGaAs heterostructures as well as at the [001] surface of $SrTiO_3$ single crystals (Nakamura et al. 2012; Nitta et al. 1997), whereas significantly larger Rashba spin splittings have been reported at the surfaces of heavy metals such as Au, Ir, and $\sqrt{3} \times \sqrt{3}$ BiAg[111] alloys ($\approx 3.7 \times 10^{-10}$ eV m in this last case) (Ast et al. 2007; LaShell et al. 1996; Varykhalov et al. 2012). Evidence of strong Rashba-like effects have been also reported for the polar bulk materials BiTeI and GeTe, where $\alpha_R \approx 4 - 5 \times 10^{-10}$ eV m (Di Sante et al. 2013b; Ishizaka et al. 2011; Liebmann et al. 2015).

An interesting consequence of the Rashba effect is the peculiar distribution of momentum-dependent spin orientations of the spin-split electronic states; such distribution can be seen as a spin texture of the electronic states associated with the set of $\mathbf{k}$-points defined for a certain energy on the $E(\mathbf{k}) = \frac{\hbar^2}{2m^*}|\mathbf{k}|^2 \pm \alpha_R|\mathbf{k}|$ surface. As depicted in Figure 3.2.1a, where spin polarization is pictorially associated with three-dimensional arrows, the spin orientations of the electronic eigenstates of the Rashba Hamiltonian Equation 3.2.1 are always tangential to the associated crystal momentum, rotating counterclockwise on one branch of the energy surface and clockwise on the other branch. A tangential spin texture is the fingerprint of a Rashba-type spin-orbit interaction, as usually detected by means of spin and angle resolved photoemission spectroscopy (spin-ARPES) measurements.

As first shown by Dresselhaus (1955), spin-orbit coupling may also produce a spin splitting proportional to $k^3$ in the bulk electronic structures of nonpolar materials lacking inversion symmetry such as zinc-blend semiconductors. This effect, arising from the loss of a reflection symmetry with respect to a plane containing at least one lattice site of the

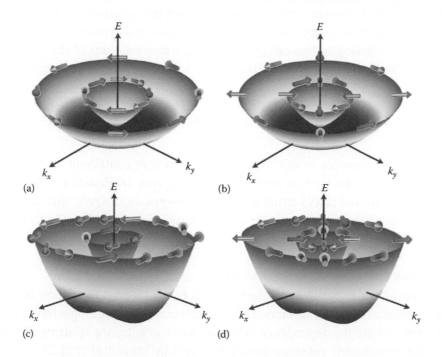

FIGURE 3.2.1 Typical "Mexican-hat" shape of the energy dispersion of electrons experiencing (a) Rashba and (b) Dresselhaus interactions. The arrows represent the spin texture: (a) tangential texture and (b) radial texture. (c, d) Energy dispersions when both interactions coexist, with a predominant Rashba (c) or Dresselhaus (d) character.

underlying crystal, is usually related to a *bulk inversion asymmetry*. As noted previously, however, the bulk inversion asymmetry may give rise also to Rashba-like spin-splitting linear in the crystal momentum when the BIA is further characterized by the presence of a polar axis. On the other hand, it has been recently pointed out that a spin splitting linear in $k$ can be realized also in noncentrosymmetric but nonpolar crystals when they belong to gyrotropic point groups (Ganichev and Golub 2014); for these systems, the spin-splitting phenomenon is generally known as a linear Dresselhaus effect, which is described by the following Hamiltonian:

$$H_D = \frac{\hbar^2}{2m^*}(k_x^2 + k_y^2)\sigma_0 + \alpha_D(-k_x\sigma_x + k_y\sigma_y) \qquad (3.2.2)$$

where $\alpha_D$ is a material-dependent coupling constant known as the Dresselhaus parameter. The energy surface is given by the same expression as for the Rashba Hamiltonian, whereas the Dresselhaus-type spin texture, arising from the different symmetry of the Dresselhaus wave functions, has a more complex form, showing also spin orientations parallel to the associated crystal momentum (Figure 3.2.1b).

It may happen that Rashba and Dresselhaus spin-momentum couplings are simultaneously present, with one coupling usually stronger than the other. This situation is in fact realized in noncentrosymmetric materials where high-symmetry $k$ points display $C_{2v}$ point group symmetry (Ganichev and Golub 2014), and—as discussed later—some hybrid perovskite halides belong to this category. The dispersion relations of the resulting spin-split electronic states become asymmetric (Figure 3.2.1c, d), with a coupling Hamiltonian given by

$$H_{RD} = \alpha_R (k_y \sigma_x - k_x \sigma_y) + \alpha_D (-k_x \sigma_x + k_y \sigma_y) \qquad (3.2.3)$$

Usually it is not possible to directly extract the relative contributions of Rashba and Dresselhaus terms to the spin-momentum coupling given in Equation 3.2.3. For semiconductor quantum well (QW) structures, such as $n$-type InAs QWs, this information is inferred from the analysis of the angular distribution of the spin-galvanic effect (Ganichev and Golub 2014).

### 3.2.2.2 Ferroelectric Rashba Semiconductors

Many theoretical and experimental efforts related to the Rashba effect have concentrated on low-dimensional systems, surfaces, and heterostructures, including a report on the Bi/BaTiO$_3$ system, where the dependence of the Bi overlayer's Rashba splitting on the polarization of the ferroelectric substrate was investigated (Mirhosseini et al. 2010). The polar semiconductor BiTeI was the first example of a three-dimensional system exhibiting a giant bulk Rashba effect (Bahramy et al. 2011; Ishizaka et al. 2011), although its characteristic layered crystalline structure makes it closer to a 2D system, or to a system in which a preferential direction can be unambiguously identified.

To modulate the magnitude of the Rashba effect, the out-of-plane electric potential gradient can be simply tuned by applying an external electric field, a route that has been followed in low-dimensional systems. Clearly, this approach implies a volatile control of the effect, as it disappears when the electric field is switched off. Interestingly, a memory effect could be achieved in heterostructures combining thin films of materials with high spin-orbit coupling and ferroelectrics, such as the aforementioned Bi/BaTiO$_3$ system, but so far single-phase materials with electrically controlled nonvolatile Rashba splitting have not been reported yet. Coming to recently proposed bulk materials displaying Rashba effects, the electrical control itself is still a big challenge. For instance, the spin-splitting modulation via an electric field in polar (but non ferroelectric) BiTeI (Ishizaka et al. 2011) is quite small, thus limiting its application in potential devices. On the other hand, ferroelectrics, that is, materials displaying long-range dipolar order below a certain critical temperature, represent an interesting class of bulk materials lacking inversion symmetry. The coexistence and coupling between ferroelectricity and Rashba SOC interaction may give rise to new functionalities with valuable applications in semiconductor spintronics. The potential relevance of this phenomenology has been demonstrated recently in ferroelectric GeTe, where the reversal of ferroelectric polarization was theoretically predicted to produce a complete switching of the spin texture (Di Sante et al. 2013b; Liebmann et al. 2015; Picozzi 2014).

### 3.2.3 Review of Evidence for Ferroelectricity in Perovskite Halides

Since the first experimental work reporting steep increase of efficiency in perovskite halides, the possible role of the ferroelectric polarization in the photovoltaic properties has been strongly debated in the literature. At first glance, it seems reasonable that these systems may have long-range dipole ordering because the MA cations are polar and the ferroelectric order has been already discussed and verified experimentally in a similar class of hybrid organic–inorganic perovskites, namely metal–organic frameworks (Di Sante et al. 2013; Jain et al. 2009; Stroppa et al. 2011; Tian et al. 2014, 2015).

The $MAPbI_3$ compound shows two structural phase transitions: from cubic ($\alpha$) to tetragonal ($\beta$) phases at 327 K and from tetragonal to orthorhombic phases at 162 K (Poglitsch and Weber 1987). Soon after the first dielectric studies, unusual hysteresis in the current density–voltage curves ($J–V$) (Juarez-Perez et al. 2014; Snaith et al. 2014) has been shown, and the nominal efficiency of the solar cells strongly depends on how the $J–V$ measurements are performed (Snaith et al. 2014). Different mechanisms have been proposed to explain this behavior, such as ion migration (Tress et al. 2015), trapping of electronic carriers at interfaces (Shao et al. 2014) and possible ferroelectric order (Wei et al. 2014). Despite the possible relevant role of electrical ordering, its experimental evidence is still under debate. The polarization in hybrid perovskites could have different contributions: polarization coming from the $BX_3$ framework, polarization coming from the displacements of the $MA^+$ cations relative to the $BX_3$ frameworks, and the dipole moments due to the MA cation orientations (Fan et al. 2015; Frost J. M. et al. 2014b; Stroppa et al. 2015).

From the experimental point of view, several studies have been published based on macroscopic polarization–electric field ($P–E$) measurements (Beilsten-Edmands et al. 2015; Sewvandi et al. 2016; Xiao et al. 2015), or microscopic probing of ferroelectric domains (Kutes et al. 2014; Xiao et al. 2015). Several factors may contribute to experimental detection of ferroelectric polarization, if it is present at all. For instance, the $MA^+$ cations are very small and the hydrogen bonding with the $BX_3$ framework is rather weak, making the organic cations mobile, such that they exhibit strongly dynamic disorder (Baikie et al. 2013; Mattoni et al. 2015; Yaffe et al. 2017). Even if a local dipole moment is formed in each unit cell, this would lead to a zero average polarization at the macroscopic scale, that is, a structure that is centrosymmetric on a space and time average. Furthermore, the $MAPbI_3$ thin films are characterized by a high density of charge carriers that contribute to a large current during polarization measurements, complicating the interpretation of remnant polarization measurements (Beilsten-Edmands et al. 2015). Furthermore, the analysis of the $P–E$ loop is made more difficult by ionic conduction and related capacitance effects. Recent piezoforce microscopy (PFM) measurements revealed the presence of submicrometer ferroelectric domains (grains nearly 100 nm in size) in $\beta$-$MAPbI_3$ samples, and the importance of ferroelectric domain walls for the photovoltaic efficiency of hybrid halide perovskites has been pointed out (Kutes et al. 2014; Liu et al. 2015). Very recently, temperature-dependent $P–E$ loops and measurements of remnant polarization were performed using positive-up–negative-down (PUND) methods and temperature-dependent impedance and dielectric spectroscopies. They concluded that, at the macroscopic level, $MAPbI_3$

thin films do not show apparent ferroelectric properties at cell operating temperatures (between 283 and 343 K). However, this does not exclude the possibility of ferroelectric domains at the nanoscale level (Hoque et al. 2016).

There have been many theoretical studies based mainly on Density Functional Theory (DFT) calculations aiming at estimating the ferroelectric polarization. Previous theoretical studies on a similar class of hybrid perovskites, $ABX_3$ metal–organic frameworks (MOFs), have shown that ferroelectric polarization in hybrid perovskite systems can range from 0.4 µC cm$^{-2}$ to 6–7 µC cm$^{-2}$ (Di Sante et al. 2013; Stroppa et al. 2011). It was theoretically predicted that apolar cations may also foster electric polarization in MOF sytems, as was later confirmed experimentally (Jain et al. 2009; Tian et al. 2014, 2015). Therefore it has been natural to investigate theoretically the possible ordering of polar cations in $ABX_3$ perovskite halides. Frost et al. performed DFT calculations on the cubic phase in which all the MA molecules are aligned in parallel but without including atomic relaxations. They estimate a ferroelectric polarization of ≈38µ C cm$^{-2}$ (Frost et al. 2014), a value as high as conventional inorganic ferroelectric oxide perovskites such as $BaTiO_3$ and $KNbO_3$. Furthermore, the same authors have also investigated the formation of possible domains with different MA molecular alignments at finite temperature and under an external applied electric field (Frost et al. 2014b). Zheng et al. have demonstrated that the expected values for the ferroelectric polarization in such a class of organohalide materials should be around 4–5 µC cm$^{-2}$ (Zheng et al. 2015). A detailed study of the ferroelectric polarization in $MAPbI_3$ combining computational DFT simulations with symmetry mode analysis has been performed by Stroppa et al. (2015). The authors disentangled the relative contribution of different functional units in the perovskite, that is, the MA cations and the $BX_3$ framework. Furthermore, the important contribution of the atomic relaxations of the $BX_3$ framework to the final value of the polarization has been highlighted. Mosconi et al. studied the tendency of tetragonal β phase of $MaPbI_3$ to show a long-range global alignment of MA molecules by DFT calculations and molecular dynamics simulations (Quarti et al. 2014). It has been demonstrated that structures showing a ferroelectric long-range alignment are more stable than the ones characterized by an antiferroelectric one. However, the energy difference is comparable to room temperature $k_BT$, making both configurations accessible at room temperature. Nevertheless, this would also suggest the possibility of favoring the ferroelectric phase in $MAPbI_3$ over the antiferroelectric one by applying an external electric field (Quarti et al. 2014).

Another promising material in the class of hybrid perovskites is the tin iodide $(NH_2)_2CHSnI_3$, hereafter referred to as $FASnI_3$, where FA is the formamidinium cation $(NH_2)_2CH^+$. It has attracted attention because of its possible polar properties and as a non-toxic alternative to Pb-based materials. The ferroelectric properties of $FASnI_3$ have been investigated by DFT calculations (Stroppa et al. 2014). Experimentally, a crystal structure belonging to a polar space group has been reported at room temperature (Stoumpos et al. 2013). Theoretically, it has been proposed that molecular dipole orientation in $FASnI_3$ can be changed easily by rotating FA around its N–N axis. The arrangements of FA cations in the cavities of the framework determine whether $FASnI_3$ shows a ferroelectric or antiferroelectric cation ordering. It is worth noting that DFT calculations suggest that the most stable structure is not the experimentally reported one. The lowest-energy structure calculated using a doubled supercell corresponds to a weak-ferroelectric (FE) configuration of molecular dipoles, where the nearby FA dipoles are mutually perpendicular. This is analogous to the case of weak-ferromagnetic spin configurations. The calculations show that the experimental polar structure (Stoumpos et al. 2013) and the AFE one are intermediate states higher in energy with respect to the weak-FE state by $\approx 100$ meV f.u$^{-1}$. In the same theoretical work it has been pointed out that the stability of the weak-FE structure is due to hydrogen bond interactions between organic cations and the perovskite framework. Because such interactions are weak, it is expected that the polarization orientation can be easily rotated by an external electric field or modified by applied strain. In the next sections, we will see how this tunability of $FASnI_3$ and the polar properties of many halide perovskites are strongly interlinked to spin-orbit coupling.

Despite the lack of conclusive experiments about the existence of FE polarization at a nanometer scale, and although it seems clear that ferroelectricity is generally absent at the macroscopic scale in the halide perovskites, the possibility remains that the global centrosymmetry of these materials may be broken locally. Molecular dynamics simulations show a "dynamical Rashba effect," which suggests that even in globally centrosymmetric structures, the dynamics of the coupled inorganic–organic degrees of freedom give rise to a spatially modulated Rasbha effect that fluctuates on the subpicosecond time scale characteristic of MA dynamics (Etienne et al. 2016). Therefore, it is extremely timely to consider the connection between polarization and spin-orbit properties.

### 3.2.4 Interplay between Spin-Orbit Coupling and Ferroelectricity in Organohalide Perovskites

As previously discussed, the interplay between ferroelectricity and spin-orbit coupling may give rise to new and interesting phenomena. The most promising effect is the so-called electric control of bulk Rashba effect, which paved the way to the class of ferroelectric Rashba semiconductors (FERSCs) (Picozzi 2014). Recent theoretical studies shed light on the possibility of acquiring the same effect in ferroelectric perovskite halides. First evidences, based on a joint model and DFT *ab initio* investigation, were reported by Kim et al. (2014) in the family of methylammonium (MA) based compounds. Their analysis showed that $S = 1/2$ and $J = 1/2$ Rashba bands emerge at the valence (VBM) and conduction (CBM) band edges, respectively (Figure 3.2.2a). These bands, directly coupled to ferroelectric polarization, host different compositions of the orbital and spin angular momentum, a distinctive feature of organohalide perovskites. The Rashba parameters, that is, the Rashba energy $E_R$, the momentum offset $k_0$, and the Rashba coefficient $\alpha_R$, for some MA-based lead and tin halides are summarized in Table 3.2.1. The Rashba parameter depends on the strength of the spin-orbit interaction, as well as on the amount of polar distortion in these materials.

From Equation 3.2.1, the momentum offset of the split bands is given by $k_0 = m^* \alpha_R / \hbar^2$ while the Rashba energy of the split band minimum is $E_R = m^* \alpha_R^2 / (2\hbar)^2$.

In contrast, $FASnI_3$ shows a more complex interplay between ferroelectricity and SOC. Focusing on the CBM, where the spin splitting is larger, the spin-split bands present nontrivial

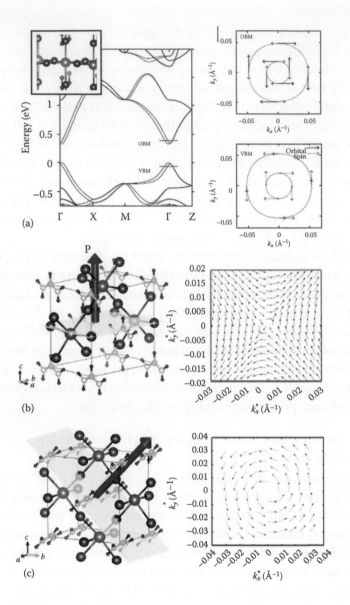

FIGURE 3.2.2 (a) Atomic and electronic structure of β-MAPbI₃ from first-principles calculations. Spin (blue arrows) and orbital (red arrows) angular momentum textures are reported in the panels for cuts above the CBM and below the VBM. (b) Atomic structure and spin texture of the conduction outer branch in FASnI₃ when the ferroelectric polarization is oriented along the [001] direction. (c) Atomic structure and spin texture of the valence outer branch in FASnI₃ when the ferroelectric polarization has been rotated and aligned along the [011] direction. (Partially reproduced with permission from Stroppa, A., D. Di Sante, P. Barone, M. Bokdam, G. Kresse, C. Franchini, M.-H. Whangbo, and S. Picozzi. 2014. Tunable ferroelectric polarization and its interplay with spin-orbit coupling in tin iodide perovskites. *Nat. Commun.* 5; 5900–5908; and Kim, M., J. Im, A. J. Freeman, J. Ihm, and H. Jin. 2014. Switchable $S = 1/2$ and $J = 1/2$ Rashba bands in ferroelectric halide perovskites. *PNAS* 111; 6900–6904.)

TABLE 3.2.1    Estimated Rashba Parameters for Some MA-Based Organohalide Perovskites

| System | Band | $E_R$ (meV) | $k_0$(Å$^{-1}$) | $\alpha_R$ (eV. Å) | Reference |
|---|---|---|---|---|---|
| β-MAPbI$_3$ | CBM | 12 | 0.015 | 1.5 | Kim et al. (2014) |
| | VBM | 11 | 0.016 | 1.4 | Kim et al. (2014) |
| β-MASnI$_3$ | CBM | 11 | 0.011 | 1.9 | Kim et al. (2014) |
| | VBM | 4 | 0.006 | 1.2 | Kim et al. (2014) |
| ortho-MASnBr$_3$ | VBM | 15 | 0.031 | 1.0 | Kim et al. (2014) |
| cubic-MAPbBr$_3$ | VBM | 240 ± 80 | 0.043 ± 0.002 | 11 ± 4 | Niesner et al. (2016) |
| ortho-MAPbBr$_3$ | VBM | 160 ± 40 | 0.031 ± 0.002 | 7 ± 1 | Niesner et al. (2016) |

and opposite spin textures, with spins orthogonal to the crystal momenta $k_x$ and $k_y$, but radial along the directions rotated by 45°, as shown in Figure 3.2.2b. Such a spin texture qualitatively differs from the one predicted by Kim et al. for β-MAPbI$_3$, β-MASnI$_3$, and ortho-MASnBr$_3$ ferroelectric halide perovskites, where spins rotate according to the circular topology of a standard Rashba-type spin splitting. A model Hamiltonian analysis shows that the $C_{2v}$ point group symmetry in FASnI$_3$ allows the SOC term to have a coexistence of Rashba as well as linear Dresselhaus effects, unlike the materials in Table 3.2.1. By fitting the *ab initio* bandstructure onto the model described by Equation 3.2.3, Stroppa et al. estimated $\alpha_D$ = 1.190 eV Å and $\alpha_R$ = 0.003 eV Å. This analysis demonstrates that a rather large Dresselhaus coupling is present and that the Dresselhaus-like term dominates over the Rashba-like, thus explaining the origin of spin texture in FASnI$_3$ that strongly resembles a pure Dresselhaus type.

Analogous to the FERSCs, a reversal of the ferroelectric polarization from P to −P causes a full switching of the momentum-dependent spin orientations (Stroppa et al. 2014). An interesting possibility in FASnI$_3$ is to change the polar axis from the [001] to the [011] crystalline direction. This can be achieved in principle by rotating molecular dipoles inside the framework cage by applying, for example, a sufficiently large external electric field. When the polar axis is parallel to [011], spin splittings in the VBM are larger than in the CBM. This is ascribed to a different response of the SnI$_3$ framework when the direction of the molecular dipole is rotated. The most evident difference is in the change of the spin-texture topology that sensibly switches from a Dresselhaus to a mostly Rashba-like, with spins rotating in a circular way on constant energy contours, without any radial component (Figure 3.2.2c).

## 3.2.5 Impact of Inversion Symmetry Breaking and the Spin Degree of Freedom on Photocurrents in the Hybrid Perovskites

### 3.2.5.1 Photocurrent Carrier Separation and Inversion Symmetry Breaking in the Hybrid Perovskites

The preceeding sections have introduced the themes of inversion symmetry breaking and strong spin-orbit interaction in the halide perovskites. Here, we explore how these affect the photovoltaic properties of this class of materials. The idea that inversion symmetry breaking (local or global) allows for efficient carrier separation has been a focus of research in the hybrid perovskites. We first review the effect of inversion symmetry breaking without spin-orbit coupling, before considering their combination in the following section. The combination of both induces a Rashba-type electronic band structure and the associated splitting of spin-up and spin-down bands, which then gives rise to several related photovoltaic signatures.

Initial explanations for the high power generation efficiencies of the hybrid perovskites involved the noncentrosymmetric nature of the polar organic cations. It was proposed that this gave rise to efficient photoexcited carrier separation in this material (Sherkar and Koster 2015). In a multidomain sample, ferroelectric domain walls between domains of different molecular orientation can act as sites for separation of electrons and holes (Brivio et al. 2013; Even et al. 2014; Liu et al. 2015). It has also been proposed that carriers diffuse along the potential minima and maxima generated by local dipole order at domain walls (Frost et al. 2014), forming a network of "ferroelectric highways" aiding carrier transport from the bulk of the material to the electrode. Even in samples with disordered molecular arrangements, large-scale DFT calculations suggest that charge separation is induced by long-range potential fluctuations associated with molecular orientational disorder (Ma and Wang 2015). The bulk photovoltaic effect (BPVE), which is the generation of photocurrents in the bulk of non-centrosymmetric materials (Belinicher and Sturman 1980; Sturman and Fridkin 1992; von Baltz and Kraut 1981), has also been studied in the hybrid perovskites (Zheng et al. 2015) using first-principles based perturbation theory (Kràl 2000; Sipe and Shkrebtii 2000; Young and Rappe 2012).

Recently, it has emerged that dynamic structural changes, including dipole alignment, may provide a more complete explanation for the photoconductivity of the hybrid perovskites. A slow increase in the photocurrent, with a time scale of seconds, was observed under illumination (Gottesman et al. 2014). The migration of methylammonium ions does not explain the observed trend of photoconductivity response in all devices, and neither does photoinduced trap formation explain the rise in current over time (Gottesman et al. 2014). It was proposed that methylammonium dipole alignment (Snaith et al. 2014) is caused by a combination of illumination and external bias, which leads to the formation of domains favoring photocurrent conduction. First-principles calculations support this claim by demonstrating a reduction in the molecule–inorganic cage binding energy in the optically excited state (Gottesman et al. 2014). More recent electron-beam–induced current experiments have also supported this hypothesis by showing that carrier diffusion lengths increase under illumination (Kedem et al. 2015). Light-induced halide redistribution (deQuilettes et al. 2016; Hoke et al. 2014; Yoon et al. 2016) and polaronic effects (Neukirch et al. 2016; Nie et al. 2016) have also been proposed as explanations for the time-dependent changes in photoconductivity and photoluminescence under illumination.

### 3.2.5.2 Photocurrents in the Presence of Spin-Orbit Coupling

Strong spin-orbit coupling and broken inversion symmetry give rise to a Rashba-type band structure. The conduction band manifold is split into two bands of opposite spin orientations, as is the valence band manifold. The conduction and valence band manifolds can be described as $J = 1/2$ and $S = 1/2$ systems with $m_j = \pm 1/2$ and $m_s = \pm 1/2$ quantum numbers (Giovanni et al. 2015). Optical transitions between states of different $m_j$ and $m_s$ respond differently to circularly polarized light, depending on selection rules (Even et al. 2014; Kim et al. 2014; Umebayashi et al. 2003). The population of spin states in the conduction and valence bands can therefore be monitored using circularly polarized pump and probe pulses. These time-resolved pump-probe experiments (Giovanni et al. 2015) have shown that the spin relaxation lifetime is 10 ps for electrons and 1 ps for holes. The observed dependence of the spin relaxation time on temperature is consistent with the $\tau \propto T^{-1/2}$ power law for the Elliot–Yafet mechanism (Elliott 1954; Yafet 1983), which involves intra-band spin-flip scattering events.

The persistence of carrier spins at these time scales gives rise to a host of magnetic field effects. In the $\Delta g$ mechanism (Zhang et al. 2015), which refers to the difference of $g$-factors of electrons and holes, the spin degree of freedom of the conduction and valence bands gives rise to a measurable dependence of the photoluminescence, electroluminescence, and photoconductivity on the applied magnetic field. In the hybrid perovskites, electrons in the $m_j = \pm 1/2$ conduction and holes in the $m_s = \pm 1/2$ valence bands combine to form singlet and triplet excited states under illumination. In a magnetic field, the electron–hole pairs precess between the singlet and triplet states, with a precession frequency dependent on the magnetic field, and on $\Delta g$. If the recombination and dissociation rates of singlet and triplet states are unequal, the magnetic field influences the luminescence and conductivity of the sample. It was found that this is indeed the case in the hybrid perovskites (Hsiao et al. 2015; Zhang et al. 2015), with the $\Delta g$ value being orders of magnitude higher than in other semiconductors.

Besides the spin degree of freedom, spin-orbit coupling can also affect the optical properties of the hybrid perovskites through the momentum space degree of freedom. As a result of the Rashba splitting, the conduction band minimum (CBM) and the valence band maximum (VBM) are not located at high-symmetry points of the Brillouin zone ($R$), but are instead shifted slightly. Because of the difference in band characters, the CBM and VBM are not located at the same points. This results in the formation of an indirect band gap, which limits the radiative recombination rate. Because the indirect band gap is only slightly lower in energy (tens of milli-electronvolts) than the direct optical transitions, the absorption spectrum is not greatly affected by the presence of the indirect gap. This combination of a low recombination rate and strong absorption has been proposed as an explanation for the high efficiencies of hybrid perovskite solar cells (Amat et al. 2014; Even et al. 2013; Motta et al. 2015).

Another way the Rashba effect changes the carrier lifetime is by modifying the spin selection rules that govern the recombination probability. This can happen either globally in a polar structure such as $FASnI_3$ in its weak-FE or FE states, or locally and dynamically in a material with polar fluctuations such as $MAPbI_3$ (Azarhoosh et al. 2016; Etienne et al. 2016). Because both the conduction and valence bands are no longer spin-degenerate, the relative spin orientations of electrons and holes will affect their recombination rate. To illustrate this, we consider the excitation, relaxation, and recombination processes of carriers in a Rashba-type band structure (Figure 3.2.3). Incident light at the band gap energy $\hbar\omega$ excites electrons from the valence band manifold to the conduction band manifold. This process preserves the spin orientations. In this example, we assume that the ordering of

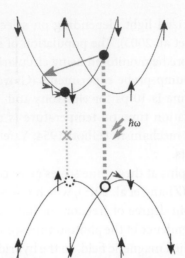

FIGURE 3.2.3 Diagram of Rashba bands and carrier excitation and relaxation processes. The black arrows indicate the directions of the spins. After absorption of photons (thick dashed, right), spin-aligned electrons and holes are created. The excited electrons relax to the CBM due to optical and acoustic phonon scattering processes, which do not flip the electron spin. Similarly, the holes relax to the VBM without flipping their spins. After this thermalization process, radiative recombination of electrons and holes at equal wavevectors is spin-forbidden due to opposite spin helicity (crossed dashed line, left).

bands is such that the CBM and VBM have opposite spin orientations. Optical excitation therefore involves bands higher than CBM or VBM.

Once generated, electrons and holes quickly relax to the band edges via inelastic (spin-conserving) scattering events with phonon modes. The time scale for this relaxation is much faster than the time scale for radiative recombination (Zheng et al. 2015b). We note that phonons of nonzero wavevector can induce interband transitions within the conduction band manifold or within the valence band manifold. For instance, the large arrow in Figure 3.2.3 connects states of the same spin orientation, and so phonon-induced transitions between them are allowed. Through a sequence of such scattering events, electron–hole pairs decohere and individually relax to the band edges, forming a thermalized distribution. The carrier lifetime in this steady state is enhanced because of the opposite spin orientations of these two bands (Zheng et al. 2015b). The enhancement of carrier lifetime therefore relies on favorable orderings of the Rashba-split conduction bands and valence bands. Because the CBM and VBM are primarily of Pb and I character, these orderings are affected by the structure of the $PbI_3$ inorganic lattice (Figure 3.2.4). In Zheng et al. (2015b), first-principles calculations show that the favorable ordering of bands commonly occurs at room temperature.

At room temperature, the molecular orientations vary spatially and change on a picosecond time scale (Poglitsch and Weber 1987; Yaffe et al. 2017). Since the inorganic $PbI_3$ lattice interacts with the molecules, it experiences dynamic disorder as well. Even though the structure, on average, may not break inversion symmetry, the dynamical inversion

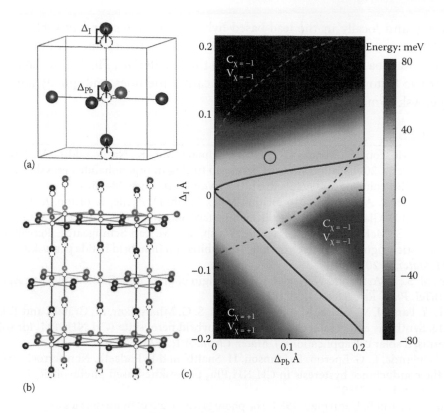

FIGURE 3.2.4 (a, b) Schematic diagram showing Pb and I displacements in $MAPbI_3$. (c) Phase diagram of Rashba splitting energy and spin texture for structures with different Pb and apical I displacements in tetragonal $MAPbI_3$ calculated from DFT. The color indicates the Rashba splitting energy of the conduction and valence bands (whichever is smaller). Positive values indicate spin textures where carrier lifetimes are enhanced, while negative values indicate spin textures where they are not. The boundaries between these two different types of spin textures are indicated by the solid red lines. The dashed lines indicate the areas with energy cost less than 25 meV per tetragonal 20-atom unit cell (below room-temperature fluctuation) to move the Pb and I atoms. The red circle marks the distortions with the lowest total energy. (Adapted from Zheng, F., L. Z. Tan, S. Liu, and A. M. Rappe. 2015. Rashba spin-orbit coupling enhanced carrier lifetime in $CH_3NH_3PbI_3$. *Nano Lett.* 15; 7794–7800.)

symmetry breaking can result in a dynamically changing indirect band gap (Motta et al. 2015), which will increase the carrier lifetime. Different regions of a bulk sample of $MAPbI_3$ could contain different local structures. Not only do these fluctuations lead to a dynamically changing indirect band gap, but they also result in spatially varying Rashba splitting. Strictly speaking, the Rashba model has to be modified to account for spatial inhomogeneity, and is applicable only when the length scale of disorder is large enough to generate local regions of Rashba splitting. MD simulations have shown that a spatially local Rashba effect arises at subpicosecond time scales (Etienne et al. 2016).

In this chapter, we have reviewed evidence for ferroelectricity and its relation to spin-orbit coupling in the hybrid perovskites. The breaking of inversion symmetry occurs dynamically and locally in the lead-based hybrid perovskites, and possibly globally in FASnI$_3$. We have seen how the combination of ferroelectricity and spin-orbit coupling gives rise to Rashba and Dresselhaus effects in these materials. Finally, we have surveyed how these two factors affect carrier separation, carrier lifetimes, and spin dynamics in the hybrid perovskite materials.

## REFERENCES

Amat, A., E. Mosconi, E. Ronca, C. Quarti, P. Umari, Md. K. Nazeeruddin, M. Grätzel, and F. De Angelis. 2014. Cation-induced band-gap tuning in organohalide perovskites: Interplay of spin-orbit coupling and octahedra tilting. *Nano Lett.* 14; 3608–3616.

Ast, C. R., J. Henk, A. Ernst, L. Moreschini, M. C. Falub, D. Pacilé, P. Bruno, K. Kern, and M. Grioni. 2007. Giant spin splitting through surface alloying. *Phys. Rev. Lett.* 98; 186807-4.

Azarhoosh, P., S. McKechnie, J. M. Frost, A. Walsh, and M. van Schilfgaarde. Research update: Relativistic origin of slow electron-hole recombination in hybrid halide perovskite solar cells. *APL Mater.* 4; 2016–2018.

Bahramy, M. S., R. Arita, and N. Nagaosa. 2011. Origin of giant bulk Rashba splitting: Application to BiTeI. *Phys. Rev. B* 84; 041202(R).

Baikie, T., Y. Fang, J. M. Kadro, M. Schreyer, F. Wei, S. G. Mhaisalkar, M. Grätzel, and T. J. White. 2013. Synthesis and crystal chemistry of the hybrid perovskite (CH$_3$NH$_3$)PbI$_3$ for solid-state sensitised solar cell applications. *J. Mater. Chem. A* 1; 5628–5641.

Beilsten-Edmands, J., G. Eperon, R. Johnson, H. Snaith, and P. Radaelli. Non-ferroelectric nature of the conductance hysteresis in CH$_3$NH$_3$PBi$_3$ perovskite-based photovoltaic devices. *Appl. Phys. Lett.* 106; 173502.

Belinicher, V. I., and B. I. Sturman. 1980. The photogalvanic effect in media lacking a center of symmetry. *Sov. Phys. Usp.* 23; 199–223.

Brivio, F., A. B.Walker, and A. Walsh. 2013. Structural and electronic properties of hybrid perovskites for high-efficiency thin-film photovoltaics from first-principles. *APL Mater.* 1; 042111–042115.

Bychkov, Y. A., and E. Rasbha. 1984. Properties of a 2d electron gas with lifted spectral degeneracy. *P. Zh. Eksp. Teor. Fiz.* 39; 66–69.

deQuilettes, D. W., W. Zhang, V. M. Burlakov, D. J. Graham, T. Leijtens, A. Osherov, V. Bulović, H. J. Snaith, D. S. Ginger, and S. D. Stranks. 2016. Photo-induced halide redistribution in organic-inorganic perovskite films. *Nature Commun.* 7; 11683–11689.

Di Sante, D., P. Barone, R. Bertacco, and S. Picozzi. 2013. Electric control of the giant Rashba effect in bulk GeTe. *Adv. Mater.* 25; 509–513.

Di Sante, D., A. Stroppa, P. Jain, and S. Picozzi. 2013. Tuning the ferroelectric polarization in a multiferroic metal-organic framework. *J. Am. Chem. Soc.* 135; 18126–18130.

Dresselhaus, G. 1955. Spin-orbit coupling effects in zinc blende structures. *Phys. Rev.* 100; 580–586.

Elliott, R. J. 1954. Theory of the effect of spin-orbit coupling on magnetic resonance in some semi-conductors. *Phys. Rev.* 96; 266–279.

Etienne, T., E. Mosconi, and F. De Angelis. 2016. Dynamical origin of the Rashba effect in organohalide lead perovskite: A key to suppressed carrier recombination in perovskite solar cells? *J. Phys. Chem. Lett.* 7; 1638–1645.

Even, J., L. Pedesseau, J.-M. Jancu, and C. Katan. 2013. Importance of spin-orbit coupling in hybrid organic/inorganic perovskites for photovoltaic applications. *J. Phys. Chem. Lett.* 4; 2999–3005.

Even, J., L. Pedesseau, and C. Katan. 2014. Analysis of multivalley and multibandgap absorption and enhancement of free carriers related to exciton screening in hybrid perovskites. *J. Phys. Chem. C* 118; 11566–11572.

Fan, Z., J. Xiao, K. Sun, L. Chen, Y. Hu, J. Ouyang, K. P. Ong, K. Zeng, and J. Wang. 2015. Ferroelectricity of $CH_3NH_3PbI_3$ perovskite. *J. Phys. Chem. Lett.* 6; 1155–1161.

Frost, J. M., K. T. Butler, F. Brivio, C. H. Hendon, M. van Schilfgaarde, and A. Walsh. 2014. Atomistic origins of high-performance in hybrid halide perovskite solar cells. *Nano Lett.* 14; 2584–2590.

Frost, J. M., K. T. Butler, and A. Walsh. 2014. Molecular ferroelectric contributions to anomalous hysteresis in hybrid perovskite solar cells. *APL Mater.* 2; 081506–081510.

Ganichev, S. D., and L. E. Golub. 2014. Interplay of Rashba/Dresselhaus spin splittings probed by photogalvanic spectroscopy: A review. *Phys. Status Solidi (b)* 251; 1801–1823.

Giovanni, D., H. Ma, J. Chua, M. Grätzel, R. Ramesh, S. Mhaisalkar, N. Mathews, and T. C. Sum. 2015. Highly spin-polarized carrier dynamics and ultralarge photoinduced magnetization in $CH_3NH_3PbI_3$ perovskite thin films. *Nano Lett.* 15; 1553–1558.

Gottesman, R., E. Haltzi, L. Gouda, S. Tirosh, Y. Bouhadna, A. Zaban, E. Mosconi, and F. De Angelis. 2014. Extremely slow photoconductivity response of $CH_3NH_3PbI_3$ perovskites suggesting structural changes under working conditions. *J. Phys. Chem. Lett.* 5; 2662–2669.

Hoffmann, A., and S. D. Bader. 2015. Opportunities at the frontiers of spintronics. *Phys. Rev. Appl.* 4; 047001.

Hoke, E. T., D. J. Slotcavage, E. R. Dohner, A. R. Bowring, H. I. Karunadasa, and M D. McGehee. 2014. Reversible photo-induced trap formation in mixed-halide hybrid perovskites for photovoltaics. *Chem. Sci.* 6; 613–617.

Hoque, M. N. F., M. Yang, Z. Li, N. Islam, X. Pan, K. Zhu, and Z. Fan. 2016. Polarization and dielectric study of methylammonium lead iodide thin film to reveal its nonferroelectric nature under solar cell operating conditions. *ACS Energy Lett.* 1; 142–149.

Hsiao, Y.-C. T. Wu, and B. Hu. 2015. Magneto-optical studies on spin-dependent charge recombination and dissociation in perovskite solar cells. *Adv. Mater.* 27; 2899–2906.

Ishizaka, K., M. S. Bahramy, H. Murakawa, M. Sakano, T. Shimojima, T. Sonobe, K. Koizumi, S. Shin, H. Miyahara, A. Kimura, K. Miyamoto et al. 2011. Giant Rashba-type spin splitting in bulk BiTeI. *Nat. Mater.* 10; 521–526.

Jain, P., V. Ramachandran, R. J. Clark, H. D. Zhou, B. H. Toby, N. S. Dalal, H. W. Kroto, and A. K. Cheetham. 2009. Multiferroic behavior associated with an order-disorder hydrogen bonding transition in metal-organic frameworks (mofs) with the perovskite $ABX_3$ architecture. *J. Am. Chem. Soc.* 131; 13625–13627.

Juarez-Perez, E.J., R. S. Sanchez, L. Badia, G. Garcia-Belmonte, Y. S. Kang, I. Mora-Sero, and J. Bisquert. 2014. Photoinduced giant dielectric constant in lead halide perovskite solar cells. *J. Phys. Chem. Lett.* 5; 2390–2394.

Kedem, N., T. M. Brenner, M. Kulbak, N. Schaefer, S. Levcenko, I. Levine, D. Abou-Ras, G. Hodes, and D. Cahen. 2015. Light-induced increase of electron diffusion length in a p-n junction type $CH_3NH_3PbBr_3$ perovskite solar cell. *J. Phys. Chem. Lett.* 6; 2469–2476.

Kim, M., J. Im, A. J. Freeman, J. Ihm, and H. Jin. 2014. Switchable $S=1/2$ and $J = 1/2$ Rashba bands in ferroelectric halide perovskites. *PNAS* 111; 6900–6904.

Král, P. 2000. Quantum kinetic theory of shift-current electron pumping in semiconductors. *J. Phys.: Condens. Matter* 12; 4851–4868.

Kutes, Y., L. Ye, Y. Zhou, S. Pang, B. D. Huey, and N. P. Padture. 2014. Direct observation of ferroelectric domains in solution-processed $CH_3NH_3PbI_3$ perovskite thin films. *J. Phys. Chem. Lett.* 5; 3335–3339.

LaShell, S., B. McDougall, and E. Jensen. 1996. Spin splitting of an au(111) surface state band observed with angle resolved photoelectron spectroscopy. *Phys. Rev. Lett.* 77; 3419.

Liebmann, M., C. Rinaldi, D. Di Sante, J. Kellner, C. Pauly, R. N. Wang, J. E. Boschker, A. Giussani, S. Bertoli, M. Cantoni, L. Baldrati et al. 2015. Giant Rashba-type spin splitting in ferroelectric GeTe(111). *Adv. Mater.* 28; 560–565.

Liu, S., F. Zheng, N. Z. Koocher, H. Takenaka, F. Wang, and A. M. Rappe. 2015. Ferroelectric domain wall induced band gap reduction and charge separation in organometal halide perovskites. *J. Phys. Chem. Lett.* 6; 693–699.

Ma, J., and L.-w. Wang. 2015. Nanoscale charge localization induced by random orientations of organic molecules in hybrid perovskite $CH_3NH_3PbI_3$. *Nano Lett.* 15; 248–253.

Mattoni, A., A. Filippetti, M. I. Saba, and P. Delugas. 2015. Methylammonium rotational dynamics in lead halide perovskite by classical molecular dynamics: The role of temperature. *J. Phys. Chem. C* 119; 17421–17428.

Mirhosseini, H., I. V. Maznichenko, S. Abdelouahed, S. Ostanin, A. Ernst, I. Mertig, and J. Henk. 2010. Toward a ferroelectric control of Rashba spin-orbit coupling: Bi on $BaTiO_3(001)$ from first principles. *Phys. Rev. B* 81; 073406-4.

Motta, C. F. El-Mellouhi, S. Kais, N. Tabet, F. Alharbi, and S. Sanvito. 2015. Revealing the role of organic cations in hybrid halide perovskite $CH_3NH_3PbI_3$. *Nature Commun.* 6; 7026–7027.

Nakamura, H., T. Koga, and T. Kimura. 2012. Experimental evidence of cubic Rashba effect in an inversion-symmetric oxide. *Phys. Rev. Lett.* 108; 206601–206604.

Neukirch, A. J., W. Nie, J.-C. Blancon, K. Appavoo, H. Tsai, M. Y. Sfeir, C. Katan, L. Pedesseau, J. Even, J. J. Crochet, G. Gupta, A. D. Mohite, and S. Tretiak. 2016. Polaron stabilization by cooperative lattice distortion and cation rotations in hybrid perovskite materials. *Nano Lett.* 16; 3809–3816.

Nie, W., J.-C. Blancon, A. J. Neukirch, K. Appavoo, H. Tsai, M. Chhowalla, M. A. Alam, M. Y. Sfeir, C. Katan, J. Even, S. Tretiak, J. J. Crochet, G. Gupta, and A. D. Mohite. 2016. Light-activated photocurrent degradation and self-healing in perovskite solar cells. *Nature Commun.* 7; 11574–11579.

Niesner, D., M. Wilhelm, I. Levchuk, A. Osvet, S. Shrestha, M. Batentschuk, C. Brabec, and T. Fauster. 2016. Giant Rashba splitting in $CH_3NH_3PbBr_3$ organic-inorganic perovskite. *Phys. Rev. Lett.* 117; 126401–126404.

Nitta, J., T. Akazaki, H. Takayanagi, and T. Enoki. 1997. Gate control of spin-orbit interaction in an inverted InGaAs/InAlAs heterostructure. *Phys. Rev. Lett.* 78; 1335–1338.

Picozzi, S. 2014. Ferroelectric Rashba semiconductors as a novel class of multifunctional materials. *Front. Physics* 2; 10–15.

Poglitsch, A., and D. Weber. 1987. Dynamic disorder in methylammoniumtrihalogenoplumbates (II) observed by millimeter-wave spectroscopy. *J. Chem. Phys.* 87; 6373–6378.

Quarti, C., E. Mosconi, and F. De Angelis. 2014. Interplay of orientational order and electronic structure in methylammonium lead iodide: Implications for solar cells operation. *Chem. Mater.* 26; 6557–6569.

Rashba, E. 1960. Properties of semiconductors with an extremum loop. 1. cyclotron and combinational resonance in a magnetic field perpendicular to the plane of the loop. *Sov. Phys. Solid State* 2; 1109–1122.

Sewvandi, G. A., K. Kodera, H. Ma, S. Nakanishi, and Q. Feng. 2016. Antiferroelectric nature of $CH_3NH_3PbI_{3-x}Cl_x$ perovskite and its implication for charge separation in perovskite solar cells. *Sci. Rep.* 6; 30680–30686.

Shao, Y., Z. Xiao, C. Bi, Y. Yuan, and J. Huang. 2014. Origin and elimination of photocurrent hysteresis by fullerene passivation in $CH_3NH_3PbI_3$ planar heterojunction solar cells. *Nat. Commun.* 5; 5784–5787.

Sherkar, T. S., and L. J. A. Koster. 2015. Can ferroelectric polarization explain the high performance of hybrid halide perovskite solar cells? *Phys. Chem. Chem. Phys.* 18; 331–338.

Sipe, J. E., and A. I. Shkrebtii. 2000. Second-order optical response in semiconductors. *Phys. Rev. B* 61; 5337–5352.

Snaith, H. J., A. Abate, J. M. Ball, G. E. Eperon, T. Leijtens, N. K. Noel, S. D. Stranks, J. T.-S. Wang, K. Wojciechowski, and W. Zhang. 2014. Anomalous hysteresis in perovskite solar cells. *J. Phys. Chem. Lett.* 5; 1511–1515.

Spanier, J. E., V. M. Fridkin, A. M. Rappe, A. R. Akbashev, A. Polemi, Y. Qi, Z. Gu, S. M. Young, C. J. Hawley, D. Imbrenda, G. Xiao, A. L. Bennett-Jackson, and C. L. Johnson. 2016. Power conversion efficiency exceeding the Shockley-Queisser limit in a ferroelectric insulator. *Nature Photonics* 10; 611–616.

Stoumpos, C. C., C. D. Malliakas, and M. G. Kanatzidis. 2013. Semiconducting tin and lead iodide perovskites with organic cations: Phase transitions, high mobilities, and near-infrared photoluminescent properties. *Inorg. Chem.* 52; 9019–9038.

Stroppa, A., D. Di Sante, P. Barone, M. Bokdam, G. Kresse, C. Franchini, M.-H. Whangbo, and S. Picozzi. 2014. Tunable ferroelectric polarization and its interplay with spin-orbit coupling in tin iodide perovskites. *Nat. Commun.* 5; 5900–5908.

Stroppa, A., P. Jain, P. Barone, M. Marsman, J. M. Perez-Mato, A. K. Cheetham, H. W. Kroto, and S. Picozzi. 2011. Electric control of magnetization and interplay between orbital ordering and ferroelectricity in a multiferroic metal–organic framework. *Angew. Chem. Int. Ed.* 26; 5847–5850.

Stroppa, A., C. Quarti, F. De Angelis, and S. Picozzi. 2015. Ferroelectric polarization of $CH_3NH_3PbI_3$: A detailed study based on density functional theory and symmetry mode analysis. *J. Phys. Chem. Lett.* 6; 2223–2231.

Sturman, B. I., and V. M. Fridkin. 1992. *The photovoltaic and photorefractive effects in noncentrosymmetric materials*. Amsterdam: Gordon and Breach Science.

Tian, Y., A. Stroppa, Y.-S. Chai, P. Barone, M. Perez-Mato, S. Picozzi, and Y. Sun. 2015. High-temperature ferroelectricity and strong magnetoelectric effects in a hybrid organic-inorganic perovskite framework. *Phys. Stat. Sol. RRL* 9; 62–67.

Tian, Y., A. Stroppa, Y.-S. Chai, L. Yan, S. Wang, P. Barone, S. Picozzi, and Y Sun. 2014. Cross coupling between electric and magnetic orders in a multiferroic metal-organic framework. *Sci. Rep.* 4; 6062–6065.

Tress, W., N. Marinova, T. Moehl, S. M. Zakeeruddin, M. K. Nazeeruddin, and M. Grätzel. 2015. Understanding the rate-dependent *J-V* hysteresis, slow time component, and aging in $CH_3NH_3PbI_3$ perovskite solar cells: The role of a compensated electric field. *Energy Environ Sci.* 8; 995–1004.

Umebayashi, T., K. Asai, T. Kondo, and A. Nakao. 2003. Electronic structures of lead iodide based low-dimensional crystals. *Phys. Rev. B* 67; 155405–155406.

Varykhalov, A., D. Marchenko, M. R. Scholz, E. D. L. Rienks, T. K. Kim, G. Bihlmayer, J. Sánchez-Barriga, and O. Rader. 2012. Ir(111) surface state with giant Rashba splitting persists under graphene in air. *Phys. Rev. Lett.* 108; 066804.

von Baltz, R., and W. Kraut. 1981. Theory of the bulk photovoltaic effect in pure crystals. *Phys. Rev. B* 23; 5590–5596.

Wei, J., Y. Zhao, H. Li, G. Li, J. Pan, D. Xu, Q. Zhao, and D. Yu. 2014. Hysteresis analysis based on the ferroelectric effect in hybrid perovskite solar cells. *J. Phys. Chem. Lett.* 5; 3937–3945.

Winkler, R. 2003. *Spin-orbit coupling effects in two-dimensional electron and hole systems*. Berlin and Heidelberg: Springer-Verlag.

Xiao, Z., Y. Yuan, Y. Shao, Q. Wang, Q. Dong, C. Bi, P. Sharma, A. Gruverman, and J. Huang. 2015. Giant switchable photovoltaic effect in organometal trihalide perovskite devices. *Nat. Mater.* 14; 193–198.

Yafet, Y. 1983. Conduction electron spin relaxation in the superconducting state. *Phys. Lett. A* 98; 287–290.

Yaffe, O., Y. Guo, L. Z. Tan, D. A. Egger, T. Hull, C. C. Stoumpos, F. Zheng, T. F. Heinz, L. Kronik, M. G. Kanatzidis, J. S. Owen, A. M. Rappe, M. A. Pimenta, and L. E. Brus. 2017. Local polar fluctuations in leadhalide perovskite crystals. *Phys. Rev. Lett.* 118; 136001–4.

Yoon, S. J., S. Draguta, J. S. Manser, O. Sharia, W. F. Schneider, M. Kuno, and P. V. Kamat. 2016. Tracking iodide and bromide ion segregation in mixed halide lead perovskites during photo-irradiation. *ACS Energy Lett.* 1; 290–296.

Young, S. M., and A. M. Rappe. 2012. First principles calculation of the shift current photovoltaic effect in ferroelectrics. *Phys. Rev. Lett.* 109; 116601–116605.

Zhang, C., D. Sun, C.-X. Sheng, Y. X. Zhai, K. Mielczarek, A. Zakhidov, and Z. V. Vardeny. 2015. Magnetic field effects in hybrid perovskite devices. *Nature Phys.* 11; 427–434.

Zheng, F., H. Takenaka, F. Wang, N. Z. Koocher, and A. M. Rappe. 2015. First-principles calculation of the bulk photovoltaic effect in $CH_3NH_3PbI_3$ and $CH_3NH_3PbI_{3-x}Cl_x$. *J. Phys. Chem. Lett.* 6; 31–37.

Zheng, F., L. Z. Tan, S. Liu, and A. M. Rappe. 2015. Rashba spin-orbit coupling enhanced carrier lifetime in $CH_3NH_3PbI_3$. *Nano Lett.* 15; 7794–7800.

# CHAPTER 4

# Alloys and Environmental Related Issues

*Toward the Computational Design of Pb-Free and Stable Hybrid Materials for Solar Cells*

Fedwa El Mellouhi, Fahhad H. Alharbi, Carlo Motta, Sergey Rashkeev, Stefano Sanvito, and Sabre Kais

## CONTENTS

## 4.1 INTRODUCTION

Since 2012, a new family of solar cell technologies based on hybrid perovskite absorbers has emerged in the world of photovoltaics and has evolved at an unprecedented pace. This new research direction was originally initiated with the use of $(CH_3NH_3)PbI_3$ as a dye replacement in dye-sensitized solar cells (DSSCs) (Im et al. 2011) and now such compounds are considered a serious competitor materials platform to silicon-based technologies. Over this short period of time the efficiency of perovskite-based solar cells has reached a certified value of >22.0% (Conings et al. 2014; NREL 2016). Most astonishing is the fact that extremely high efficiencies are achieved regardless of the device design as long as the light-absorbing medium is one of the Pb-based hybrid organic–inorganic halide perovskites (NREL 2016; Stoumpos et al. 2013; Zhao and Zhu 2013). This fact illustrates how rich the field is and similarly reveals that we are just at the beginning of a scientific revolution. Owing to these rapid developments and remarkable facts, the editors of *Science* magazine listed perovskite solar cells (PSCs) among the top 10 scientific breakthroughs of the year 2013 (Science Focus 2013). It is furthermore expected that the rapid development rate and extreme simplicity of PSCs fabrication will lead to commercialization in the very near future (Hodes and Cahen 2014; Kojima et al. 2009; Leitens et al. 2015; Nayak et al. 2012; NREL 2016; Peplow 2014; Science Focus 2013; Service 2014; Snaith 2013).

### 4.1.1 Environment-Related Issues in PSC Deployment

At the present time, questions related to mass production of PSCs and their commercialization are rapidly emerging, largely anticipating the typical decennial research and development (R&D)-to-production process (Babayigit et al. 2016; Hodes and Cahen 2014). This further testifies to the high potential of such novel materials in solar devices and the expectations that they raise. Yet, many practical difficulties remain that prevent commercialization of perovskite cells. The two main problems are the presence of the toxic element lead [many countries prohibit the use of Pb for residential applications (Chen et al. 2014; NREL 2016)] and the mechanical instability of the cell. In principle, the aforementioned fact that many efficient PSCs are made using a wide range of Pb-based hybrid perovskites (NREL 2016; Stoumpos et al. 2013; Zhao and Zhu 2013) and with tens of device concepts and designs (Conings et al. 2014; Kojima et al. 2009) leads to two important conclusions. On the one hand, it illustrates how rich the field is even at this infant stage. On the other hand, it indicates that PSCs are far from being optimized. Thus, the efficiency can be improved even further (up to 32% corresponding to the Shockley–Quiesser limit). It is worth noting that a maximum efficiency of 28% (Nayak et al. 2012; Snaith 2013) was estimated based on the voltage at maximum power point ($V_{mpp}$), which depends on the structural disorder often observed in the hybrid PSCs (Nayak et al. 2012).

In reality, it is now established that the main problem regarding PSCs is not their efficiency, but rather their stability (Leitens et al. 2015). The best reported lifetime for a PSC is about 10,000 hours, which is obtained for a PSC device fabricated under 20% relative humidity (Mei et al. 2014). The lifetime is reduced to 1,000 hours for 50% relative humidity (Zhao and Zhu 2013), as PSCs degrade rapidly if exposed to moisture. Both of these reported devices have been encapsulated after fabrication. Research efforts have taken several directions. The main trends are toward stability improvement, developing Pb-free PSCs cells, device optimization, particular applications, and fundamental understanding of the high performance of PSCs. Importantly, the PSC industry can take advantage of the well-established knowledge concerning DSSC and organic light emitting diodes (OLED) technology (NREL 2016). However, there are other important challenges that should be addressed beforehand (Hodes and Cahen 2014; Kim et al. 2014; Osedach et al. 2013; Park 2013; Pellet et al. 2014; Peplow 2014; Science Focus 2013; Service 2014). To tackle the instability issue, many approaches have been suggested (Pellet et al. 2014); however, the improvement is still marginal with a considerable increase in the cost. For example, encapsulation needs the use of relatively expensive processes. Therefore, alternative cost-efficient solutions of the stability problem are needed.

## 4.1.2 Perovskite Structure Formability and the Concept of Tolerance Factor

The perovskite crystal structure, $AMX_3$, is characterized by $[MX_6]^{4-}$ octahedra (see Figure 4.1), which share corners in all three orthogonal directions to generate infinite three-dimensional (3D) $[MX_3]^-$ frameworks. The $A^+$ cations function as structural

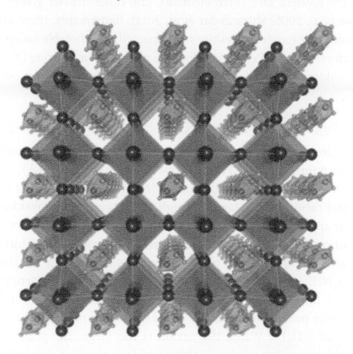

FIGURE 4.1    Representation of prototypical $(CH_3NH_3)PbI_3$ hybrid perovskite crystal structure.

templates and their shape, size, and charge distribution are crucial factors for the properties and the stabilization of the perovskite structure. The Goldschmidt's tolerance factor (TF) (Goldschmidt 1926) can help us in understanding which specific structure forms with the $AMX_3$ composition. This is defined as

$$t = (r_A + r_X)/\sqrt{2}(r_B + r_X), \tag{4.1}$$

where $r_A$, $r_B$, and $r_X$ are the ionic radii of the corresponding ions in the $AMX_3$ formula. For values of $t$ in the range 0.9–1.0, mostly cubic perovskites are found, whereas values of 0.80–0.89 predominantly lead to distorted perovskites that can be further classified by using the concept of octahedral tilting (Kieslich et al. 2014). Below 0.80, other structures such as the ilmenite-type ($FeTiO_3$) are more stable because of the similar sizes of the cations A and B. Values of $t$ larger than 1 lead to hexagonal structures in which layers of face-sharing octahedra are introduced into the structure.

Halide perovskites, however, have several additional specific features (Stoumpos and Kanatzidis 2015). Unlike their oxide counterparts, the $A^I M^{II} X_3^-$ composition can be formed only with specific combinations of elements. This is because the halide ($X^-$) anions (X = Cl, Br, I) impose two major differences in comparison to the oxide ($O^{2-}$) anion: (1) they bear a smaller negative charge, which is sufficient only to compensate metal ions in lower oxidation states, and (2) they have a much larger ionic radii, which prevents much smaller $M^{II}$ metal anions to be incorporated in an octahedral coordination geometry. Because of these restrictions, the $M^{II}$ metal anion can be selected only from a narrow set of elements including the alkaline earths, the bivalent rare earth elements, and other heavier group elements ($Ge^{2+}$, $Sn^{2+}$, $Pb^{2+}$) (Hesse et al. 2006; Shirwakdar et al. 2011). Remarkably, there are no transition metals that can adopt the halide perovskite structure, with few notable exceptions (Horowitz et al. 1982). The aforementioned considerations do not apply for fluorine (X = $F^-$), which can stabilize the perovskite structure for virtually all the bivalent metal ions owing to its small size.

### 4.1.3 Revised Tolerance Factor for Organo-Metallic (Hybrid) Perovskites

Only three A cations known to date are able to stabilize the perovskite structure in heavy halides: namely, $Cs^+$, $CH_3NH_3^+$ (methylammonium, MA), and $HC(NH_2)_2^+$ (formamidinium, FA) (Stoumpos and Kanatzidis 2015). Whereas it is clear that $Cs^+$ is the only elemental cation that is large enough to sustain the perovskite structure, for organic cations it appears that it is not only size and structure that are important, but also the distribution of the net positive charge (dipole moment). Thus, $CH_3NH_3^+$ and $HC(NH_2)_2^+$ are able to stabilize the perovskite, but cations with similar size, such as $HONH_3^+$, $CH_3CH_2NH_3^+$, or $(CH_3)_2NH_2^+$ form lower dimension 2D structures with edge sharing octahedra, for example (Im et al. 2012; Thiele and Serr 1996). When the cation is too small, the preferred structure is $NH_4CdCl_3$-type, which can be described as double chains of $[MI_5]^{3-}$. When the cation is too large, the preferred structure is $CsNiBr_3$-type, which consists of single chains of face-sharing octahedra. In contrast, very large, densely packed cations can give rise to perovskite-like structures albeit with a lower dimensionality (Cao et al. 2015; Wang et al. 1995).

Recently, Goldschmidt's concept of the tolerance factor was extended (Kieslich et al. 2014, 2015) to account for the nonspherical cations. Similar to Goldschmidt's initial approach, (effective) ionic radii were used for calculating the tolerance factor of hybrid perovskites. Protonated amines including $[NH_4]^+$ (ammonium), $[NH_3OH]^+$ (hydroxylammonium), $[CH_3NH_3]^+$ (methylammonium), $[NH_3NH_2]^+$ (hydrazinium), $[(CH_2)_3NH_2]^+$ (azetidinium), $[CH(NH_2)_2]^+$ (formamidinium), $[(C_3N_2H_5]^+$ (imidazolium), $[(CH_3)_2NH_2]^+$ (dimethylammonium), $[(CH_3CH_2)NH_3]^+$ (ethylammonium), $[(NH_2)_3C]^+$ (guanidinium), $[(CH_3)_4N]^+$ (tetramethylammonium), $[C_3H_4NS]^+$ (thiazolium), and $[NC_4H_8]^+$ (3-pyrollinium) were treated as spheres with an effective radius $r_{Aeff}$. Highly asymmetric molecular anions such as $HCOO^-$, $CN^-$, and $N_3^-$, were treated as cylinders with effective height $h_{Xeff}$ and effective radius $r_{Xeff}$ for the cylinder, respectively. TFs of hybrid perovskites were calculated according to the generalized Goldschmidt equation,

$$t = (r_{Aeff} + r_{Xeff})/\sqrt{2}(r_B + 0.5h_{Xeff}) \tag{4.2}$$

Considering a wide variety of divalent metal ions ($Be^{2+}$, $Mg^{2+}$, $Ca^{2+}$, $Sr^{2+}$, $Ba^{2+}$, $Mn^{2+}$, $Fe^{2+}$, $Co^{2+}$, $Ni^{2+}$, $Pd^{2+}$, $Pt^{2+}$, $Cu^{2+}$, $Zn^{2+}$, $Cd^{2+}$, $Hg^{2+}$, $Ge^{2+}$, $Sn^{2+}$, $Pb^{2+}$, $Eu^{2+}$, $Tm^{2+}$, $Yb^{2+}$) as cation B and adding some "symmetric" ions such as halides ($F^-$, $Cl^-$, $Br^-$, $I^-$) and $BH_4^-$, Kieslich et al. (2015) calculated TF for more than 2500 amine–metal–anion permutations. Their results suggested the potential existence of more than 600 undiscovered hybrid perovskites (with $0.8 < t < 1.0$) including alkaline earth metal– and lanthanide-based materials. Although the limitations of tolerance factors for hybrid materials are not yet clear, one could expect that this study is a great initial step for selection of materials for certain applications. Also, these results show the large tuning adaptability of the perovskite structure, which is encouraging for further search on hybrid perovskites for photovoltaic applications.

Along this direction, several computational and experimental efforts have recently been conducted toward the quest for new materials for solar cells inspired by the current understanding of the prototypical $(CH_3NH_3)PbI_3$. The search for Pb-free alternatives either consists of steric engineering of hybrid organic–inorganic materials, or the design or prediction of new inorganic compounds that might be good candidates to host organic moieties. Current efforts and future directions are discussed in the sections to follow.

## 4.2 COMPUTATIONAL MATERIALS SCREENING AND DESIGN

The discovery and design of new compounds with a desired set of properties is one of the main challenges of modern materials science and engineering. Traditionally, the characterization of a new material can take years of research and involve theoretical modeling at different time and length scales and also a large set of experimental measurements. Although such an approach will always be needed to achieve a deep and complete understanding of materials, it is limited in speed and also requires expertise in many different fields. Furthermore, at the computational level, different models should be developed for the material at different scales, creating a bottleneck between the initial material model (at an atomistic level) and the application, which is typically at a mesoscopic level. This section discusses major materials screening and discovery initiatives active today followed by a discussion of important components critical for the success of the screening procedure. This includes structural prediction and minima search, descriptor development for a fast and efficient screening of solar cell materials, and finally, data analytics.

### 4.2.1 Major Computational Material Discovery Initiatives

In the last few years, a new protocol for materials discovery has emerged. This is the so-called high-throughput electronic structure theory (HTEST) scheme, which is based on combining advanced electronic structure methods, database construction, intelligent machine-learning–based database search, and experimental validation. Such research protocol is at the core of the Materials Genome Initiative (MGI 2016), an initiative announced by the US White House in 2011, whose goal involves accelerating the materials-to-applications development time. Following the MGI a number of consortia have been formed, mainly in the United States, with the scope of building the necessary computational and database infrastructure, and of exploring different classes of materials for specific targeted applications. The Center for Materials Genomics (CMG 2016), based at Duke University, develops and maintains the Automatic - Flow for Materials Discovery (AFLOW) code (Curtarolo et al. 2012) and the AFLOWLIB.org library (Curtarolo et al. 2012b). The center hosts a database containing electronic structure data for about 800,000 inorganic compounds (either experimentally synthesized or hypothetical) and it is specialized in the search for intermetallic alloys, mostly binary. Success of the consortium includes the mapping of many novel binary alloys from the platinum group (Hart et al. 2013) and the discovery of a number of half-Heusler alloys displaying exceptionally low thermal conductivity (Carrete et al. 2014).

The Materials Project (MP 2016), hosted jointly at MIT and the Lawrence Berkeley National Laboratory, has developed a range of screening tools namely Python Materials Genomics (PYMATGEN) and a web portal building a database of about 80,000 compounds. While the main focus of the initiative is to engineer novel materials for Li-ion batteries (Chen et al. 2012), a task that has already shown several successes, their database covers other classes of materials as well.

The Harvard Clean Energy Project (HCEP 2016), in collaboration with IBM, is a virtual high-throughput discovery and design effort for the next generation of plastic solar cell materials. The target of the project is the design of novel organic materials for both solar energy harvesting and organic electronics. The project maintains a database for existing and hypothetical organic materials comprising several million entries. Successes of the initiative include the discovery and synthesis of a novel high-mobility organic single crystal compound (Sokolov et al. 2011). A novel major initiative has been recently funded in Switzerland [The National Centre of Competence in Research (NCCR) "MARVEL"] with the aim of constructing a large computational and infrastructural platform for materials discovery. Although this initiative has a large breadth, there is a significant component dedicated to HTEST, in particular the development of the Automated Interactive Infrastructure and Database for Computational Science (AiiDA) platform for atomistic simulations (AiiDA 2016).

The Novel Materials Discovery repository (NoMaD 2016) very recently invited the computational materials community to upload their computed materials data for sharing with users worldwide. Sharing data from computational materials science aims to enable the confirmatory analysis of materials data, their reuse, and repurposing. The Open Quantum Materials Database is a database of Density Functional Theory (DFT) calculated thermodynamic and structural properties containing a large amount of data and prototype structures (OQMD 2016). Also, the Computational Materials Repository (CMR) is one of the first databases focused on perovskite structures for visible light absorption (CMR 2016).

The CRAQSolar project, recently funded by the Qatar National Research fund (CRAQSolar 2016), a collaboration between Qatar Environment and Energy Research Institute (QEERI 2016) and Centre for Research on Adaptive Nanostructures and Nanodevices (CRANN 2016), avails of the computational infrastructure of the Center for Materials Genomics to design a new class of light harvesting materials, currently unexplored by any other HTEST initiatives, namely that of hybrid organic/inorganic perovskites. This is an area with huge potential for developing a novel generation of toxic-element–free light harvesting devices. The research methodology of the project shall be elaborated further in the next sections.

## 4.2.2 Global Minima and Stability Search Methods

High-throughput computational materials design and discovery is an emerging area of materials science combining advanced crystal structure prediction, thermodynamic and electronic-structure calculation, as well as machine learning and database construction (Curtarolo et al. 2013). The first step in materials discovery or in the design of materials with specific properties consists in determining their structural properties. Structural information is critical to the understanding and predicting materials properties and their functionalities. This information can be obtained by using computer simulation methods to map the high-dimensional potential free energy surface and find the global minimum among the large number of local minima (Dixon and Szego 1975; Gavezzotti 1994). It is well known, based on heuristic estimates that the number of local minima grows exponentially with increasing system size (Stillinger 1999). Minimization methods can be mainly classified into two groups, deterministic and stochastic. Deterministic methods, variations on Newton's method, have the strength of being extremely fast, but have the weakness of often being trapped in local minima. Conversely, a stochastic method is far less likely to be trapped in local minima, but it can be shown that no stochastic method has the potential of converging to the global minimum in a finite number of steps. In principle, there is a need of a full visit of all local minima on the potential energy surface. To find a global minimum in a complex potential energy surface, it is not feasible to use deterministic methods and there is a need for adapting stochastic search strategies (Stillinger 1999).

In the field of optimization for solving the structure prediction problem the efforts have been focused mainly on enhancing the sampling efficiency and on reducing the configuration space of the potential energy surface. This led to the development of many efficient structure prediction methods, including genetic algorithm (Abraham and Probert 2006; Deaven and Ho 1995; Kolmogorov et al. 2010; Lonie and Zurek 2011; Oganov and Glass 2006; Trimarchi and Zunger 2007; Woodley et al. 1999), simulated annealing (Brooks and Morgan 1995; Kirkpatrick 1984; Schön and Jansen 1996), pivot methods (Serra et al. 1997), group leader optimization (Daskin and Kais 2011), data mining (Fischer et al. 2006), minimal hopping (Goedecker 2004), basin hopping (Wales and Doye 1997), the activation relaxation technique (El Mellouhi et al. 2008; Mousseau et al. 2012), quantum annealing (Finnila et al. 1994), J-walking (Frantz et al. 1990), tabu search (Cyijovic and Klinowski 1995), random sampling (Pickard and Needs 2011), and the Particle Swarm Optimization method (Wang et al. 2010). These and other computational methods allow scientists to achieve a predictive capability sufficient for guiding experimentalists to discover or design materials with specific properties (Ceder 2010; Lyakhov et al. 2013; Wang and Ma 2014; Wang et al. 2012; Woodley and Catlow 2008).

### 4.2.3 Advanced Minima Search in High-Throughput Materials Discovery

Recently, Trimarchi and co-workers demonstrated that standard high-throughput materials discovery methods have limitation in predicting the correct crystal structure of RbCuS and RbCuSe (Trimarchi et al. 2015) ternary alloys. They pointed out that structural analogies often employed in rapid materials screening using high-throughput methods might not always be sufficient. This is because specific crystallographic classes of formal structure types, each corresponding to a series of chemically distinct "daughter structure types" (DSTs), have the same space group but possess totally different local bonding configurations, including coordination types. Hence, the importance of global minima search methods and their inclusion in the materials screening procedure have been emphasized. Thus, they propose to first perform a high-throughput screening of a set of structure types followed by a global optimization search, for instance, using an evolutionary algorithm or a random search. This additional step will allow exploring parts of the surface energy landscape that will be otherwise left unexplored.

### 4.2.4 Data Handling, Post Processing, and Machine Learning

When searching for novel materials, regardless of the modeling strategy applied, one needs to handle large datasets. By exploiting the power of current supercomputer architectures, scientists generate, manage, and analyze enormous data repositories using intelligent data mining and database construction (Curtarolo et al. 2013). One key bottleneck for any screening project in this regard is the lack of availability of machine-readable output for virtually all quantum chemistry codes, and the option to use databases for the analysis and exchange of the data. For example, in the InfoMol project, Lüthi et al. (2016) extended the TURBOMOLE and PSI4 program packages to generate Extensible Markup Language (XML) output files that can be imported in an eXist database. The focus of "InfoMol" is not only on virtual screening, but even more on the establishment of relationships between the (molecular) properties of light harvesting compounds and their chemical activity. Information obtained from these quantum chemical computations can serve as input for machine learning and other data-centric modeling tools. From this, one expects to find the chemical information contained in the molecule, which is responsible for a particular property. These so called "descriptors" also enter the scoring functions to screen an array of prospects against a set of target properties.

### 4.2.5 Progress in Developing New Screening Descriptors for Solar Cell Materials

The search for promising materials for photovoltaic applications usually proceeds through the definition of *descriptors*. These are quantities, either rigorous or empirical, easily accessible from the database (e.g., effective mass, atomization energy, enthalpy of formation, etc.). For example, the effective mass can give information on the potential of a material as charge conductor. Clearly such descriptor is not the material's mobility, but may provide a first estimate on whether this can be large in a real device. As such, the descriptor represents a compromise between the need of evaluating given quantities accurately and the need to perform the calculation over an enormous number of prototypes. The descriptors will provide a first rapid screening tool for searching the database and allow one to match the desired properties with a range of potential materials. Generally, the screening space can be very large and hence not all the details of the screened materials can be calculated. Thus, the screening is performed in multistages where only a smaller set can reach to the later stages of the workflow. To determine this "small set," simple "descriptor" models within a minimal number of parameters are needed at each stage in which only the materials with high potential for the desired effect or functionality are qualified (Alharbi et al. 2015; Curtarolo et al. 2013; Greeley et al. 2006; Potyrailo et al. 2011; Scharber et al. 2006).

For solar cells, the most commonly used descriptor is the Scharber model (Scharber et al. 2006), which was designed for organic photovoltaics (OPV). This model is very simplistic, where all the device performance parameters depend only on one parameter (the energy gap, $E_g$). Such level of simplicity needs inevitably rough assumptions. The main two shortages of the Scharber model are the assumption that all the photons above $E_g$ are absorbed and that the transport is efficient. Therefore, other improved models were suggested to correct those shortages and to cover other types of absorbers (Alharbi et al. 2015; Dou et al. 2013; Castelli et al. 2012; Martsinovich and Troisi 2011; Pastore et al. 2010; Yu and Zunger 2012). The most recent model (Alharbi et al. 2015) includes the absorption spectrum $\alpha(E)$ of the materials and empirically models the transport based on the materials' family and its electronic structure. $\alpha(E)$ is calculated from the same electronic structure calculations used to determine $E_g$ while the transport is characterized by a given diffusion length $L_d$ to avoid further extensive calculations. Later, the performances of the main suggested descriptors are compared.

Before the comparison, the general aspects of solar cells performance are presented. Figure 4.2 shows solar photon flux density as extracted from American Society for Testing and Materials standard (ASTM G173-03 (ASTM 2016)) for AM0, AM1.5g, and AM1.5d. AM0 is the density according to solar radiation just outside Earth's atmosphere. It is due to black-body radiation of the sun surface which temperature is about 5800 K. Once the radiation enters Earth's atmosphere, some bands are lost due to absorption by the atmosphere constituents. After entering the atmosphere, the radiation travels different distances until it reaches different regions on the earth. This is quantified by air mass coefficient (AM). AM1.5 corresponds to a zenith angle of 48.2°. Based on the

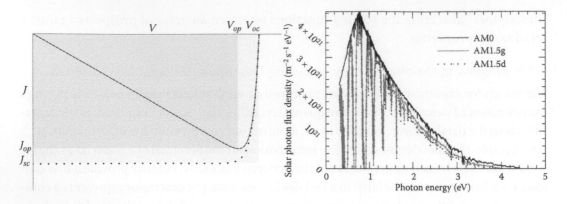

FIGURE 4.2 (Left) The output current (dotted) and power (solid) versus the voltage. (Right) Solar photon flux density for AM0, AM1.5g, and AM1.5d as extracted from American Society for Testing and Materials Standard [ASTM G173-03 (ASTM 2016)].

measurements apparatus and solar cell system, there are two defined AM1.5 standards: the global (AM1.5g) and direct spectrum (AM1.5d). For flat panels, AM1.5g is more appropriate as it is the routinely used one, while AM1.5d is more appropriate for concentrated solar cell systems.

The performance is characterized by the fraction (efficiency) of the input solar power that can be converted to utilizable electric power:

$$\eta = \frac{V_{op} J_{op}}{P_{in}} \tag{4.3}$$

where $\eta$ is the efficiency, $V_{op}$ and $J_{op}$ are the optimum voltage and current density (see Figure 4.2), and $P_{in}$ is the input solar power density (around 100 mW cm$^{-2}$ for the commonly used AM1.5g solar spectrum). This is commonly represented in the following alternative form:

$$\eta = \frac{V_{oc} J_{sc} FF}{P_{in}} \tag{4.4}$$

where $V_{oc}$ is the open circuit voltage, $J_{sc}$ is the short circuit current density, and $FF$ is the fill factor (see Figure 4.2). In general, the relations between these three parameters are tightly coupled, complicated, and depend heavily on the materials used, the cell design, and the fabrication quality. The ideal case is when the system is governed by Shockley ideal diode (with unit ideality factor), i.e.,

$$J(V) = J_0(e^{V/kT} - 1) - J_{sc} \tag{4.5}$$

where $J_0$ is the reverse saturation current and $kT$ is the thermal voltage (around 25 mV at room temperature). In this case,

$$J_{sc} = -J(0), \tag{4.6}$$

$$V_{oc} = kT \ln\left(\frac{J_{sc}}{J_0} + 1\right), \tag{4.7}$$

$V_{op}$, $J_{op}$, and $FF$ are obtained by maximizing $V_{op}J(V_{op})$.

Due to the coupling, it is inevitable to apply some approximations for any descriptor model. Furthermore, summing all the incident photons above a given $E_g$ is routinely needed as it is related to the maximum obtainable current $J_{ph}(E_g)$ (in case of no carrier multiplicity), where

$$J_{ph}(E_g) = q \int_{E_g} \phi_{ph}(E) dE \tag{4.8}$$

where $q$ is the electron charge and $\phi_{ph}(E)$ is the solar photon flux density as a function of photon energy. The resulting $J_{ph}(E_g)$ from AM1.5g is shown in Figure 4.3 and can be approximated by

$$\tilde{J}_{ph} = 73.5 \, \exp\left(-0.44 E_g^{1.861}\right) \tag{4.9}$$

The difference between the calculated and approximated $J_{ph}(E_g)$ for the desired $E_g$ (0.5 – 2.5 eV) is in most cases less than 1%. As aforementioned, many descriptors were suggested and they use generally different *scoring metrics*.

To illustrate the differences, four descriptors (Table 4.1) are compared, where the metric is the efficiency. The first two descriptors are the original Scharber model (Scharber et al. 2006) and the proposed improvement by Dou et al. (2013), where $E_g$

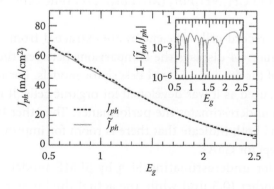

FIGURE 4.3  The exact and approximated $J$ versus $E$. In the inset, the error of the approximated $J$ is plotted in logarithmic scale.

TABLE 4.1 Comparison of Descriptors for Solar Cell Materials Screening

| Model | Descriptor | Set Parameters |
|---|---|---|
| Scharber et al. (2006) | $\eta = \dfrac{V_{oc} J_{sc} FF}{P_{in}}$ where $V_{oc} = \dfrac{1}{q} E_g - 0.3 - \Delta V$, $J_{sc} = 0.65 J_{ph}(E_g)$, and $FF = 0.65$ | $\Delta V$ is the band offset |
| Dou et al. (2013) | $\eta = \dfrac{V_{oc} J_{sc} FF}{P_{in}}$ where $V_{oc} = \dfrac{1}{q} E_g - V_L$, $J_{sc} = 0.80 J_{ph}(E_g)$, and $FF = 0.8$ | $V_L$ is voltage loss: 0.4–0.5 V for inorganic cells 0.6–0.7 V for organic cells |
| Alharbi et al. (2015) | $\eta = \dfrac{V_{oc} J_{sc} FF}{P_{in}}$ where $V_{oc} = (1-0.057)E_g - V_L - 0.0114 E_g^{1.861}$, $J_{sc} = q \displaystyle\int_{E_g} \phi_{ph}(E)\left[1 - \exp\left(\dfrac{-\alpha(E)L_d}{\cos(\theta_{eff})}\right)\right] dE$, and $FF = \dfrac{V_{oc}}{V_{oc} + akT}$ where $L_d$ is the diffusion length. | For nonexcitonic cells: $V_L$ is 0.2 V, $\theta_{eff} = \pi/2.75$, $a = 6$, $L_d$ is 200, 10, 0.6 μm for indirect, direct, and hybrid semiconductors respectively. For excitonic cells: $V_L$ is 0.5 V, $\theta_{eff} = \pi/4$, $a = 12$, $L_d$ is 0.1 μm. |
| SLME (Yu and Zunger 2012) | $\eta = \dfrac{1}{P_{in}} \max_{V_{op}}\left(V_{op} J_{op}(V_{op})\right)$ where $J_{op} = q \displaystyle\int_{E_g} \phi_{ph}(E)\left[1 - e^{-2\alpha(E)L}\right] dE - \dfrac{1}{f_r} qb \int_{E_g} \dfrac{E^2}{\exp\left(\dfrac{E - V_{op}}{kT}\right) - 1} dE$ | $f_r = e^{-\Delta/kT}$ where $\Delta$ is the difference between the 1st dipole allowed direct transition and the energy gap. $L$ is 0.5 μm $b = \dfrac{2\pi q^3}{c^2 h^3}$ |

is the only calculated parameter. The last two approaches (Alharbi et al. 2015; Yu and Zunger 2012) use the calculated $\alpha(E)$ and set some constraints on the device thickness. The comparison is based on the best experimentally reported solar cells where two sets of materials are used; one for nonexcitonic cells (inorganic and hybrid) and one for excitonic cells (organic). The used materials are: nonexcitonic cells: Si, GaAs, InP, GaInP, CdTe, CIGS, and $CH_3NH_3PbI_3$ ($MAPbI_3$); excitonic cells: Sq, DTS, CuPc, ZnPc, DBP, P3HT, PTB7.

The properties and record cells efficiencies are extracted from (NREL 2016) and the references within. Figure 4.4 displays the comparisons between the reported efficiencies and the estimations of the four considered descriptor models. For nonexcitonic cells, the Scharber model is not used as it was developed for organic cells. It is clear that Dou et al. (2013) correction still underestimates the performance. The other two models agree in $\eta$ (except for Si case) and both indicate that there is room for improvement for most of the studied materials except for GaAs.

The main reason for underestimating Si $\eta$ by SLME model is the small assumed thickness for the absorber (0.5 μm) while the actual thickness ranges for Si solar cell is between 100 and 200 μm. For excitonic cells, the SLME model is not used as the assumed thickness is much larger than the actual diffusion length, which limits the

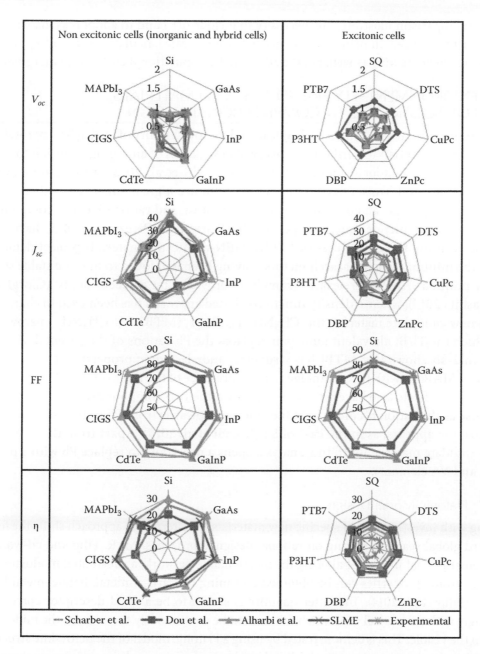

FIGURE 4.4 Comparison between the predictions of the four descriptor models with respect to experimental data for the best reported solar cell materials.

absorber layer thickness. Clearly, the Scharber model underestimates *FF* and exaggerates $V_{oc}$. The Dou et al. correction improves the estimations of the original Schaber model. However, its estimated efficiencies are impractically large. Among the three, the third descriptor results in the most reasonable estimation. The main reason is the inclusion of $\alpha(E)$. The SLME model should result in comparable estimations by using more practical device thickness.

The comparison illustrates the necessity to include $\alpha(E)$ and to adopt reasonable thickness for the devices. In principle, the needed atomic scale calculations to find $E_g$ are sufficient to calculate $\alpha(E)$ as well. So, the additional computational cost is very marginal.

## 4.3 ENVIRONMENTALLY FRIENDLY HYBRID MATERIALS FOR SOLAR CELLS BY COMPUTATIONAL DESIGN

The presence of a toxic element such as lead and the low stability of lead hybrid perovskites are pushing scientists toward the discovery of new classes of emerging materials for photovoltaics. Computational modeling studies (Castelli et al. 2012, 2015a, 2015b, 2015c; Martsinovich and Troisi 2011; Pastore et al. 2010) are a great support to this research area, as they provide an efficient route to design, scan, and rationalize new compounds. Recently, stable nitride inorganic perovskites (Sarmiento-Pèrez et al. 2015) have been predicted using the capabilities of PYMATGEN within the materials project database, with the minima hopping search method having been employed to find the stable structures minima. Another relevant example is provided by a recent study (Giorgi and Yamashita 2015), where a density functional-based approach has been used to characterize a new candidate material, the $CH_3NH_3^+ Tl_{0.5}Bi_{0.5}I_3$ (MTBI, M: $CH_3NH_3^+$) perovskite, in which the Tl/Bi aliovalent ionic pair replaces the Pb cations of the parental MAPbI₃. The analysis shows that MTBI has structural and electronic properties comparable to those of MAPbI₃, with the advantage that the presence of the toxic element (Tl) is halved. Relativistic effects also have a similar impact on the two compounds. This is important because as the cation becomes light (i.e., moving upwards from Pb in the periodic table) oxidization appears, as in the case of Sn (Brandt et al. 2015). Apart from the interest of the candidate material itself, this analysis opens a new route to replace Pb with a pair of IIIA and VA elements.

### 4.3.1 Platonic Model Design

Along with investigations of specific new materials, an alternative approach shifts the focus toward global searches relying on rational design approaches. M. R. Filip and co-workers demonstrate that instead of altering the metal–halide network, a substantial modulation of the electronic properties can be obtained by tuning the largest metal–halide–metal bond angle (Filip et al. 2014). The latter quantity is shown to be a good descriptor correlating strongly with the optical gap, which in turn can be continuously tuned from the mid-infrared to the visible. This is demonstrated by using a Platonic model of the perovskite structure in which the apical and equatorial bond angles are continuously varied. In combination with this, DFT is used to calculate the band gap, which spans almost 1 eV by varying the descriptor. The findings suggest that to make low-gap perovskites for optimum photovoltaic efficiency, structures should be engineered with minimal octahedral tilt. Practical alteration of the Pb–I–Pb angle can be obtained by a careful choice of the cation. In fact, this work also shows that the descriptor correlates strongly with the steric size of the cation. Therefore, precise band gap engineering is achieved by controlling the bond angles through the steric size of the molecular cation. This study represents a significant advance from a methodological point of view. In fact, the theory is based on general considerations and its predictions are not linked to the underlying calculations, which are often very sensitive to the details. The same predictions done for the MAPbI$_3$ can be then carried across other families of metal–halide perovskites. They subsequently performed a systematic computational screening of potential divalent metal atoms that could replace Pb in MAPbI$_3$ (Filip and Giustino 2016).

### 4.3.2 The CRAQ-Solar Hybrid Materials Design Project

In parallel, the CRAQ project for hybrid solar cell materials screening and design proposes first adopting an "ad hoc" intuitive approach by using "known" molecular cations with +1 oxidation number for the A site and metals with +5 oxidation number for the B site. A rigorous, yet computationally expensive, approach is employed to have more freedom and innovation in the molecular cation side. The research strategy is based on three main parts: (1) crystal structure prediction; (2) band gap estimation using DFT calculations; (3) band gap correction using hybrid functional and spin-orbit coupling (SOC) as summarized in Figure 4.5.

All the calculations are performed initially at the level of DFT in the generalized gradient approximation (GGA–PBE), which allows us to maintain an adequate accuracy concerning the structural properties. Dispersion corrections will be then included at the level of DFT–van der Waals (vdW), which according to Motta et al. (2015) is sufficient, with little computational cost, to reproduce the experimental volumes otherwise estimated with PBE alone (see Figure 4.6). It is also crucial to describe the vdW forces governing the free energy landscape of hybrid organic inorganic perovskites with respect to the molecular cation orientation (Filip et al. 2014). Hence, benchmarking and tailoring vdW forces must

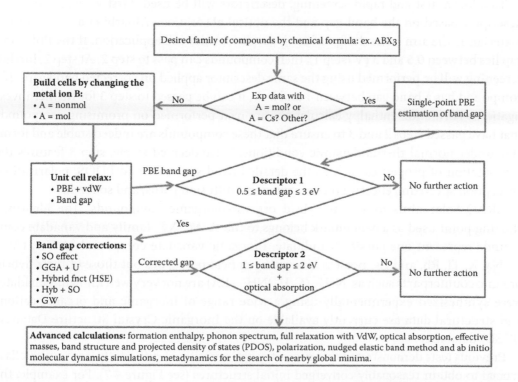

FIGURE 4.5 Flowchart of the CRAQ Solar Materials screening project (El Mellouhi et al. 2017).

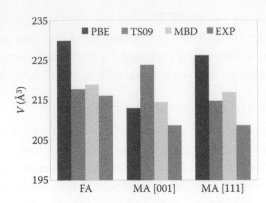

FIGURE 4.6 Calculated unit cell volume of $CH_3NH_3PbI_3$ perovskite after relaxation with PBE without vdW, with TS-vdW, and with many-body dispersion interactions. The nominal experimental value is also reported. (Reproduced under a Creative Commons CC-BY 4.0 license from Motta, C., F. El Mellouhi, S. Kais, N. Tabet, F. Alharbi, and S. Sanvito. 2015. Revealing the role of organic cations in hybrid halide perovskites $CH_3NH_3PbI_3$. *Nat. Commun.* 6; 7026–7027.)

be undertaken when moving from a class of materials to the other. A key part of the characterization is the refined calculation of the electronic properties of the promising compounds, with higher levels (and more computationally expensive) of theory such as hybrid functionals and GW.

To allow a first and rapid screening, descriptors will be used. First we will consider a descriptor based on the band gap and the optical absorption (Alharbi et al. 2015) of the material, as the aim is the design of compounds for solar cell application. If the PBE band gap lies between 0.5 and 3 eV (step 1), then compounds can pass to step 2. At step 2, further screening will be performed using the same descriptor applied on corrected band gaps. If a compound has a band gap between 1 and 2 eV, it can be passed to step 3 for further investigations. Formation enthalpy calculations are next performed on promising compounds that have passed steps 2 and 3 to ensure that these compounds are indeed stable and formable under normal thermodynamic conditions. Once deemed stable, step 3 features the computation of properties such as the detailed band structure and PDOS, polarization, absorption spectrum, effective masses, phonons, diffusion paths, and so forth.

CRAQ-Solar aims to screen hybrid organic-inorganic compounds. The identified starting point used as a benchmark belongs to the $A^+M^{5+}X_3^{2-}$ family and vanadate compounds represent one possible candidate. Inorganic vanadate compounds $A^+VO_3$ ($A^+$ = Li, Na, K, Tl, Rb, and Cs) possess interesting optical properties but those of their hybrid organic counterparts such as $NH_4^+$, $(C_2H_5)NH_3^+$ (EA) are not very well known. Vanadates were synthesized experimentally using a wide range of inorganic and organic cations and structural data are currently available on the Inorganic Crystal Structure Database (ICSD) database.

Previous tests demonstrated that geometry optimization of sufficiently large supercells is crucial to obtain reasonably converged initial structures (see Figure 4.7). For example, the five-atom $CsVO_3$ perovskite unit cell is unstable after geometry optimization and relaxes to

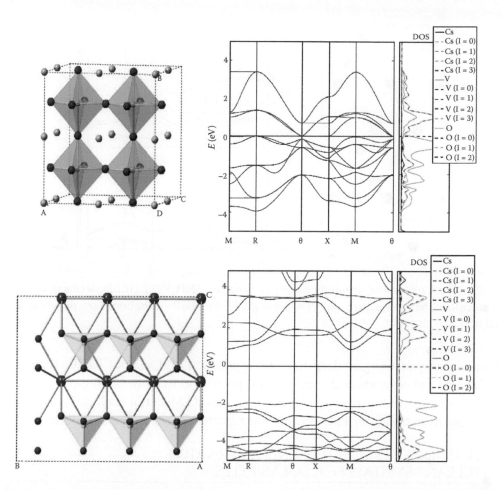

FIGURE 4.7 Our calculated band structure and PDOS for CsVO3 in different geometries. (Top) Unrelaxed simple cubic perovskite containing five atoms. (Bottom) Fully optimized structure starting form experimentally published orthorhombic unit cell containing 20 atoms CsVO3 (ICSD database code 1489) adapting the orthorhombic structure with lattice parameters a = 5.393 b = 12.249 c = 5.786 Å (El Mellouhi et al. 2017).

the pyroxene structure opening the band gap to approximately 3 eV. This is a clear indication that substitution of Cs with organic cation with different symmetries and dipole moment might require careful consideration of the size of the cell used during the screening. For small organic cations such as ammonium $(NH_4)^+$, $2 \times 2 \times 1$ unit cells are sufficient. However, larger cations are also desirable because of their dipole moment but fitting them into small cells would alter the tetrahedral ordering and break the symmetry. Hence, larger unit cells as in the case of the ethylammonium vanadates $(EAVO_3)$ might be needed.

Starting from the experimentally reported hybrid oxovanadate structure, electronic structure calculation of for ammonium-catena vanadate $(NH_4VO_3)$ and ethylammonium vanadate $((C_2H_5)NH_3VO_3)$ reveal that the band gap remains between 2.7 and 3 eV, a value, which is comparable to the Cs inorganic counterpart (see Figure 4.8). Also, analysis of the band structure and DOS demonstrates that the molecule-derived energy levels have spectral

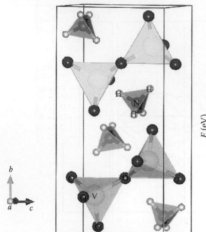
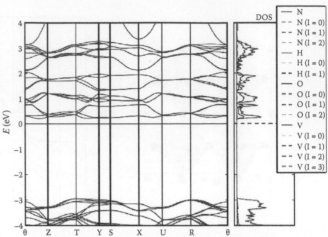

FIGURE 4.8 Unit cell of ammonium-catena vanadate ($NH_4VO_3$) (ICSD database code 1487), our calculated band structure and PDOS using DFT-PBE+vdW functional are shown as well (El Mellouhi et al. 2017).

amplitude only at energies deep into the conduction and valence bands. Interestingly, El-Mellouhi and co-workers showed that it is possible to tune the band gap of this class of materials on substitution of some organic molecules (El Mellouhi et al. 2017). For instance, a band gap reduction from 3 eV to 1.7 eV was reported for the sulfonium cation.

## 4.4 LEAD-FREE PEROVSKITES AND ALTERNATIVE "CLEAN" SYSTEMS FOR SOLAR CELLS

In this section, we summarize and discuss some previous experimental synthesis efforts performed for several groups of some environmentally friendly materials.

### 4.4.1 Organo-Sn Halides

Solid solutions of halide perovskites $AMX_3$ can be prepared on both the M and X sites, but only as long as the M and X are immediate neighbors in the periodic table (Stoumpos and Kanatzidis 2015). For example, for a true $AMX'_{3-x}X_x$ solid solution, such as $CH_3NH_3SnI_{3-x}Br_x$, the absorption edge can be tuned between the two parent compounds (for x = 0 and x = 3, respectively) as a function of $x$ following a simple linear relationship (Vegard's law). This systematic behavior is not observed in the $CH_3NH_3SnI_{3-x}Cl_x$ composition, where instead of a continuous change in the absorption edge, a constant value of the band gap is observed, which corresponds to that of $CH_3NH_3SnI_3$. This behavior, coupled with the x-ray diffraction data shows that this material is a phase mixture of $CH_3NH_3SnI_3$ and $CH_3NH_3SnCl_3$ (Figure 4.9).

An attempt to use the lead-free methylammonium tin halide perovskites ($CH_3NH_3SnI_{3-x}Br_x$) as light-absorbing materials in solid-state photovoltaic devices fabricated by solution-processed approach was reported in (Hao et al. 2014). Lead-free $CH_3NH_3SnI_3$ shows an ideal optical bandgap of 1.3 eV, which is lower than the 1.55 eV band gap of $CH_3NH_3PbI_3$. Also, devices with $CH_3NH_3SnI_3$ in conjunction with an organic spiro-OMeTAD hole-transport

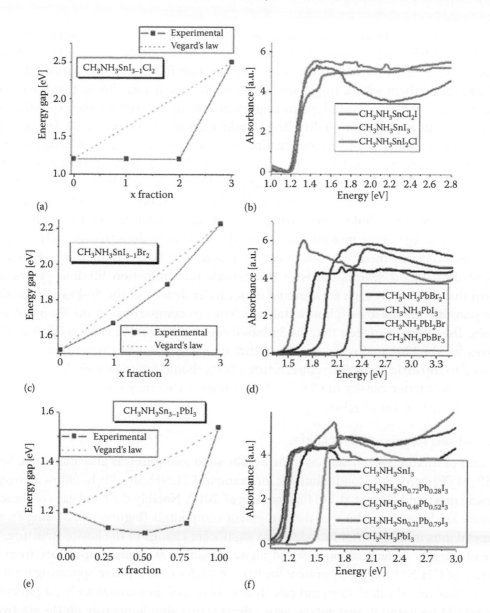

FIGURE 4.9 Dependence of the band gap (a, c, e) and the corresponding diffuse reflectance spectra (b, d, f) of the $MASnI_{3-x}Cl_x$ phase mixtures and of the regular $MAPbI_{3-x}Br_x$ and irregular $MAPb_{1-x}Sn_xI_3$ solid solutions, respectively . (Reprinted with permission from C.C. Stoumpos and M.G. Kanatzidis. *The Renaissance of Halide Perovskites and Their Evolution as Emerging Semiconductors.* Copyright 2015 American Chemical Society.)

layer show an absorption onset of 950 nm, which is significantly red-shifted compared to $CH_3NH_3PbI_3$-based devices. For a wider class of materials ($CH_3NH_3SnI_{3-x}Br_x$ perovskites) it was shown that the band gap can be controllably tuned to cover much of the visible spectrum, thus enabling the realization of lead-free, colorful solar cells.

Although the initial power conversion efficiency (PCE) of 5.73% under simulated full sunlight was lower than that reached in $MAPbI_3$, possibilities for further efficiency enhancements were recently investigated by the same group at Northwestern University (Hao et al. 2015). In particular, they investigated solvent effects on the crystallization processes of lead-free methylammonium tin triiodide ($CH_3NH_3SnI_3$) perovskite films and compared them with a single-step solution process. It was shown that the solvent induced intermediate phase plays an important role in controlling the crystallization process and final thin film quality. The stabilization of the intermediate $SnI_2$ solvates prior to the formation of the perovskite films was found to promote homogeneous nucleation and enabled an adjustable perovskite film growth rate. High-quality, pinhole-free $CH_3NH_3SnI_3$ films were achieved using dimethyl sulfoxide (DMSO) and $N$-methyl-2-pyrrolidone (NMP) as solvents and afforded lead-free perovskite heterojunction depleted solar cells without a hole-transporting material layer. Charge extraction and transient photovoltage decay measurements reveal high carrier densities in a $CH_3NH_3SnI_3$ perovskite device. These are one order of magnitude larger than that achieved in $CH_3NH_3PbI_3$-based devices but with comparable recombination lifetime. It was also shown that a relatively high background dark carrier density of the Sn-based perovskite is responsible for the lower photovoltaic efficiency in comparison to the Pb-based analogues. These results suggest that the Sn-based compounds can be used as efficient alternatives for the Pb-based ones, providing that the "background" carrier density can be reduced to intrinsic levels. Therefore, future efforts should focus on ways to decrease the background carrier density in $CH_3NH_3SnI_3$, a feature that may increase the device efficiencies to 15% or even higher.

### 4.4.2 Mixed Organo-Pb/Sn Halides

The idea of mixing Pb and Sn ions in perovskite solid solutions was also attempted. Some $AM'_{1-x}M_xX_3$ perovskite solid solutions, for example, $CH_3NH_3Sn_{1-x}Pb_xI_3$, follow an irregular band gap trend (Hao et al. 2014b; Ogomi et al. 2014). Notably the band gaps of the solid solution are lower than those of the two parent compounds (Figure 4.9). Mixing Sn and Pb metal ions itself does not introduce any significant change in the lattice structure, but instead triggers a composition-directed phase transition at room temperature from the α-phase of $CH_3NH_3Sn_{1-x}Pb_xI_3$ present for $0.0 < x < 0.5$ to the β-phase appearing for $0.5 < x < 1.0$. Theoretical calculations indicate that the origin of the anomalous band gap evolution should be found in two antagonistic effects acting simultaneously on the electronic structure: (1) a bandgap reduction caused by strong spin-orbit coupling (SOC), which is dominant in the $0.0 < x < 0.5$ range; and (2) a bandwidth decrease (which increases bandgap) caused by the lattice distortion as a function of $x$, which is dominant for $0.5 < x < 1.0$ (Im et al. 2015). As a result of this anomalous trend, it became possible to fabricate functional photovoltaic devices that are able to absorb light up to approximately 1050 nm (lowest energy achieved yet) and at the same time display superior current characteristics with respect to $CH_3NH_3PbI_3$ (Hao et al. 2014b).

Although Pb/Sn mixing does not solve the problem of developing lead-free, environmentally friendly materials for solar cells, it provides a great example of how the properties of perovskite halide materials could be modified by changing their chemical composition. In particular, further pathways for property-tuning based on unconventional trends in electronic structure of perovskites may open new strategies toward the development of panchromatic absorbers for light harvesting in solar devices, that is, materials that could absorb at all wavelengths of the solar spectrum. This is discussed in the next section.

### 4.4.3 Organo-Germanium Iodides

The idea to use organo-germanium iodides also sounds attractive. Recently, materials with methylammonium ($CH_3NH_3GeI_3$), formamidinium ($HC-(NH_2)_2GeI_3$), (acetamidinium, $CH_3C(NH_2)_2GeI_3$), guanidinium ($C-(NH_2)_3GeI_3$), trimethylammonium (($CH_3)_3NHGeI_3$), and isopropylammonium (($CH_3)_2C(H)NH_3GeI_3$) organic cations were synthesized (Stoumpos et al. 2015). There are however, some problems with using these compounds in photovoltaics. First, $CH_3NH_3GeI_3$ and $HC-(NH_2)_2GeI_3$ have a direct band gap respectively of 1.9 eV and 2.2 eV, which is much higher that the band gaps of their lead analogues (1.57 eV for $CH_3NH_3PbI_3$ and 1.48 eV for $HC-(NH_2)_2PbI_3$, see (Eperon et al. 2014)). Therefore, their ability to absorb light from the solar spectrum is rather restricted. Second, the three-dimensional crystal structure of these two compounds is assembled through single Ge–I–Ge bridges, which form $[GeI_6]^{4-}$ corner-sharing octahedra. The octahedra adopt a trigonally distorted $CaTiO_3$ structure crystallizing in the polar R3m space group. The trigonal distortion causes the loss of the fourfold symmetry axes present in the cubic perovskites. At the same time, organo-germanium iodides exhibit strong, type I phase matchable second harmonic generation (SHG) response and high laser-induced damage thresholds (up to ~3 GW cm$^{-2}$), which makes these materials attractive for nonlinear optical applications in the visible and near-infrared regions (Stoumpos et al. 2015). It would be interesting to mix Ge and Sn in organo-metallic iodide perovskite solid solutions. However, to the best of our knowledge, such investigation has not been done yet.

### 4.4.4 Other Synthesized Materials

We should also mention two recent synthesis efforts performed for double perovskite systems (McClure et al. 2016; Wei et al. 2016). $Cs_2AgBiBr_6$ and $Cs_2AgBiCl_6$ have been synthesized from both solid state and solution routes (McClure et al. 2016). These materials adopt the cubic double perovskite structure, and reveal band gaps of 2.19 eV (X = Br) and 2.77 eV (X = Cl) which are slightly smaller than the band gaps of the analogous lead halide perovskites (2.26 eV for $CH_3NH_3PbBr_3$ and 3.00 eV for $CH_3NH_3PbCl_3$). Band structure calculations indicate that the interaction between the Ag 4$d$- orbitals and the $p$- orbitals of the halide ion modifies the valence band leading to an indirect band gap. Both compounds are stable when exposed to air, but $Cs_2AgBiBr_6$ degrades over a period of weeks when exposed to both ambient air and light. Another hybrid double halide perovskite, $(MA)_2KBiCl_6$, also showed several striking similarities to the lead analogues, which was recently confirmed by a combination of spectroscopic measurements, nanoindentation studies, and density functional calculations (Wei et al. 2016).

Although some parameters of these systems still need to be tuned up (e.g., the band gaps of $Cs_2AgBiBr_6$ and $Cs_2AgBiCl_6$ are too high so a significant fraction of visible light spectrum will not be absorbed by these materials), these results clearly indicate that halide double perovskite semiconductors could potentially be an environmentally friendly alternative to the lead halide perovskite semiconductors.

### 4.4.5 Other Possible Alternative "Clean" Systems

The recent success of methylammonium lead halide ($MAPbX_3$) perovskites has motivated the identification of the unique properties of this material set, giving rise to the good bulk transport characteristics and identifying materials and systems with similar properties. It was suggested that a general "defect tolerance" emerges from the fundamental electronic-structure properties, including the orbital character of the conduction and valence band edges, the charge carrier effective masses, and the static dielectric constant (Eperon et al. 2014). For this purpose, the Materials Project (MP 2016) database was used and details of electronic structure obtained from first-principles calculations were analyzed. In particular, it was demonstrated that materials other than $MAPbX_3$ may also have similar or even superior properties, making them reasonable candidates to serve as light absorbers in solar cells.

Many authors have proposed explanations for the success of $MAPbX_3$ as photovoltaic material. These include its large absorption coefficient, long electron and hole diffusion lengths, low exciton-binding energies, low effective masses and high mobilities, and the presence of only shallow defects in the band gap (Brandt et al. 2015; Green et al. 2014; Stoumpos et al. 2013; Stranks et al. 2013). The existence of only shallow defects, and the disperse valence band, have both been tied to the presence of filled Pb $6s$ orbitals, deriving from the partial oxidation of Pb relative to its $Pb^{4+}$ oxidation state. This orbital character has been identified by several authors, and explains both the shallow binding energy of defects and the typical dependence of band gap on strain or temperature in $MAPbI_3$ (Yamada et al. 2014; Yin et al. 2014a, 2014b). In addition, device-level observations support the claims that $MAPbI_3$ has excellent transport properties, including long carrier diffusion lengths up to 175 μm in single crystals (Shi et al. 2015). Low nonradiative minority-carrier recombination rates are also supported by measurements of high photoluminescence quantum yield, long carrier lifetimes, and the high open-circuit voltage ($V_{OC}$) demonstrated by many devices.

This combination of properties, however, is not completely unique and can be frequently found in metal–nonmetal systems with partially oxidized cations such as binary group III halides or group IV chalcogenides. By filtering the materials data drawn from the Materials Project database, Brandt et al. (2015) focused on several interesting classes of compounds that may share the beneficial properties identified previously. This search first identified a specific class of halide perovskites ($CsPbI_3$, $CsSnI_3$, $RbPbBr_3$, etc.), a result that is encouraging since prior optoelectronic screening efforts always neglect such compounds. Other classes of materials include

1. Binary chalcogenides and halides ($BiI_3$, $Bi_2S_3$, $SbI_3$, $Sb_2S_3$, $SnI_2$, $InI$, etc.): There are many binary iodides, sulfides, and selenides formed with $ns^2np^0$ cations, several of which have band gaps in the range of interest for PV.

2. Binary halides, sulfides, and selenides stabilized in cubic structures such as NaCl and CsCl: Given the anisotropic transport properties that result from stereochemically active lone pairs in materials such as SnS, more promising materials may result by stabilizing them in higher symmetry phases.

3. Chalcohalides (BiOI, BiSI, BiSeI, SbSI, SbSeI, BiSBr, BiSeBr, etc.): The $ns^2np^0$ chalcohalides (V–VI–VII compounds) band gaps range over the visible to the UV, and many compounds demonstrate ferroelectric behavior at lower temperatures as well.

4. Ternary alkali chalcogenides (LiBiS$_2$, NaBiS$_2$, KBiS$_2$, RbBiS$_2$, CsBiS$_2$, etc.).

5. Ternary halides: (In$_3$SnI$_5$, InAlI$_4$, etc.): In$^+$ and Tl$^+$ may be stabilized by several non-coordinating molecular anions such as AlI$^{4-}$.

6. In$^+$–II–VII$^3$ ternary halides (CdInBr$_3$, CaInBr$_3$, etc.): In$^+$ is also stabilized in a number of In–II–VII$^3$ compounds, which appear to have band gaps outside of the range for PV, but that may exhibit similar defect tolerant properties.

7. Bismides and antimonides (e.g., KSnSb): Sn$^{2+}$ and other partially oxidized cations are stabilized in several bismides and antimonides, most of which have band gaps too small for PV absorbers under 1-Sun illumination.

8. Octahedrally coordinated metal halides (Cs$_3$Bi$_2$I$_9$, Rb$_3$Bi$_2$I$_9$, K$_3$Bi$_2$I$_9$, Cs$_3$Sb$_2$I$_9$, etc.): Bi$^{3+}$ and Sb$^{3+}$ form III$^2$X$_9^{3-}$ anions that consist of edge or face-sharing octahedra, bonding with alkali metals to form a variety of materials with different band gaps, crystal structures, and ferroelectric phase transitions.

9. Cs-containing compounds substituting 1+ molecular cations [e.g., (MA)$_3$Bi$_2$I$_9$, (FA)$_3$Bi$_2$I$_9$, etc.]: Lastly, one can identify a variety of Cs-containing materials, which could be converted to hybrid materials by the substitution of the Cs with an alkyl-ammonium or other molecular cation (Plackowski et al. 1995). A simple example is the family containing (MA)$_3$Bi$_2$I$_9$ or (FA)$_3$Bi$_2$I$_9$ (with formamidinium as the organic cation). A number of other molecular cations may be substituted as well based on size, analogous to the substitution explored in MAPbX$_3$ (Brandt et al. 2015; Plackowski et al. 1995), to tune the band gap or to form lower-dimensional structures.

For some of these materials, the electronic structure was calculated from the first-principles, and this information was analyzed together with available experimental data. Based on this analysis, several conclusions were made (Brandt et al. 2015). First, many of these materials, such as MAPbI$_3$, are relatively ionic, so that the band gaps are usually large. Second, anisotropic crystal structures with stereochemically active lone pairs often lead to heavy holes. Third, most of the materials investigated display large ionic dielectric constants, independent of whether or not they have a polar space group.

## 4.5 CONCLUDING REMARKS AND FUTURE OUTLOOK

Organo-lead halide perovskite solar cells have gained enormous significance and have now achieved power conversion efficiencies of approximately 20% in $CH_3NH_3PbI_3$. However, the potential toxicity of lead in these systems raises environmental concerns for widespread deployment. As a result, a search for organometallic lead-free halide perovskites that could be used as solar absorbers in PV devices becomes an important practical issue. The prototypical hybrid halide perovskite family such as $(CH_3NH_3)PbX_3$ compounds offers an unusual combination of ionicity, high dielectric constant, low effective masses, and an ideal band gap. Some semiconductors can claim good performance across all these categories and it is expected that the transport characteristics of some of them may exhibit defect tolerance similar to that of $(CH_3NH_3)PbX_3$. Therefore, $(CH_3NH_3)PbX_3$ materials may not be unique and one may identify a broad class of semiconductors containing partially oxidized cations, as well as several specific instances where compounds may share the same properties of $(CH_3NH_3)PbX_3$. Hence, a search based on these principles may yield materials with similar performance, but enhanced stability and lower toxicity. This is a field so far dominated by empirical trial-and-error approaches, so that progress has been slow. In general the discovery of a radically new compound is usually a serendipitous process and the improvement of the solar harvesting efficiency is usually fast in the first few years from the discovery but then proceeds extremely slowly. There is a need to establish a research protocol for the rapid design of new high-performing materials for photovoltaics. So far very few materials screening and discovery initiatives have taken the challenge to design hybrid materials, probably because it would require a revolution in the traditional materials screening procedure. For instance, extensive structural prediction initiatives (El Mellouhi et al. 2017; Trimarchi and Zunger 2007) emerged that use genetic algorithms for the search of stable structure not limited to very small cell sizes commonly used in high-throughput screening methods. Combined theoretical, computational, and experimental efforts can provide a way for producing a significant palette of new promising compounds for further experimental characterization.

## REFERENCES

Abraham, N., and M. Probert. 2006. A periodic genetic algorithm with real-space representation for crystal structure and polymorph prediction. *Phys. Rev. B* 73; 224104–224106.

AiiDA (Automated Interactive Infrastructure and Database for Computational Science). 2016. http://www.aiida.net/ (Accessed February, 2016).

Alharbi, F. H., S. N. Rashkeev, F. El-Mellouhi, H. P. Lüthi, N. Tabet, and S. Kais. 2015. An efficient descriptor model for designing materials for solar cells. *NPJ Comput. Mater.* 1; 15003–15009.

ASTM (American Society for Testing and Materials Standard). 2016. http://www.astm.org/ (Accessed February, 2016).

Babayigit, A., A. Ethirajan, M. Muller, and B. Conings. 2016. Toxicity of organometal halide perovskite solar cells. *Nat. Mater.* 15; 247–251.

Brandt, R, E., V. Stevanović, D. S. Ginley, and T. Buonassisi. 2015. Identifying defect-tolerant semiconductors with high minority-carrier lifetimes: Beyond hybrid lead halide perovskites. *Mater. Res. Soc. Commun.* 5; 265–275.

Brooks, S., and B. Morgan. 1995. Optimization using simulated annealing. *Statistician* 44; 241–257.

Cao, D. H., C. C. Stoumpos, O. K. Farha, J. T. Hupp, and M. G. Kanatzidis. 2015. 2D Homologous perovskites as light-absorbing materials for solar cell applications. *J. Am. Chem. Soc.* 137; 7843–7850.

Carrete, J., W. Li, N. Mingo, and S. Curtarolo. 2014. Finding unprecedentedly low-thermal-conductivity half-Heusler semiconductors via high-throughput materials modeling. *Phys. Rev. X* 4; 011019.

Castelli, I. E., T. Olsen, S. Datta, D. D. Landis, S. Dahl, K. S. Thygesen, and K. W. Jacobsen. 2012. Computational screening of perovskite metal oxides for optimal solar light capture. *Energy Environ. Sci.* 5; 5814–5819.

Castelli, I. E., M. Pandey, K.S. Thygesen, and K.W. Jacobsen. 2015a. Band-gap engineering of functional perovskites through quantum confinement and tunneling. *Phys. Rev. B* 91; 165309, 6 pp.

Castelli, I. E., K. S. Thygesen, and K. W. Jacobsen. 2015b. Calculated optical absorption of different perovskite phases. *J. Mater. Chem. A* 3; 12343–12349.

Castelli, I. E., F. Hüser, M. Pandey, H. Li, K. S. Thygesen, B. Seger, A. Jain, K. A. Persson, G. Ceder, and K. W. Jacobsen 2015c. New light-harvesting materials using accurate and efficient band-gap calculations. *Adv. Energy Mater.* 5; 1400915–1400917.

Ceder, G. 2010. Opportunities and challenges for first-principles materials design and applications to Li battery materials. *Mater. Res. Soc. Bull.* 35; 693–701.

Chen, H., G. Hautier, A. Jain, C. Moore, B. Kang, R. Doe, L. Wu, Y. Zhu, Y. Tang, and G. Ceder. 2012. Carbonophosphates: A new family of cathode materials for Li-ion batteries identified computationally. *Chem. Mater.* 24; 2009–2016.

Chen, P.-Y., J. Qi, M. T. Klug, X. Dang, P. T. Hammond, and A. M. Belcher. 2014. Environmentally-responsible fabrication of efficient perovskite solar cells from recycled car batteries. *Energy Environ. Sci.* 7; 3659–3965.

CMG (Center for Materials Genomics). 2016. http://materials.duke.edu (Accessed February, 2016).

CMR (Computational Materials Repository). 2016. https://cmr.fysik.dtu.dk/organometal/organometal .html (Accessed February, 2016).

Conings, B., L. Baeten, T. Jacobs, R. Dera, J. D'Haen, J. Manca, and H.-G. Boyen. 2014. An easy-to-fabricate low-temperature $TiO_2$ electron collection layer for high efficiency planar heterojunction perovskite solar cells. *APL Mater.* 2; 081505–081508.

CRANN (Centre for Research on Adaptive Nanostructures and Nanodevices). 2016. http://crann .tcd.ie/ (Accessed February, 2016).

CRAQSolar. 2016. CRAQSolar project http://qnrf.org.qa (Accessed February, 2016).

Curtarolo, S., W. Setyawan, G. L. W. Hart, M. Jahnatek, R. V. Chepulskii, R. H. Taylor, S. Wang, J. Xue, K. Yang, O. Levy, M. J. Mehl, H. T. Stokes, D. O. Demchenko, and D. Morgan. 2012. AFLOW: An automatic framework for high-throughput materials discovery. *Comp. Mater. Sci.* 58; 218–226.

Curtarolo, S., G. L. W. Hart, M. Buongiorno Nardelli, N. Mingo, S. Sanvito, and O. Levy. 2013. The high-throughput highway to computational. *Nat. Mater.* 12; 191–201.

Curtarolo, S., W. Setyawan, S. Wang, J. Xue, K. Yang, R. H. Taylor, L. J. Nelson, G. L. W. Hart, S. Sanvito, M. Buongiorno-Nardelli, N. Mingo, and O. Levy. 2012. AFLOWLIB.ORG: A distributed materials properties repository from high-throughput *ab-initio* calculations. *Comp. Mater. Sci.* 58; 227–235.

Cvijovic, D., and J. Klinowski. 1995. Taboo search: An approach to the multiple minima problem. *Science* 267; 664–666.

Daskin, A., and S. Kais 2011. Group leaders optimization algorithm. *Mol. Phys.* 109; 761–772.

Deaven, D., and K. Ho. 1995. Molecular geometry optimization with a genetic algorithm. *Phys. Rev. Lett.* 75; 288–291.

Dixon, L. C. W., and G. P. Szego. 1975. *Towards global optimization*, Vols. 1 and 2. New York: North Holland.

Dou, L., J. You, Z. Hong, Z. Xu, G. Li, R. A. Street, and Y. Yang. 2013. 25th anniversary article: A decade of organic/polymeric photovoltaic research. *Adv. Mater.* 25; 6642–6671.

El-Mellouhi, F., A. Akande, C. Motta, S. Rashkeev, G. Berdiyorov, M. El-Amine Madjet, A. Marzouk, E. T. Bentria, S. Sanvito, S. Kais, and F. Hussain AlHarbi. 2017. Solar cells materials by design: Hybrid pyroxene corner-sharing $VO_4$ tetrahedral chains. *Chem Sus Chem* http://dx.doi.org/10.1002/cssc.201700121.

El Mellouhi, F., L. J. Lewis, and N. Mousseau. 2008. The kinetic activation-relaxation technique: A powerful off-lattice on-the-fly kinetic Monte Carlo algorithm. *Phys. Rev. B* 78; 153202–153204.

Eperon, G. E., S. D. Stranks, C. Menelaou, M. B. Johnston, L. M. Herz, and H. J. Snaith. 2014. Formamidinium lead trihalide: A broadly tunable perovskite for efficient planar heterojunction solar cells. *Energy Environ. Sci.* 7; 982–988.

Filip, M. R., G. E. Eperon, H. J. Snaith, and F. Giustino. 2014. Steric engineering of metal-halide perovskites with tunable optical band gaps. *Nat. Commun.* 5; 5757–5759.

Filip, M. R., and F. Giustino. 2016. Computational screening of homovalent lead substitution in organic–inorganic halide perovskites. *J. Phys. Chem. C* 120; 166–173.

Finnila, A. B., M.A. Gomez, C. Sebenik, C. Stenson, and J.D. Doll. 1994. Quantum annealing: A new method for minimizing multidimensional functions. *Chem. Phys. Lett.* 219; 343–348.

Fischer, C. C., K. J. Tibbetts, D. Morgan, and G. Ceder. 2006. Predicting crystal structure by merging data mining with quantum mechanics. *Nat. Mater.* 5; 641–643.

Frantz, D. D., D. L. Freeman, D. L., and J. D. Doll. 1990. Reducing quasi-ergodic behavior in Monte Carlo simulations by *J*-walking: Applications to atomic clusters. *J. Chem. Phys.* 93; 2769.

Gavezzotti, A. 1994. Are crystal structures predictable? *Acc. Chem. Res.* 27; 309–314.

Giorgi, G., and K. Yamashita 2015. Alternative, lead-free, hybrid organic-inorganic perovskites for solar applications: A DFT analysis. *Chem. Lett.* 44; 826–828.

Goedecker, S. 2004. Minima hopping: An efficient search method for the global minimum of the potential energy surface of complex molecular systems. *J. Chem. Phys.* 120; 9911–9917.

Goldschmidt, V. M. 1926. Die Gesetze der Krystallochemie. *Naturwissenschaften* 14; 477–485.

Greeley, J., T. F. Jaramillo, J. Bonde, I. Chorkendorff, and J. K. Nørskov. 2006. Computational high-throughput screening of electrocatalytic materials for hydrogen evolution. *Nat. Mater.* 5; 909–913.

Green, M. A., A. Ho-Baillie, and H. J. Snaith. 2014. The emergence of perovskite solar cells. *Nat. Photonics* 8; 506–514.

Hao, F., C. C. Stoumpos, D. H. Cao, R. P. H. Chang, and M. G. Kanatzidis. 2014. Lead-free solid-state organic–inorganic halide perovskite solar cells. *Nat. Photonics* 8; 489–494.

Hao, F., C. C. Stoumpos, R. P. H. Chang, and M. G. Kanatzidis. 2014. Anomalous band gap behavior in mixed Sn and Pb perovskites enables broadening of absorption spectrum in solar cells. *J. Am. Chem. Soc.* 136; 8094–8099.

Hao, F., C. C. Stoumpos, P. Guo, N. Zhou, T. J. Marks, R. P. H. Chang, and M. G. Kanatzidis. 2015. Solvent-mediated crystallization of $CH_3NH_3SnI_3$ films for heterojunction depleted perovskite solar cells. *J. Am. Chem. Soc.* 137; 11445–11452.

Hart, G. L. W., S. Curtarolo, T. B. Massalski, and O. Levy. 2013. Comprehensive search for new phases and compounds in binary alloy systems based on platinum-group metals, using a computational first-principles approach. *Phys. Rev. X* 3; 041035, 33 pp. DOI:10.1103/Phys RevX.3.041035.

Harvard Clean Energy Project http://cleanenergy.molecularspace.org/ (Accessed February, 2016).

Hesse, S., J. Zimmermann, H. von Seggern, H. Ehrenberg, H. Fuess, C. Fasel, and R. Riedel. 2006. $CsEuBr_3$: Crystal structure and its role in the photostimulation of CsBr: $Eu^{2+}$. *J. Appl. Phys.* 100; 083506.

Hodes, G., and D. Cahen. 2014. Photovoltaics: Perovskite cells roll forward. *Nat. Photonics* 8; 87–88.

Horowitz, A., M. Amit, J. Makovsky, L.Ben Dor, and Z.H. Kalman. 1982. Structure types and phase transformations in $KMnCl_3$ and $TlMnCl_3$. *J. Solid State Chem.* 43; 107–125.

Im, J.-H., J. Chung, S.-J. Kim, and N.-G. Park. 2012. Synthesis, structure, and photovoltaic property of a nanocrystalline 2H perovskite-type novel sensitizer $(CH_3CH_2NH_3)PbI_3$. *Nanoscale Res. Lett.* 7; 353–357.

Im, J.-H., C.-R. Lee, J.-W. Lee, S.-W. Park, and N.-G. Park. 2011. 6.5% efficient perovskite quantum-dot-sensitized solar cell. *Nanoscale* 3; 4088–4093.

Im, J., C. C. Stoumpos, H. Jin, A. J. Freeman, and M. G. Kanatzidis. 2015. Antagonism between spin-orbit coupling and steric effects causes anomalous band gap suppression in the perovskite photovoltaic materials $CH_3NH_3Sn_{1-x}Pb_xI_3$. *J. Phys. Chem. Lett.* 6; 3503–3509.

Kieslich, G., S. Sun, and A. K. Cheetham. 2014. Solid-state principles applied to organic-inorganic perovskites: New tricks for an old dog. *Chem. Sci.* 5; 4712–4715.

Kieslich, G., S. Sun, and A. K. Cheetham. 2015. An extended tolerance factor approach for organic-inorganic perovskites. *Chem. Sci.* 6; 3430–3433.

Kim, H.-S., S. H. Im, and N.-G. Park. 2014. Organolead halide perovskite: New horizons in solar cell research. *J. Phys. Chem. C* 118; 5615–5625.

Kirkpatrick, S. 1984. Optimization by simulated annealing: Quantitative studies. *J. Statis. Phys.* 34; 975–986.

Kojima, A., K. Teshima, Y. Shirai, and T. Miyasaka. 2009. Organometal halide perovskites as visible-light sensitizers for photovoltaic cells. *J. Am. Chem. Soc.* 131; 6050–6051.

Kolmogorov, A., S. Shah, E. R. Margine, A. F. Bialon, T. Hammerschmidt, and R. Drautz. 2010. New superconducting and semiconducting Fe–B compounds predicted with an *ab-initio* evolutionary search. *Phys. Rev. Lett.* 105; 217003–217004.

Leijtens, T., G. E. Eperon, N. K. Noel, S. N. Habisreutinger, A. Petrozza, and H. J. Snaith. 2015. Stability of metal halide perovskite solar cells. *Adv. Energy Mater.* 5; 1500963.

Lonie, D. C., and E. Zurek. 2011. XtalOpt: An open-source evolutionary algorithm for crystal structure prediction. *Comp. Phys. Commun.* 182; 372–387.

Lüthi, H. P., S- Heinen, G. Schneider, A. Glöss, M. P. Brändle, R. A. King, E. Pyzer-Knapp, F. H. Alharbi, and S. Kais. 2016. The quantum chemical search for novel Materials and the issue of data processing: The InfoMol Project. *J. Comput. Sci.* 15; 65–73.

Lyakhov, A. O., A. R. Oganov, H. T. Stokes, and Q. Zhu. 2013. New developments in evolutionary structure prediction algorithm USPEX. *Comp. Phys. Commun.* 184; 1172–1182.

Martsinovich, N., and A. Troisi. 2011. High-throughput computational screening of chromophores for dye-sensitized solar cells. *J. Phys. Chem. C* 115; 11781–11792.

McClure, E. T., M. R. Ball, W. Windl, and P. M. Woodward. 2016. $Cs_2AgBiX_6$ (X = Br, Cl): New visible light absorbing, lead-free halide perovskite semiconductors. *Chem. Mater.* 28; 1348–1354.

Mei, A., X. Li, L. Liu, Z. Ku, T. Liu, Y. Rong, M. Xu, M. Hu, J. Chen, Y. Yang, M. Grätzel, and H. Han. 2014. A hole-conductor–free, fully printable mesoscopic perovskite solar cell with high stability. *Science* 345; 295–298.

MGI (Materials Genome Initiative). 2016. http://www.whitehouse.gov/mgi (Accessed February, 2016).

Motta, C., F. El Mellouhi, S. Kais, N. Tabet, F. Alharbi, and S. Sanvito. 2015. Revealing the role of organic cations in hybrid halide perovskites $CH_3NH_3PbI_3$. *Nat. Commun.* 6; 7026–7027.

Mousseau, N., L. K. Bèland, P. Brommer, J.-F. Joly, F. El-Mellouhi, E. Machado-Charry, M.-C. Marinica, and P. Pochet. 2012. The activation-relaxation technique: ART nouveau and kinetic ART. *J. At. Mol. Opt. Phys.* 2012; 925278–9252714.

MP (Materials Project). 2016. https://www.materialsproject.org/ (Accessed February, 2016).

Nayak, P. K., G. Garcia-Belmonte, A. Kahn, J. Bisquert, and D. Cahen. 2012. Photovoltaic efficiency limits and material disorder. *Energy Environ. Sci.* 5; 6022–6039.

NoMaD (Novel Materials Discovery). 2016. http://NoMaD-Repository.eu/ (Accessed February, 2016).

NREL. 2016. NREL Research Cell Efficiency Records. http://www.nrel.gov/ncpv/images/efficiency _chart.jpg (Accessed February, 2016).

Oganov, A. R., and C. W. Glass. 2006. Crystal structure prediction using *ab initio* evolutionary techniques: Principles and applications. *J. Chem. Phys.* 124; 244704.

Ogomi, Y., A. Morita, S. Tsukamoto, T. Saitho, N. Fujikawa, Q. Shen, T. Toyoda, K. Yoshino, S. S. Pandey, T. Ma, and S. Hayase. 2014. $CH_3NH_3Sn_xPb_{1-x}I_3$ perovskite solar cells covering up to 1060 nm. *J. Phys. Chem. Lett.* 5; 1004–1011.

Open Quantum Materials Database (OQMD). 2016. http://oqmd.org/ (Accessed February, 2016).

Osedach, T. P., T. L. Andrew, and V. Bulovic. 2013. Effect of synthetic accessibility on the commercial viability of organic photovoltaics. *Energy Environ. Sci.* 6; 711–718.

Park, N.-G. 2013. Organometal perovskite light absorbers toward a 20% efficiency low-cost solidstate mesoscopic solar cell. *J. Phys. Chem. Lett.* 4; 2423–2429.

Pastore, M., E. Mosconi, F. De Angelis, and M. Grätzel. 2010. A computational investigation of organic dyes for dye-sensitized solar cells: Benchmark, strategies, and open issues. *J. Phys. Chem. C* 114; 7205–7212.

Pellet, N., P. Gao, G. Gregori, T.-Y. Yang, M. K. Nazeeruddin, J. Maier, and M. Grätzel. 2014. Mixedorganic-cation perovskite photovoltaics for enhanced solar-light harvesting. *Angew. Chem. Int. Ed.* 53; 3151–3158.

Peplow, M. 2014. The perovskite revolution. *Spectrum, IEEE* 51; 16–17.

Pickard, C. J., and R. Needs. 2011. *Ab initio* random structure searching. *J. Phys.: Condens. Matter* 23; 053201–053223.

Plackowski, T., D. Włosewicz, P. E. Tomaszewzki, J. Baran, and M. K. Marchewka. 1995. Specific heat of $(NH_2(CH_3)_2)_3Bi_2I_9$. *Acta Phys. Pol. A* 87; 635–641.

Potyrailo, R., K. Rajan, K. Stoewe, I. Takeuchi, B. Chisholm, and H. Lam. 2011. Combinatorial and highthroughput screening of materials libraries: Review of state of the art. *ACS Comb. Sci.* 13; 579–633.

QEERI (Qatar Environment and Energy Research Institute). 2016. http://www.qeeri.org.qa/, Hamad Bin Khalifa University http://www.hbku.edu.qa (Accessed February, 2016).

Sarmiento-Pérez, R., T. F. T. Cerqueira, S. Körbel, S. Botti, and M. A. L. Marques. 2015. Prediction of stable nitride perovskites. *Chem. Mater.* 27; 5957–5963.

Scharber, M. C., D. Mühlbacher, M. Koppe, P. Denk, C. Waldauf, A. J. Heeger, and C. J. Brabec. 2006. Design rules for donors in bulk-heterojunction solar cells: Towards 10% energy conversion efficiency. *Adv. Mater.* 18; 789–794.

Schön, J. C., and M. Jansen. 1996. First step towards planning of syntheses in solid-state chemistry: Determination of promising structure candidates by global optimization. *Angew. Chem. Int. Ed.* 35; 1286–1304.

Science Focus. 2013. Newcomer juices up the race to harness sunlight. *Science* 342; 1438–1439.

Service, R. F. 2014. Perovskite solar cells keep on surging. *Science* 344; 458.

Serra, P., A. F. Stanton, and S. Kais. 1997. Comparison study of pivot methods for global optimization. *J. Chem. Phys.* 106; 7170–7177.

Shi, D., V. Adinolfi, R. Comin, M. Yuan, E. Alarousu, A. Buin, Y. Chen, S. Hoogland, A. Rothenberger, K. Katsiev, Y. Losovyj, X. Zhang et al. 2015. Low trap-state density and long carrier diffusion in organolead trihalide perovskite single crystals. *Science* 347; 519–522.

Shirwadkar, U., E. V. D. van Loef, R. Hawrami, S. Mukhopadhyay, J. Glodo, and K. S. Shah. 2011. New promising scintillators for gamma-ray spectroscopy: Cs(Ba,Sr) (Br,I)$_3$. In: *IEEE: Nucl. Sci. Symp. Conf. Rec.* 1583–1585.

Snaith, H. J. 2013. Perovskites: The emergence of a new era for low-cost, high-efficiency solar cells. *J. Phys. Chem. Lett.* 4; 3623–3630.

Sokolov, A. N., S. Atahan-Evrenk, R. Mondal, H. B. Akkerman, R. S. Sánchez-Carrera, S. GranadosFocil, J. Schrier, S. C. B. Mannsfeld, A. P. Zoombelt, Z. Bao, and A. Aspuru-Guzik. 2011. From computational discovery to experimental characterization of a high hole mobility organic crystal. *Nat. Commun.* 2; 437–438.

Stillinger, F. H. 1999. Exponential multiplicity of inherent structures. *Phys. Rev. E* 59; 48–51.

Stoumpos, C. C., L. Frazer, D. J. Clark, Y. S. Kim, S. H. Rhim, A. J. Freeman, J. B. Ketterson, J. I. Jang, and M. G. Kanatzidis. 2015. Hybrid germanium iodide perovskite semiconductors: Active lone pairs, structural distortions, direct and indirect energy gaps, and strong nonlinear optical properties. *J. Am. Chem. Soc.* 137; 6804–6819.

Stoumpos, C. C., and M. G. Kanatzidis. 2015. The renaissance of halide perovskites and their evolution as emerging semiconductors. *Acc. Chem. Res.* 48; 2791–2802.

Stoumpos, C. C., C. D. Malliakas, and M. G. Kanatzidis. 2013. Semiconducting tin and lead iodide perovskites with organic cations: Phase transitions, high mobilities, and near-infrared photoluminescent properties. *Inorg. Chem.* 52; 9019–9038.

Stranks, S. D., G. E. Eperon, G. Grancini, C. Menelaou, M. J. P. Alcocer, T. Leijtens, L. M. Herz, A. Petrozza, and H. J. Snaith. 2013. Electron-hole diffusion lengths exceeding 1 micrometer in an organometal trihalide perovskite absorber. *Science* 342; 341–344.

Thiele, G., and B. R. Serr. 2010. Crystal structure of dimethylammonium triiodostannate(II), $(CH_3)_2NH_2SnI_3$. *Z. Kristallogr. Crystall. Mater.* 211; 48.

Trimarchi, G., X. Zhang, M. J. DeVries Vermeer, J. Cantwell, K. R. Poeppelmeier, and A. Zunger. 2015. Emergence of a few distinct structures from a single formal structure type during high-throughput screening for stable compounds: The case of RbCuS and RbCuSe. *Phys. Rev. B* 92; 165103–165109.

Trimarchi, G., and A. Zunger. 2007. Global space-group optimization problem: Finding the stablest crystal structure without constraints. *Phys. Rev. B* 75; 104113–104118.

Wales, D. J., and J. P. Doye. 1997. Global optimization by basin-hopping and the lowest energy structures of Lennard–Jones clusters containing up to 110 atoms. *J. Phys. Chem. A* 101; 5111–5116.

Wang, Y., J. Lv, L. Zhu, and Y. Ma. 2010. Crystal structure prediction via particle-swarm optimization. *Phys. Rev. B* 82; 094116–094118.

Wang, Y., J. Lv, L. Zhu, and Y. Ma. 2012. CALYPSO: A method for crystal structure prediction. *Comp. Phys. Commun.* 183; 2063–2070.

Wang, Y., and Y. Ma. 2014. Perspective: Crystal structure prediction at high pressures. *J. Chem. Phys.* 140; 040901.

Wang, S., D. B. Mitzi, C. A. Feild, and A. Guloy. 1995. Synthesis and characterization of $[NH_2C(I) = NH_2]_3MI_5$ (M = Sn, Pb): Stereochemical activity in divalent tin and lead halides containing single <110> perovskite sheets. *J. Am. Chem. Soc.* 117; 5297–5302.

Wei, F., Z. Deng, S. Sun, F. Xie, G. Kieslich, D. M. Evans, M. A. Carpenter, P. D. Bristowe, and A. K. Cheetham. 2016. The synthesis, structure and electronic properties of a lead-free hybrid inorganic organic double perovskite $(MA)_2KBiCl_6$ (MA = methylammonium). *Mater. Horiz.* 3; 328–332.

Woodley, S. M., P. D. Battle, J. D. Gale, and C. R. A. Catlow. 1999. The prediction of inorganic crystal structures using a genetic algorithm and energy minimisation. *Phys. Chem. Chem. Phys.* 1; 2535–2542.

Woodley, S. M., and R. Catlow. 2008. Crystal structure prediction from first principles. *Nat. Mater.* 7; 937–946.

Yamada, Y., T. Nakamura, M. Endo, A. Wakamiya, and Y. Kanemitsu. 2014. Near-band-edge optical responses of solution-processed organic-inorganic hybrid perovskite $CH_3NH_3PbI_3$ on mesoporous $TiO_2$ electrodes. *Appl. Phys. Express* 7; 032302–032304.

Yin, W.-J., T. Shi, and Y. Yan. 2014a. Unusual defect physics in $CH_3NH_3PbI_3$ perovskite solar cell absorber. *Appl. Phys. Lett.* 104; 063903.

Yin, W.-J., T. Shi, and Y. Yan. 2014b. Unique properties of halide perovskites as possible origins of the superior solar cell performance. *Adv. Mater.* 26:4653–4658.

Yu, L., and A. Zunger. 2012. Identification of potential photovoltaic absorbers based on first-principles spectroscopic screening of materials. *Phys. Rev. Lett.* 108; 068701–068705.

Zhao, Y., and K. Zhu. 2013. Charge transport and recombination in perovskite $(CH_3NH_3)PbI_3$ sensitized $TiO_2$ solar cells. *J. Phys. Chem. Lett.* 4; 2880–2884.

Stillinger, F. 1999. Exponential multiplicity of inherent structures. *Phys. Rev.* E, 59, 48–51.

Stoumpos, C. C., L. Frazer, D. J. Clark, Y. S. Kim, S. H. Rhim, A. J. Freeman, J. B. Ketterson, J. I. Jang, and M. G. Kanatzidis. 2015. Hybrid germanium iodide perovskite semiconductors: Active lone pairs, structural distortions, direct and indirect energy gaps, and strong nonlinear optical properties. *J. Am. Chem. Soc.* 137, 6804–6819.

Stoumpos, C. C. and M. G. Kanatzidis. 2015. The renaissance of halide perovskites and their evolution as emerging semiconductors. *Acc. Chem. Res.* 48, 2791–2802.

Stoumpos, C. C., C. D. Malliakas, and M. G. Kanatzidis. 2013. Semiconducting tin and lead iodide perovskites with organic cations: Phase transitions, high mobilities, and near-infrared photoluminescent properties. *Inorg. Chem.* 52, 9019–9038.

Strauss, S. J., G. F. Epenscheidt-Jaegel, C. Mangelsen, M. E. A. Stocker, J. Teilhorst, L. M. Herz, A. Petrozza, and H. J. Snaith. 2013. Electron-hole diffusion lengths exceeding 1 micrometer in an organometal trihalide perovskite absorber. *Science*, 342, 341–344.

Thiele, G., and B. R. Serr. 2010. Crystal structure of dimethylammonium triiodoarsenate(III). (VIII). *Z. Kristallogr. Cryst. Mater.* 217, 633.

Tranchemi, G., X. Zhang, M. J. Berrler-Vannous, J. Carswell, L. X. Bergensen, and A. Zunger. 2015. Emergence of a new distinct structure from a single formal structure in perovskite halides through screening metastable compounds: The case of KBiO₃ and HBr. *Inorg. Rev. B* 92, 16105–16105.

Trimarchi, G. and A. Zunger. 2007. Global space-group optimization problem: Finding the stablest crystal structure without constraints. *Phys. Rev. B* 75, 104113–104115.

Wales, D. J. and J. P. Doye. 1997. Global optimization by basin-hopping and the lowest energy structures of Lennard-Jones clusters containing up to 110 atoms. *J. Phys. Chem. A* 101, 5111–5116.

Wang, Y., J. Lv, L. Zhu, and Y. Ma. 2010. Crystal structure prediction via particle-swarm optimization. *Phys. Rev. B* 82, 094116–094118.

Wang, Y., J. Lv, L. Zhu, and Y. Ma. 2012. CALYPSO: A method for crystal structure prediction. *Comp. Phys. Commun.* 183, 2063–2070.

Wang, Y. and Y. Ma. 2014. Perspective: Crystal structure prediction at high pressures. *J. Chem. Phys.* 140, 040901.

Weber, D. 1978. CH₃NH₃PbX₃, ein Pb(II)-System mit kubischer Perowskitstruktur / CH₃NH₃PbX₃, a Pb(II)-system with cubic perovskite structure. *Z. Naturforsch. B* 33, 1443–1445.

Wei, F., Z. Deng, S. Sun, F. Xie, G. Kieslich, A. A. Carpenter, P. D. Bristowe, and A. K. Cheetham. 2016. The synthesis, structure and electronic properties of a lead-free hybrid inorganic-organic double perovskite (MA)₂KBiCl₆ (MA = methylammonium). *Mater. Horiz.* 3, 328–332.

Woodley, S. et al., P. D. Battle, J. D. Gale, and C. R. A. Catlow. 1999. The prediction of inorganic crystal structures using a genetic algorithm and energy minimisation. *Phys. Chem. Chem. Phys.* 1, 2535–2542.

Woodley, S. M. and R. Catlow. 2008. Crystal structure prediction from first principles. *Nat. Mater.* 7, 937–946.

Yamada, Y., T. Nakamura, M. Endo, A. Wakamiya, and Y. Kanemitsu. 2014. Near-band-edge optical responses of solution-processed organic-inorganic hybrid perovskite CH₃NH₃PbI₃ on mesoporous TiO₂ electrodes. *Appl. Phys. Express* 7, 032302–032304.

Yin, W. J., T. Shi, and Y. Yan. 2014. Unusual defect physics in CH₃NH₃PbI₃ perovskite solar cell absorber. *Appl. Phys. Lett.* 104, 063903.

Yin, W. J., T. Shi, and Y. Yan. 2014. Unique properties of halide perovskites as possible origins of the superior solar cell performance. *Adv. Mater.* 26, 4653–4658.

Yu, L. and A. Zunger. 2012. Identification of potential photovoltaic absorbers based on first-principles spectroscopic screening of materials. *Phys. Rev. Lett.* 108, 068701–068705.

Zhou, Y. and K. Zhu. 2016. Perovskite solar cells shine in the "valley of the sun". *ACS Energy Lett.* 1, 64–67.

Zhao, Y., and K. Zhu. 2013. Charge transport and recombination in perovskite (CH₃NH₃)PbI₃ sensitized TiO₂ solar cells. *J. Phys. Chem. Lett.* 4, 2880–2884.

# CHAPTER 5

# Atomic Structures and Electronic States at the Surfaces and Interfaces of CH₃NH₃PbI₃ Perovskite

Jun Haruyama, Keitaro Sodeyama, and Yoshitaka Tateyama

## CONTENTS

## 5.1 INTRODUCTION

### 5.1.1 History of Perovskite Solar Cells

Conversion of solar energy to electrical power using photovoltaic (PV) devices is one of the most important sources of renewable energy. Solar cells based on organic–inorganic perovskites, referred to as perovskite solar cells (PSCs), have recently drawn global interest because their power conversion efficiencies (PCEs) have increased dramatically, from 3.8% to more than 20%, over the course of only five to six years (NREL 2016). Specifically, in 2009, Miyasaka et al. used methylammonium lead halide, $CH_3NH_3PbX_3$ (MAPbX$_3$, where X = Br or I), as a light-absorbing layer in liquid-electrolyte–based dye-sensitized solar cells (DSSCs), and reported a PCE of 3.8% (Kojima et al. 2009). However, this type of PSC had a drawback: instability of the deposited MAPbI$_3$ in the liquid electrolyte. This problem was solved by replacing the liquid electrolyte with solid-state, hole-transporting materials (HTMs), leading to an efficiency as high as 9.7% and devices exhibiting long-term stability (Kim et al. 2012). Up to that point, the PSC architectures were identical to those of conventional DSSCs. Light-harvesting dyes or organic–inorganic perovskite nanoparticles were regarded to play the role of sensitizer, which injects the excited electrons into a mesoporous-TiO$_2$ (mp-TiO$_2$) scaffold or the holes into the HTM.

In 2012, Lee et al. made a discovery that changed the general concept. They found that the use of a mesoporous scaffold made of insulating $Al_2O_3$ instead of TiO$_2$ resulted in a PCE of 10.9%, which was similar or superior to what had already been achieved with conventional architectures. These researchers proposed that electrons are transported in the perovskite phase and that mp-TiO$_2$ is thus unnecessary. These results also indicated that electrons and holes are allowed to move along the surfaces of the $Al_2O_3$ nanoparticles through a thin coating layer of MAPbI$_3$. However, the role of the mp-$Al_2O_3$ scaffold in these devices remains an open question (Lee et al. 2012).

Initially, single-step deposition of the perovskite from a solution of PbX$_2$ and MAX (X = Cl, Br, or I) in the same solvent, referred to as a one-step method, was often employed. However, this method produced perovskites with variable morphologies and particle sizes, which resulted in poor reproducibility of the PV performance. In 2013, Burschka et al. reported that a two-step method involving spin-coating of PbI$_2$ onto an mp-TiO$_2$ film and subsequent exposure of the coated film to a solution of MAI was effective for the fabrication of high-performance cells (Burschka et al. 2013). This method led to the perovskite particles completely covering the mp-TiO$_2$ layer and allowed much better control over the perovskite morphology, which greatly improved the reproducibility of the solar cell performance and increased the PCE to 15.0%. High-performance cells, prepared by Sang Il Seok and his group, show a certified PCE of 17.9% (Jeon et al. 2015). These cells combine formamidinium lead iodide, $HC(NH_2)_2PbI_3$ (FAPbI$_3$), with MAPbBr$_3$ as the light-harvesting unit in a bilayer architecture. That combination stabilizes the perovskite phase and improves the overall performance.

### 5.1.2 Bulk Properties of PSCs

The excellent efficiency of PSCs can be attributed to the remarkable intrinsic optical and electronic properties of organolead halide perovskites (Park 2013; Snaith 2013). First, they strongly absorb sunlight over a broader wavelength spectra covering the entire visible light range than other dye sensitizers. Representative MAPbI$_3$ has a direct band gap of 1.55 eV (Baikie et al. 2013) and an optical absorption coefficient greater than $10^5$ cm$^{-1}$ (comparable to or higher than that of crystalline GaAs) in the short-wavelength region (De Wolf et al. 2014). The excitons produced by light absorption have a weak binding energy [(~0.030 eV

(Ponseca et al. 2014)], indicating that most excitons can dissociate rapidly into free carriers at room temperature. This behavior of the photoexcited carriers is in contrast to that of organic PV solar cells. Femtosecond transient absorption spectroscopy measurements have shown that the dominant relaxation path is recombination of free electrons and holes (Christians et al. 2015; Manser and Kamat 2014; Yamada et al. 2014). The high mobility and slow recombination rate of the charge carriers are responsible for the remarkable intrinsic properties of organolead halide perovskites (Oga et al. 2014; Savenije et al. 2014). Also interesting is the fact that the perovskites show considerably long diffusion lengths, approaching and sometimes exceeding 1 μm for both holes and electrons (Dong et al. 2015a; Eperon et al. 2014; Stranks et al. 2013; Xing et al. 2013).

These remarkable optoelectronic properties can be elucidated by their electronic band structures (Yin et al. 2014c). Perovskite materials typically have a primitive cell that can be represented as $ABX_3$ in a cubic structure (Figure 5.1a), where A and B correspond to the cations and X is the anion. In the case of $MAPbI_3$, lower-symmetry structures consisting of tetragonal and orthorhombic phases have been reported to be stable around and much below room temperature, respectively (Baikie et al. 2013; Kawamura et al. 2002; Stoumpos et al. 2013). The unit cell of the tetragonal phase can be depicted as a $\sqrt{2} \times \sqrt{2} \times 2$ $ABX_3$ cubic unit cell as shown in Figure 5.1a. A number of Density Functional Theory (DFT) calculation studies have shown the band structure and density of states (DOS) of $MAPbI_3$ (Baikie et al. 2013; Mosconi et al. 2013; Umebayashi et al. 2003). In all three phases (cubic, tetragonal, and orthorhombic), the valence band maximum (VBM) has strong Pb-$s$ and I-$p$ antibonding character, whereas the conduction band minimum (CBM) consists almost entirely of a Pb-$p$ state (Umebayashi et al. 2003; Yin et al. 2014c). The antibonding character of the VBM is particularly important for the prominent optoelectronic properties of organolead halide perovskites.

Meanwhile, the organic moieties play no significant role in the electronic structures, because their orbitals are located far from the band edges. An important function of the organic moieties is to stabilize the perovskite structure and change the lattice constant

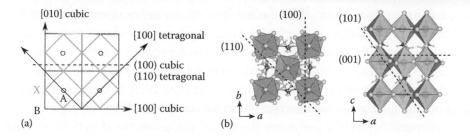

FIGURE 5.1  (a) Top view of the relationship between the conventional cubic perovskite and the tetragonal cell. A, B, and X sites are also shown. (b) Unit cell of tetragonal $MAPbI_3$. The H, C, N, I, and Pb atoms are depicted as small white spheres, small black spheres, small gray spheres, large white spheres, and large black spheres, respectively. VESTA was used to visualize the crystal structures (Momma and Izumi 2008). Gray octahedrons represent the $PbI_6$ units. Dashed lines indicate typcial planes. (Reprinted with permission from Haruyama, J., K. Sodeyama, L. Han, and Y. Tateyama. 2014. Termination dependence of tetragonal $CH_3NH_3PbI_3$ surfaces for perovskite solar cells. *J. Phys. Chem. Lett.* 5; 2903–2909. Copyright © 2014 American Chemical Society.)

(Giorgi et al. 2014). Other possible effects of organic moieties (e.g., dipole polarization or ionic diffusion) have been discussed recently (Azpiroz et al. 2015; Eames et al. 2015; Giorgi et al. 2015; Haruyama et al. 2015).

DFT calculations of energy gaps of lead halide perovskites with the Perdew–Burke–Ernzerhof (PBE) functional (Perdew et al. 1996) have reproduced the experimental values accidentally, because the energy gap underestimation owing to the generalized gradient approximations is canceled out by neglect of the effects of spin–orbit coupling (SOC) (Even et al. 2013). Investigation of the band structures of lead halide perovskites requires advanced computational methods such as GW+SOC (Umari et al. 2014) and HSE+SOC (Feng and Xiao 2014). However, the high optical absorption coefficient from the transition dipole moment between the VBM and the CBM can be reproduced with the PBE functional (Yin et al. 2014a). SOC is important for calculating the effective masses of the carriers, mainly electrons at the CBM. Giorgi and Yamashita have pointed out that the estimated effective mass of electrons is substantially decreased by SOC due to the CBM character of the Pb-$p$ orbitals (Giorgi et al. 2013). The results suggest that lead halide perovskites have long-range ambipolar transport properties.

Other essential properties are intrinsic defect states (Agiourgosis et al. 2014; Buin et al. 2014; Kim et al. 2014; Yin et al 2014b). Yin et al. showed that all the possible point defects such as $V_{MA}$, $V_I$, $V_{Pb}$, and $MA_i$ have shallow defect levels (Yin et al. 2014b). In particular, the absence of additional midgap states leads to few carrier traps and few nonradiative recombination centers, which can account for the experimentally observed long diffusion length. Similar properties of additional states are expected even at surfaces (Buin et al. 2014; Haruyama et al. 2014) or grain boundaries (Yin et al. 2014a), which is discussed in the text that follows.

### 5.1.3 Surface and Interface Properties of PSCs

The properties of the interfaces between the perovskites and the adjacent layers are also essential for smooth transport of the carriers. Experimental observations of ultrafast events such as electron and hole injection to charge-separated layers and recombination processes at interfaces have been reported (Marchioro et al. 2014; Ogomi et al. 2014). For further optimization of PSCs, a clear understanding of the mechanisms of carrier-transport dynamics is crucial. The first principles calculation studies by De Angelis and his co-workers are regarded as a first step toward understanding and controlling charge-transfer processes at the PSC interfaces (De Angelis 2014; Mosconi et al. 2014; Roiati et al. 2014). Although there are some other first principles studies of the perovskite interfaces as well (Feng et al. 2015; Nemnes et al. 2015), little is known about the surface and the interface properties compared to the bulk ones.

In this chapter, we describe our recent theoretical studies on the structural and electronic properties of MAPbI$_3$ perovskites, with an emphasis on the surfaces and the interfaces. In Sections 5.3.1 and 5.3.2, we focus on the termination effects of representative tetragonal MAPbI$_3$ surfaces at various chemical potentials and DFT geometry optimizations. In particular, the properties of interfaces between MAPbI$_3$ and organic HTMs such as spiro-OMeTAD (2,2′,7,7′-tetrakis-($N$,$N$-di-$p$-methoxyphenylamine)-9,9′-bifluorene) are discussed because the calculated vacuum surfaces can be taken as reasonable interface models owing to their weak interactions with the organic HTMs. Desirable properties for good PSC performance are suggested on the basis of the resultant stable surface terminations and their electronic states.

In Section 5.3.3, we describe our recent calculations of $TiO_2/MAPbI_3$ and $Al_2O_3/MAPbI_3$ interfaces, comparing them with the other reports. Exploring interfacial matching between the $TiO_2$ ($Al_2O_3$) scaffold and $MAPbI_3$ perovskites was an important issue. We then carried out a systematic search for the stable matching under stoichiometric conditions, and examined the geometries, binding energies, and band alignments at the stable interfaces. Like the band gap estimation, we found that the band alignments at the interfaces were satisfactorily reproduced by the PBE or van der Waals (vdW) functionals. These results are highly encouraging because such calculations may allow computer-aided interface engineering in the near future.

## 5.2 METHODS

### 5.2.1 DFT Calculation

The Quantum ESPRESSO code was used to carry out all DFT calculations within a spin-unpolarized treatment (Giannozzi et al. 2009). Both the PBE (Perdew et al. 1996) and the vdW-DF2-B86R (Hamada 2014) exchange-correlation functionals were adopted with a plane-wave basis set. The cutoff energies were set to 40 Ry for the wave functions and 320 Ry for the augmented charge with the ultrasoft pseudopotential framework (Rappe et al. 1990; Vanderbilt 1990). The electronic configurations of the pseudopotentials were $1s^1$ for H, $2s^2 2p^2$ for C with nonlinear core correction (NCC) (Louie et al. 1982), $2s^2 2p^3$ for N with NCC, $2s^2 2p^4$ for O, $3s^1 3p^2$ for Al, $3s^2 3p^6 4s^2 4d^2$ for Ti, $5s^2 5p^5$ for I with NCC, and $5d^{10} 6s^2 6p^2$ for Pb with NCC. Convergent $k$-point sampling was adopted for surface systems whereas only a gamma point was adopted in examining the interface systems. The relaxed atomic structures showed residual forces of less than 0.001 Ry/bohr. The occupation number was determined by the Gaussian smearing technique with a smearing parameter of 0.01 Ry. SOC coupling was treated with pseudopotentials tailored to reproduce the solutions of fully relativistic equations (Bachelet and Schlüter 1982; Dal Corso and Mosca Conte 2005; Kleinman 1980). The SOC-DFT calculations were used only for surface projected density of states (PDOS), in which we used the geometries optimized under the non-SOC condition.

### 5.2.2 Computational Models for Perovskite Surfaces

The [010], [001], [110], and [111] directions in a conventional cubic $ABX_3$ unit correspond to the [110], [001], [100], and [101] directions of the tetragonal unit, respectively (Figure 5.1a). The tetragonal (110) and (001) planes are flat nonpolar surfaces that consist of alternately piled-up $[MAI]^0$ and $[PbI_2]^0$ layers. The tetragonal (100) plane is constructed with charged layers of $[MAPbI]^{2+}$ and $[I_2]^{2-}$; and the charged layers of $[MAI_3]^{2-}$ and $[Pb]^{2+}$ compose the tetragonal (101) plane (Figure 5.1b).

X-ray diffraction experiments have indicated presence of the (110) and (001) planes of tetragonal $MAPbI_3$ on mp-$TiO_2$ films (Heo et al. 2013), whereas a (110) peak has been observed in planar type cells including $MAPbI_{3-x}Cl_x$ (Liu et al. 2013). A recent experiment indicated that Cl is depleted in the surface region (Starr et al. 2015), and computational studies have suggested that the concentration of Cl in the final $MAPbI_3$ crystal is relatively low (Colella et al. 2015). However, whether Cl improves morphological control, decreases the defect concentration, or plays some other role is still not clear (Grätzel 2014). After analyzing the structural stability, we dealt with the probable (110) and (001) surfaces of $MAPbI_3$.

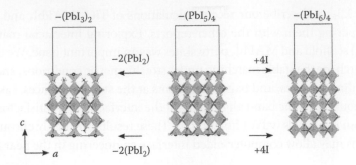

FIGURE 5.2 Schematic of the construction of various terminations.

On the four target planes of the tetragonal phase, all of the surface structures are composed of two or four $PbI_6$ octahedrons in the primitive cell. Various terminations, including flat terminations, were investigated. To distinguish the termination types, the outermost $PbI_x$ polyhedron layers were used as labels. For example, a flat termination of the (001) surface was expressed as $-(PbI_5)_4$ as shown in Figure 5.2. A surface attaching four I atoms to the $-(PbI_5)_4$ termination was represented as $-(PbI_6)_4$, and a flat plane missing $2(PbI_2)$ was represented as $-(PbI_3)_2$. Thus, more than five terminations were investigated on each target plane. All of the terminations investigated are presented in Haruyama et al. (2014). The polarizations emerged from defect formations or polar layer alternations sometimes cause supercell calculations to fail. To prevent such a polarization appearance in the supercell, we introduced the same termination compositions on both surfaces of the calculation slab.

In a previous experiment with the two-step synthesis, a small understoichiometry of N and I atoms of the $TiO_2/MAPbI_3$ interface are reported (Lindblad et al. 2014). The observed composition ratio of N/I = 1:2.8 may be regarded that the number of MA molecules attached on $MAPbI_3$ surfaces is lower than the number of apical I atoms sticking on the surfaces. Therefore, MA terminations except for apical I atoms were not considered in this study. The initial MA configuration was determined as described previously by Mosconi et al. (2013). To avoid an unreasonably large dipole in the calculation supercells, we adjusted the MA configurations and canceled out the surface polarizations.

### 5.2.3 Phase Diagrams of MAPbI₃

In this section, we describe the grand potential approach, which can determine the structural stabilities of different terminations. We calculated the grand potential $\Omega$ as follows:

$$\Omega(\Delta\mu_{Pb}, \Delta\mu_I, \Delta\mu_{MAPbI_3}) \approx E_{tot}\left[Pb_\alpha I_\beta (MAPbI_3)_\gamma\right]$$

$$-\alpha\mu_{Pb}^{metal} - \frac{\beta}{2}\mu_{I_2}^{gas} - \gamma\mu_{MAPbI_3}^{tetragonal} + -\alpha\Delta\mu_{Pb} - \beta\Delta\mu_I - \gamma\Delta\mu_{MAPbI_3}, \quad (5.1)$$

$$\Delta\mu_{Pb} = \mu_{Pb} - \mu_{Pb}^{metal}, \ \Delta\mu_I = \mu_I - \frac{1}{2}\mu_{I_2}^{gas}, \ \Delta\mu_{MAPbI_3} = \mu_{MAPbI_3} - \mu_{MAPbI_3}^{tetragonal} \quad (5.2)$$

Here $E_{tot}[Pb_\alpha I_\beta(MAPbI_3)_\gamma]$ is the total energy of a surface slab constructed by $\alpha$ Pb atoms, $\beta$ I atoms, and $\gamma$ MAPbI$_3$ complexes; and $\mu_i$ ($i$ = Pb, I, MAPbI$_3$) are the chemical potentials of the environmental elements. Reference chemical potentials for Pb metal, I$_2$ gas, and MAPbI$_3$ tetragonal phases are assigned to $\mu_{Pb}^{metal}$, $\mu_{I_2}^{gas}$, and $\mu_{MAPbI_3}^{tetragonal}$, respectively. All components in Equation 5.1 are approximated by the DFT total energies per formula unit. Variations of chemical potentials from the reference values are designated as $\Delta\mu_i$, and these variations define the environmental conditions. As in conventional surface DFT studies, the entropy terms in Equation 5.1 are ignored (Qian et al. 1988). The system is assumed to always be in equilibrium with tetragonal MAPbI$_3$, that is, $\Delta\mu_{MAPbI_3} = 0$. For phase-equilibrium conditions with Pb metal, I$_2$ gas, MAI, PbI$_2$, and MAPbI$_3$, thermodynamically stable regions are derived as follows:

$$\Delta H_{form}[MAPbI_3] \leq \Delta\mu_{Pb} \leq 0, \ \Delta H_{form}[MAPbI_3] \leq \Delta\mu_I \leq 0, \quad (5.3)$$

$$\Delta H_{form}[MAPbI_3] - \Delta H_{form}[MAI] \leq \Delta\mu_{Pb} + 2\Delta\mu_I \leq \Delta H_{form}[PbI_2] \quad (5.4)$$

A heat of formation $\Delta H_{form}$ is the difference among the total energy of the chemical compounds and their constituents:

$$\Delta H_{form}[AB] = E_{tot}[AB] - E_{tot}[A] - E_{tot}[B] \quad (5.5)$$

The values of $\Delta H_{form}$(MAPbI$_3$, MAI, PbI$_2$) calculated from our PBE (Perdew et al. 1996) (or vdW (Hamada 2014))-based first principles calculations were –5.49 (–5.86), –3.02 (–3.02), and –2.39 (–2.80) eV, respectively (Haruyama et al. 2014). The range of chemical potentials described by Equations 5.3 and 5.4 was almost the same as the range reported in the other studies (Figure 5.3) (Agiourgosis et al. 2014; Yin et al 2014b). In this case, the narrow

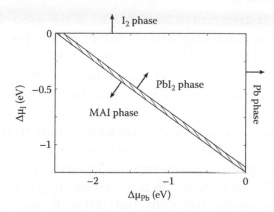

FIGURE 5.3  Range of chemical potentials (Pb and I) described by Equations 5.3 and 5.4, based on the DFT-PBE calculations. The shaded area is the thermodynamically stable range for equilibrium growth of MAPbI$_3$. (Reprinted with permission from Haruyama, J., K. Sodeyama, L. Han, and Y. Tateyama. 2014. Termination dependence of tetragonal CH$_3$NH$_3$PbI$_3$ surfaces for perovskite solar cells. *J. Phys. Chem. Lett.* 5; 2903–2909. Copyright © 2014 American Chemical Society.)

diagonal area indicates that the dissociation energy of $MAPbI_3$ into $PbI_2$ and MAI is quite small; the value is about 0.1 eV/formula. In the metal Pb-, gas $I_2$-, $PbI_2$-, and MAI-rich conditions of this narrow region, $MAPbI_3$ decomposes into the constituent substances. These findings are consistent with the experimentally observed instability of lead iodide materials.

### 5.2.4 Interface Construction

Anatase-$TiO_2$(001) and $\alpha$-$Al_2O_3$(0001) were used as model mesoporous scaffolds. The relaxed structures and electronic states of their interfaces with $MAPbI_3$ were examined. The slab thicknesses were about 1 nm, which is thick enough to show the converged energy gaps for each slab. We multiplied the surface slab in the lateral directions for the supercell to reduce the lattice mismatch between the scaffolds and $MAPbI_3$. Considering the high elastic moduli of oxide materials, we fit the $MAPbI_3$ lattice constants as much as possible to those of anatase-$TiO_2$(001) (3.784Å × 3.784Å) and $\alpha$-$Al_2O_3$(0001) (4.760Å × 8.244Å). The $MAPbI_3$[110] (12.446 × 2)Å × (12.686 × 2)Å plane was combined with the anatase-$TiO_2$[001] (3.784 × 7)Å × (3.784 × 7)Å and $\alpha$-$Al_2O_3$[0001] (4.760 × 5)Å × (8.244 × 3)Å surfaces. Based on these experimental lattice parameters, misfit parameters for the constructed interfaces were 5.3% and 3.5%, respectively. Then we slightly adjusted the lateral lattice constants of the attaching surfaces, and carried out the relaxation of the interface supercell. At this stage, we used a systematic lateral slide of one surface slab with respect to the other to prepare possible initial configurations for the relaxation. We eventually optimized four ($TiO_2$/$MAPbI_3$) and six ($Al_2O_3$/$MAPbI_3$) supercells of the interface structures by using a vdW functional, taking the surface symmetry into account.

## 5.3 RESULTS

### 5.3.1 Stable Surface Structures

The grand potentials $\Omega$ described in Section 5.2.3 of the selected terminations allow us to draw surface termination diagrams of the four planes (Figure 5.4). The four diagrams consist mainly of the $(MAPbI_3)_\gamma$ and $(PbI_2)_\alpha(MAPbI_3)_\gamma$ stoichiometric regions (dark and light gray areas, respectively). Except for the (100) surface, the $(MAPbI_3)_\gamma$ and $(PbI_2)_\alpha(MAPbI_3)_\gamma$ terminations can be classified as vacant- and flat-type terminations, respectively. Hereafter, the terms stable-vacant terminations and $PbI_2$-rich-flat terminations (or vacant and flat, simply) are used to refer to them. The thermodynamically stable range is located mostly in the stable-vacant region, implying the predominance of this type of surface, although the flat termination region is still close to the thermodynamically stable range. In fact, surface energy differences between the two major terminations on the (110), (001), and (101) surfaces are small; the differences are within 0.3 eV $nm^{-2}$ [see values of the grand potentials in Haruyama et al. (2014)]. This result suggests that the stable-vacant and $PbI_2$-rich-flat terminations coexist on various $MAPbI_3$ planes and that the latter may predominate under $PbI_2$-rich growth conditions. We confirmed that these discussions are qualitatively applicable to larger slabs and even using other (e.g., vdW) functional.

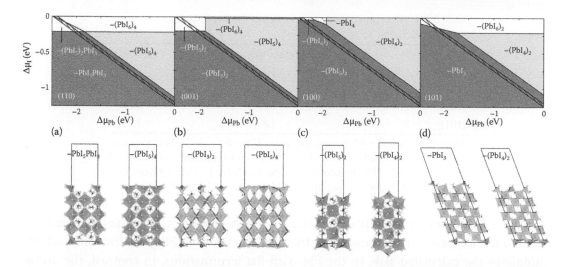

FIGURE 5.4  Surface termination diagrams with respect to the chemical potentials of Pb and I for different MAPbI$_3$ growth conditions: (a) (110), (b) (001), (c) (100), and (d) (101) surfaces. The dark and light gray regions indicate stable-vacant and PbI$_2$-rich-flat terminations, respectively. The relaxed surface structures of the two major terminations are shown in the lower panel. (Adapted with permission from Haruyama, J., K. Sodeyama, L. Han, and Y. Tateyama. 2014. Termination dependence of tetragonal CH$_3$NH$_3$PbI$_3$ surfaces for perovskite solar cells. *J. Phys. Chem. Lett.* 5; 2903–2909. Copyright © 2014 American Chemical Society.)

Relaxed structures of the stable-vacant and PbI$_2$-rich-flat terminations are obtained with the geometry optimizations (lower panel of Figure 5.4). Almost all the terminations retained their initial outermost PbI$_x$ polyhedrons; however, only the relaxed structure of the –(PbI$_5$)(PbI$_3$) termination on the (110) plane formed PbI$_4$ tetrahedrons at the outermost layers. The vacant surfaces showed large symmetry breaking with interior lattice distortions, especially on the (110) and (001) surfaces. In contrast, all the PbI$_2$-rich-flat terminations exhibited little structural deviations from the tetragonal crystal phase of MAPbI$_3$.

The change of the surface structure affects the energy gap, the energy difference between the occupied and unoccupied states (Table 5.1). The energy gaps of stable-vacant terminations are approximately 0.2 eV larger than those of the PbI$_2$-rich-flat terminations, and this difference can be attributed to a wide range of deformations of the PbI$_6$ octahedron (e.g., increase of Pb–I distances) in the interior of the calculated slab. The absence of PbI$_4$ polyhedrons on the stable-vacant surfaces is the origin of these distorted structures. These lower-symmetry structures lead to larger energy gaps. The band gap difference between tetragonal and orthorhombic MAPbI$_3$ can be understood in a similar way in terms of the symmetry breaking. In fact, our DFT calculations confirmed that the phase transformation from the tetragonal phase to the orthorhombic increases the energy gap from 1.64 to 1.70 eV (Haruyama et al. 2014). Increase of energy gaps corresponding to dimensional change (Umebayashi et al. 2003) and structural modifications (Filip et al. 2014; Kim et al. 2015) are analogous to the preceding discussion.

TABLE 5.1   Calculated Energy Gaps (eV) of the Stable-Vacant
and PbI$_2$-Rich-Flat Terminations

|  | Stable-Vacant | PbI$_2$-Rich Flat |
|---|---|---|
| (110) | 1.91 | 1.63 |
| (001) | 1.83 | 1.56 |
| (100) | 1.87 | 1.77 |
| (101) | 1.87 | 1.77 |

*Source:*  Reprinted with permission from Haruyama, J., K. Sodeyama, L.
Han, and Y. Tateyama. 2014. Termination dependence of tetragonal
CH$_3$NH$_3$PbI$_3$ surfaces for perovskite solar cells. *J. Phys. Chem. Lett.*
5; 2903–2909. Copyright © 2014 American Chemical Society.

Note that the energy gaps listed in Table 5.1 reflect dependences of the calculated slab size. In the stable-vacant terminations, the distortions of the PbI$_6$ octahedron reach the middle of the calculated slab. In the PbI$_2$-rich-flat terminations, in contrast, the distortions are rather small and almost vanish at the subsurface region. Thus, the energy gap of flat terminations converges at a small slab thickness [the data for energy gaps at larger slab sizes are available in Haruyama et al. (2014)]. However, the surface electronic properties discussed in Section 5.3.2 were not sensitive to the size of the calculation slab.

### 5.3.2  Electronic Properties and Surface States

For simple analysis of electronic states, we introduced the PDOSs of the two major phases on the probable (110) and (001) surfaces, that is, the stable-vacant and the PbI$_2$-rich-flat terminations (Figure 5.5). All of PDOSs have properties similar to those of the bulk phase: the valence and conduction bands characterized by I-5$p$ (partially Pb-6$s$) and Pb-6$p$ orbitals, respectively (Umebayashi et al. 2003). The PDOSs of the MA molecules are located far from their energy gaps (Baikie et al. 2013). We stress that there are no midgap surface states on any of the PDOSs in Figure 5.5. In addition, Buin et al. calculated the DOSs for MAI- and PbI$_2$-terminated surfaces and found that there were no states in their gaps (Buin et al. 2014). These results revealed that charge recombination unlikely happen even at surfaces, consistent with the observed long lifetimes and long diffusion lengths of the excited carriers in organolead halide perovskites (Christians et al. 2015; Dong et al. 2015a; Eperon et al. 2014; Manser and Kamat 2014; Oga et al. 2014; Ponseca et al. 2014; Savenije et al. 2014; Stranks et al. 2013; Xing et al. 2013; Yamada et al. 2014).

Both terminations have shallow surface states composed of I-5$p$ and Pb-6$s$ orbitals near the VBM, which show strong antibonding character. In the stable-vacant terminations, the absence of PbI$_x$ polyhedrons at the outermost layer makes the surface states of these terminations stable. As a result, the surface states of the vacant terminations are located 0.2 eV below the highest occupied molecular orbital (HOMO) states. Namely, we can regard the strong antibonding character within the I-5$p$ bonds as the origin of the long hole-diffusion lengths.

Charge distributions of HOMOs are important for discussing the hole-transfer abilities through MAPbI$_3$ surfaces to HTMs. The HOMOs of the vacant terminations are distributed inside the surface slab, whereas those of the flat terminations concentrate at the surface layers (insets of Figure 5.5). The surface-localized HOMOs in the latter

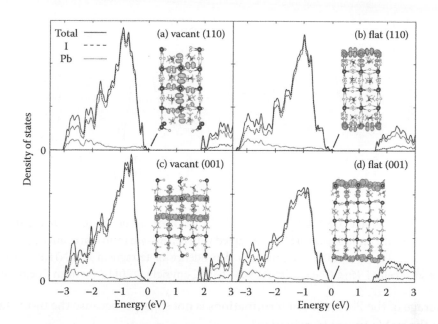

FIGURE 5.5 PDOSs of selected MAPbI$_3$ surfaces. Solid lines represent total DOSs. Dashed and dotted lines represent PDOSs of I (5$s$+5$p$) and Pb (5$d$+6$s$+6$p$), respectively. Energy origins are set to the levels of the HOMOs. The corresponding charge distributions of the HOMOs are also shown. (Adapted with permission from Haruyama, J., K. Sodeyama, L. Han, and Y. Tateyama. 2014. Termination dependence of tetragonal CH$_3$NH$_3$PbI$_3$ surfaces for perovskite solar cells. *J. Phys. Chem. Lett.* 5; 2903–2909. Copyright © 2014 American Chemical Society.)

case permit smooth charge transition at the interface. In addition, the energy levels of the surface states are located just above the VBM of the bulk MAPbI$_3$, indicating that energy loss associated with charge transfer through the surface states is expected to be very small or negligible. Conversely, the HOMOs in the bulk in the former case are not beneficial to charge transfer because of the small electronic coupling between the HOMOs and HTM.

Because a number of studies have reported the splitting of degenerate conduction bands consisting of Pb-6$p$ orbitals by inclusion of SOC (Even et al. 2013; Feng and Xiao 2014; Giorgi et al. 2013; Umari et al. 2014), the surface energy gaps can be expected to be decreased by the SOC effect. The PDOSs and the lowest unoccupied molecular orbitals (LUMOs) of the PbI$_2$-rich-flat terminations on the (110) and (001) surfaces with SOC were calculated (Figure 5.6). SOC correction results in the lower energy levels of the LUMOs by about 1 eV on both surfaces, whereas the valence bands hardly change. Compared with the charge distributions of the LUMOs without SOC, those with SOC correction have little difference; they are spread over the entire slab. Because the large SOC effects are due to the Pb-6$p$ character of the conduction bands, we can expect energy shifts for all of the surfaces.

Consequently, the surface state of the flat termination is more advantageous for the hole transfer to HTMs. The calculated phase diagrams predict that the formation of the stable-vacant terminations usually occur by one- or two-step methods. However, the PbI$_2$-rich-flat surfaces can be prepared by the two-step method because PbI$_2$-rich conditions may be realized at this synthesized procedure. Unfortunately, thermodynamic analysis indicates

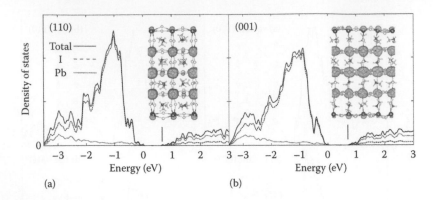

FIGURE 5.6 PDOSs and LUMOs of the $PbI_2$-rich-flat terminations on the (a) (110) and (b) (001) surfaces calculated with the inclusion of SOC. (Adapted with permission from Haruyama, J., K. Sodeyama, L. Han, and Y. Tateyama. 2014. Termination dependence of tetragonal $CH_3NH_3PbI_3$ surfaces for perovskite solar cells. *J. Phys. Chem. Lett.* 5; 2903–2909. Copyright © 2014 American Chemical Society.)

that coverage of the $PbI_2$-rich-flat terminations is not so large, because the most stable surfaces are the stable-vacant terminations.

Solar cell efficiency is known to be sensitive to material synthesis and processing conditions. A recent study showed that adsorption of a Pb dimer on a $PbI_2$-terminated (001) surface causes the formation of electron trap states (Dong et al. 2015b); and another study showed evidence for the presence of hole traps on $MAPbI_3$ surfaces (Wu et al. 2015), and these traps can be detrimental to PSC performance. In addition, there have been many reports about the instability of $MAPbI_3$/HTM interfaces, and an HTM-free layer or an alternative spiro-OMeTAD layer can improve cell performance (Liu et al. 2014; Mei et al. 2014). Recently, Mosconi et al. reported that $PbI_2$-rich-flat terminations act as a protective layer against degradation by water, but $PbI_2$ defects triggered rapid dissolution (Mosconi et al. 2015). Therefore, formation of a larger area of $PbI_2$-rich-flat surface significantly contributes to improvements in all the measures of PSC performance.

### 5.3.3 Atomic Structures and Electronic States at $TiO_2/MAPbI_3$ and $Al_2O_3/MAPbI_3$ Interfaces

In addition to examining stable surfaces consisting of exposed cut planes of $PbI_6$ octahedrons, we examined the relaxed structures and electronic states of interfaces of $MAPbI_3$ with anatase-$TiO_2(001)$ and with α-$Al_2O_3(0001)$, as model mesoporous scaffolds (Figure 5.7). Note that MA is not present at these interfaces; that is, flat terminations of $MAPbI_3$ were selected and almost stoichiometric conditions were retained. The atoms on the $MAPbI_3$ surfaces were found to be slightly deformed to fit the scaffolds. Strong Pb–O interactions at the $Al_2O_3/MAPbI_3$ interface likely attracted Pb atoms to the $Al_2O_3(0001)$ surface. As a result, the $MAPbI_3$ surface could not retain its flat structure or its outermost $PbI_5$ polyhedrons.

The binding energies per unit area for the $TiO_2/MAPbI_3$ and $Al_2O_3/MAPbI_3$ interfaces were calculated to be 6.6 and 5.8 eV $nm^{-2}$, respectively. Mosconi et al. previously reported a binding energy of 4.2 eV $nm^{-2}$ for an anatase-$TiO_2(101)$/tetragonal-$MAPbI_3(110)$ interface

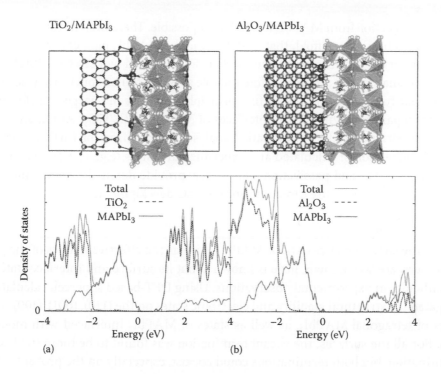

FIGURE 5.7 Relaxed structures (upper panels) and calculated PDOSs (lower panels) of (a) $TiO_2$/ $MAPbI_3$ and (b) $Al_2O_3$/$MAPbI_3$ interfaces. Solid, dashed, and dotted lines represent PDOSs of total atoms, Ti(Al)+O, and MA+Pb+I, respectively. Energy origins are set to the top of the valence band. Only the results obtained with the vdW functional are depicted.

(Mosconi et al. 2014). The binding energy of our constructed interfaces was larger than that value. The difference might be attributed to a lattice mismatch between $TiO_2(101)$ and $MAPbI_3(110)$ or to different terminations for the $TiO_2(101)$, $TiO_2(001)$, MA-terminated, and flat $MAPbI_3(110)$ surfaces. $TiO_2$ nanosheets exposing (001) facets have been experimentally shown to have higher PCEs than those exposing (101) facets (Etgar et al. 2012), implying that the small lattice distortion of the $MAPbI_3$ structure to match with $TiO_2(001)$ improves cell efficiency. However, it is difficult to draw conclusions about the efficiency of $TiO_2$ surfaces on the basis of the binding energies only. Trap states, electron injection times, and recombination rates at each $TiO_2$ interface should be investigated.

The PDOSs of the two interfaces are also shown in Figure 5.7. In spite of the large structural breaking and deformation at the interfaces, associated midgap states did not appear. Consequently, simple band alignments of the two semiconductors were obtained. As expected, the VBMs of $TiO_2$ and $Al_2O_3$ were much lower (ca. 2 eV) than the VBM of $MAPbI_3$, whereas the CBMs of $TiO_2$ and $Al_2O_3$ were lower and higher, respectively, than the CBM of $MAPbI_3$. Under our interfacial conditions, there were sufficient band gaps, consistent with previously reported energy diagrams obtained experimentally (Heo et al. 2013; Kim et al. 2012; Lindblad et al. 2014; Xing et al. 2013). Note that the $Al_2O_3$ states located in the 3–4 eV region were (0001) surface states, whereas the bulk CBM was not visible in this range because of the large band gap of $Al_2O_3$. The calculated PDOSs at the $Al_2O_3$/$MAPbI_3$ interface confirm that electron injection from $MAPbI_3$ to $Al_2O_3$ is impossible. Therefore, the role of $Al_2O_3$ might be limited to acting as a scaffold, that is, stabilizing the $MAPbI_3$ nanocrystals. These results seem to support the surface percolation mechanism discussed by Lee et al. (2012), but more sophisticated analyses are required before the role of $Al_2O_3$ can be conclusively determined.

In the near future, we will report first principles analyses of charge-transfer processes and charge separation dynamics at interfaces of $MAPbI_3$ with $TiO_2$, $Al_2O_3$, and the other carrier transport materials. Such detailed analyses are required for further development of PSCs. Comparison of calculated and experimental results (e.g., charge injection times, recombination rates, and passivation effects) may provide detailed insights into interface roles and may point to future directions for the PSC development.

## 5.4 CONCLUSIONS

Although the characteristics of bulk $MAPbI_3$, such as the effective masses of the photoexcited carriers, are well known, little is known about its surface and interface states owing to the difficulty in experimental observations. Using DFT-based supercell calculations, we investigated the structural stability and electronic states on the (110), (001), (100), and (101) surfaces of tetragonal $MAPbI_3$, as well as states of $MAPbI_3$ interfaced with model oxide anodes. For all the surfaces, the vacant termination was found to be more stable than the flat termination, but both terminations could coexist, especially on the probable (110) and (001) surfaces. The electronic states of the two types of terminations on the two surfaces were almost the same as the states of bulk $MAPbI_3$ without midgap levels. The absence of deep midgap states well explains the long carrier diffusion lengths. Furthermore, flat terminations on the (110) and (001) surfaces can facilitate efficient charge transfer to the neighboring HTM. The stable interfacial matching between a $TiO_2$ anode (or an $Al_2O_3$ scaffold) and light-absorptive $MAPbI_3$ was also discussed. The atomic structures and band alignments at the interfaces were determined by vdW-level DFT calculations. This information about the surfaces and interfaces will provide a basic understanding of charge dynamics, and allow design of an optimized interface for maximizing PSC performance.

## REFERENCES

Agiorgousis, M. L., Y.-Y. Sun, H. Zeng, and S. Zhang. 2014. Strong covalency-induced recombination centers in perovskite solar cell material $CH_3NH_3PbI_3$. *J. Am. Chem. Soc.* 136; 14570–14575.

Azpiroz, J. M., E. Mosconi, J. Bisquert, and F. De Angelis. 2015. Defect migration in methylammonium lead iodide and its role in perovskite solar cell operation. *Energy Environ. Sci.* 8; 2118–2127.

Bachelet, G. B., and M. Schlüter. 1982. Relativistic norm-conserving pseudopotentials. *Phys. Rev. B* 25; 2103–2108.

Baikie, T., Y. Fang, J. M. Kadro, M. Schreyer, F. Wei, S. G. Mhaisalkar, M. Grätzel, and T. J. White. 2013. Synthesis and crystal chemistry of the hybrid perovskite $(CH_3NH_3)PbI_3$ for solid-state sensitised solar cell applications. *J. Mater. Chem. A* 1; 5628–5641.

Buin, A., P. Pietsch, J. Xu, O. Voznyy, A. H. Ip, R. Comin, and E. H. Sargent. 2014. Materials processing routes to trap-free halide perovskites. *Nano Lett.* 14; 6281–6286.

Burschka, J., N. Pellet, S.-J. Moon, R. Humphry-Baker, P. Gao, Md. K. Nazeeruddin, and M. Grätzel.2013. Sequential deposition as a route to high-performance perovskite-sensitised solar cells. *Nature* (London) 499; 316–319.

Christians, J. A., J. S. Manser, and P. V. Kamat. 2015. Multifaceted excited state of $CH_3NH_3PbI_3$. Charge separation, recombination, and trapping. *J. Phys. Chem. Lett.* 6; 2086–2095.

Colella, S., E. Mosconi, P. Fedeli, A. Listorti, F. Gazza, F. Orlandi, P. Ferro, T. Besagni, A. Rizzo, G. Calestani, G. Gigli, F. De Angelis, and R. Mosca. 2013. $MAPbI_{3-x}Cl_x$ mixed halide perovskite for hybrid solar cells: The role of chloride as dopant on the transport and structural properties. *Chem. Mater.* 25; 4613–4618.

Dal Corso, A., and A. Mosca Conte. 2005. Spin-orbit coupling with ultrasoft pseudopotentials: Application to Au and Pt. *Phys. Rev. B* 71; 115106.

De Angelis, F. 2014. Modeling materials and processes in hybrid/organic photovoltaics: From dye-sensitized to perovskite solar cells. *Acc. Chem. Res.* 47; 3349–3360.

De Wolf, S., J. Holovsky, S.-J. Moon, P. Loeper, B. Niesen, M. Ledinsky, F.-J. Haug, J.-H. Yum, and C. Ballif. 2014. Organometallic halide perovskites: Sharp optical absorption edge and its relation to photovoltaic performance. *J. Phys. Chem Lett.* 5; 1035–1039.

Dong, Q., Y. Fang, Y. Shao, P. Mulligan, J. Qiu, L. Cao, and J. Huang. 2015a. Electron-hole diffusion lengths > 175 mm in solution-grown $CH_3NH_3PbI_3$ single crystals. *Science* 347; 967–970.

Dong, R., Y. Fang, J. Chae, J. Dai, Z. Xiao, Q. Dong, Y. Yuan, A. Centrone, X. C. Zeng, and J. Huang. 2015b. High-grain and low-driving-voltage photodetectors based on organolead triiodide perovskites. *Adv. Mater.* 27; 1912–1918.

Eames, C., J. M. Frost, P. R. F. Barnes, B. C. O'Regan, A. Walsh, and M. S. Islam. 2015. Ionic transport in hybrid lead iodide perovskite solar cells. *Nat. Commun.* 6; 7497.

Eperon, G. E., S. D. Stranks, C. Menelaou, M. B. Johnston, L. M. Herz, and H. J. Snaith. 2014. Formamidinium lead trihalide: A broadly tunable perovskite for efficient planar heterojunction solar cells. *Energy Environ. Sci.* 7; 982–988.

Etgar, L., P. Gao, Z. Xue, Q. Peng, A. K. Chandiran, B. Liu, Md. K. Nazeeruddin, and M. Grätzel. 2012. Mesoscopic $CH_3NH_3PbI_3/TiO_2$ heterojunction solar cells. *J. Am. Chem. Soc.* 134; 17396–17399.

Even, J., L. Pedesseau, J.-M. Jancu, and C. Katan. 2013. Importance of spin–orbit coupling in hybrid organic/inorganic perovskites for photovoltaic applications. *J. Phys. Chem. Lett.* 4; 2999–3005.

Feng, J., and B. Xiao. 2014. Crystal structures, optical properties, and effective mass tensors of $CH_3NH_3PbX_3$ (X = I and Br) phases predicted from HSE06. *J. Phys. Chem. Lett.* 5; 1278–1282.

Feng, H.-J., T. R. Paudel, E. Y. Tsymbal, and X. C. Zeng. 2015. Tunable optical properties and charge separation in $CH_3NH_3Sn_xPb_{1-x}I_3/TiO_2$–based planar perovskites cells. *J. Am. Chem. Soc.* 137; 8227–8236.

Filip, M. R., G. E. Eperon, H. J. Snaith, and F. Giustino. 2014. Steric engineering of metal-halide perovskites with tunable optical band gaps. *Nat. Commun.* 5; 5757–5759.

Giannozzi, P., S. Baroni, N. Bonini, M. Calandra, R. Car, C. Cavazzoni, D. Ceresoli, G. L. Chiarotti, M. Cococcioni, I. Dabo, A. Dal Corso et al. 2009. Ab-initio study of the effects induced by the electron-phonon scattering in carbon based nanostructures *J. Phys.: Condens. Matter* 21, 395502–395519.

Giorgi, G., J.-I. Fujisawa, H. Segawa, and K. Yamashita. 2013. Small photocarrier effective masses featuring ambipolar transport in methylammonium lead iodide perovskite: A density functional analysis. *J. Phys. Chem. Lett.* 4; 4213–4216.

Giorgi, G., J.-I. Fujisawa, H. Segawa, and K. Yamashita. 2014. Cation role in structural and electronic properties of 3D organic–inorganic halide perovskites: A DFT analysis. *J. Phys. Chem. C* 118; 12176–12183.

Giorgi, G., J.-I. Fujisawa, H. Segawa, and K. Yamashita. 2015. Organic–inorganic hybrid lead iodide perovskite featuring zero dipole moment guanidinum cations: A theoretical analysis. *J. Phys. Chem. C* 119; 4694–4701.

Grätzel, M. 2014. The light and shade of perovskite solar cells. *Nat. Mater.* 13; 838–842.

Hamada, I. 2014. van der Waals density functional made accurate. *Phys. Rev. B* 89; 121103(R).

Haruyama, J., K. Sodeyama, L. Han, and Y. Tateyama. 2014. Termination dependence of tetragonal $CH_3NH_3PbI_3$ surfaces for perovskite solar cells. *J. Phys. Chem. Lett.* 5; 2903–2909.

Haruyama, J., K. Sodeyama, L. Han, and Y. Tateyama. 2015. First-principles study of ion diffusion in perovskite solar cell sensitizers. *J. Am. Chem. Soc.* 137; 10048–10051.

Heo, J. H., S. H. Im, J. H. Noh, T. N. Mandal, C.-S. Lim, J. A. Chang, Y. H. Lee, H.-j. Kim, A. Sarkar, Md. K. Nazeeruddin, M. Grätzel, and S. I. Seok. 2013. Efficient inorganic–organic hybrid heterojunction solar cells containing perovskite compound and polymeric hole conductors. *Nat. Photon.* 7; 486–491.

Jeon, N. J., J. H. Noh, W. S. Yang, Y. C. Kim, S. Ryu, J. Seo, and S. I. Seok. 2015. Compositional engineering of perovskite materials for high-performance solar cells. *Nature* 517; 476–480.

Kawamura, Y., H. Mashiyama, and K. Hasebe. 2002. Structural study on cubic–tetragonal transition of $CH_3NH_3PbI_3$. *J. Phys. Soc. Jpn.* 71; 1694–1697.

Kim, H. S., C. R. Lee, J. H. Im, K. B. Lee, T. Moehl, A. Marchioro, S. J. Moon, R. Humphry-Baker, J. H. Yum, J. E. Moser, M. Grätzel, and N. G. Park. 2012. Lead iodide perovskite sensitized all-solid state submicron thin film mesoscopic solar cell with efficiency exceeding 9%. *Sci. Rep.* 2; 591.

Kim, J., S.-H. Lee, J. H. Lee, and K.-H. Hong. 2014. The role of intrinsic defects in methylammonium lead iodide perovskite. *J. Phys. Chem. Lett.* 5; 1312–1317.

Kim, J., S.-C. Lee, S.-H. Lee, and K.-H. Hong. 2015. Importance of orbital interactions in determining electronic band structures of organo-lead iodide. *J. Phys. Chem. C* 119; 4627–4634.

Kleinman, L. 1980. Relativistic norm-conserving pseudopotential. *Phys. Rev. B* 21; 2630–2631.

Kojima, A., K. Teshima, Y. Shirai, and T. Miyasaka. 2009. Organometal halide perovskites as visible-light sensitizers for photovoltaic cells. *J. Am. Chem. Soc.* 131; 6050–6051.

Lee, M. M., J. Teuscher, T. Miyasaka, T. N. Murakami, and H. J. Snaith. 2012. Efficient hybrid solar cells based on meso-superstructured organometal halide perovskites. *Science* 338; 643–647.

Lindblad, R., D. Bi, B.-W. Park, J. Oscarsson, M. Gorgoi, H. Siegbahn, M. Odelius, E. M. J. Johansson, and H. Rensmo. 2014. Electronic structure of $TiO_2/CH_3NH_3PbI_3$ perovskite solar cell interfaces. *J. Phys. Chem. Lett.* 5; 648–653.

Liu, M., M. B. Johnston, and H. J. Snaith. 2013. Efficient planar heterojunction perovskite solar cells by vapour deposition. *Nature* 501; 395–398.

Liu, J., Y. Wu, C. Qin, X. Yang, T. Yasuda, A. Islam, K. Zhang, W. Peng, W. Chen, and L. Han. 2014. A dopant-free hole-transporting material for efficient and stable perovskite solar cells. *Energy Environ. Sci.* 7; 2963–2967.

Louie, S. G., S. Froyen, and M. L. Cohen. 1982. Nonlinear ionic pseudopotentials in spin-density-functional calculations. *Phys. Rev. B* 26; 1738–1742.

Manser, J. S., and P. V. Kamat. 2014. Band filling with free charge carriers in organometal halide perovskites. *Nat. Photonics* 8; 737–743.

Marchioro, A., J. Teuscher, D. Friedrich, Marinus Kunst, R. van de Krol, T. Moehl, M. Grätzel, and J.-E. Moser. 2014. Unravelling the mechanism of photoinduced charge transfer processes in lead iodide perovskite solar cells. *Nat. Photonics* 8; 250–255.

Mei, A., X. Li, L. Liu, Z. Ku, T. Liu, Y. Rong, M. Xu, M. Hu, J. Chen, Y. Yang, M. Grätzel, and H. Han. 2014. A hole-conductor-free, fully printable mesoscopic perovskite solar cell with high stability. *Science* 345; 295–298.

Momma, K., and F. Izumi. 2008. *VESTA*: A three-dimensional visualization system for electronic and structural analysis. *J. Appl. Crystallogr.* 41; 653–658.

Mosconi, E., A. Amat, M. K. Nazeeruddin, M. Grätzel, and F. De Angelis. 2013. First-principles modeling of mixed halide organometal perovskites for photovoltaic applications. *J. Phys. Chem. C* 117; 13902–13913.

Mosconi, E., J. M. Azpiroz, and F. De Angelis. 2015. *Ab initio* molecular dynamics simulations of methylammonium lead iodide perovskite degradation by water. *Chem. Mater.* 27; 4885–4892.

Mosconi, E., E. Ronca, and F. De Angelis. 2014. First-principles investigation of the $TiO_2$/organohalide perovskites interface: The role of interfacial chlorine. *J. Phys. Chem. Lett.* 5; 2619–2625.

Nemnes, G. A., C. Goehry, T. L. Mitran, Adela Nicolaev, L. Ion, S. Antohe, N. Plugaru, and A. Manolescu. 2015. Band alignment and charge transfer in rutile-$TiO_2$/$CH_3NH_3PbI_{3-x}Cl_x$ interfaces. *Phys. Chem. Chem. Phys.* 17; 30417–30423.

NREL. 2016. Best research-cell efficiencies, National Renewable Energy Laboratory. http://www.nrel.gov/ncpv/images/efficiency_chart.jpg (Accessed February, 2016).

Oga, H., A. Saeki, Y. Ogomi, S. Hayase, and S. Seki. 2014. Improved understanding of the electronic and energetic landscapes of perovskite solar cells: High local charge carrier mobility, reduced recombination, and extremely shallow traps. *J. Am. Chem. Soc.* 136; 13818–13825.

Ogomi, Y., K. Kukihara, S. Qing, T. Toyoda, K. Yoshino, S. Pandey, H. Momose, and S. Hayase. 2014. Control of charge dynamics through a charge-separation interface for all-solid perovskite-sensitized solar cells. *Chem. Phys. Chem.* 15; 1062–1069.

Park, N.-G. 2013. Organometal perovskite light absorbers toward a 20% efficiency low-cost solid-state mesoscopic solar cell. *J. Phys. Chem. Lett.* 4; 2423–2429.

Perdew, J. P., K. Burke, and M. Ernzerhof. 1996. Generalized gradient approximation made simple. *Phys. Rev. Lett.* 77; 3865–3868.

Ponseca, C. S., Jr., T. J. Savenije, M. Abdellah, K. Zheng, A. Yartsev, T. Pascher, T. Harlang, P. Chabera, T. Pullerits, A. Stepanov, J.-P. Wolf, and V. Sundström 2014. Organometal halide perovskite solar cell materials rationalized: Ultrafast charge generation, high and microsecond-long balanced mobilities, and slow recombination. *J. Am. Chem. Soc.* 136; 5189–5192.

Qian, G.-X., R. M. Martin, and D. J. Chadi. 1988. Stoichiometry and surface reconstruction: An *ab initio* study of GaAs(100) surfaces. *Phys. Rev. Lett.* 60; 1962–1965.

Rappe, A. M., K. M. Rabe, E. Kaxiras, and J. D. Joannopoulos. 1990. Optimized pseudopotentials. *Phys. Rev. B* 41; 1227–1230.

Roiati, V., E. Mosconi, A. Listorti, S. Colella, G. Gigli, and F. De Angelis. 2014. Stark effect in perovskite/$TiO_2$ solar cells: Evidence of local interfacial order. *Nano Lett.* 14; 2168–2174.

Savenije, T. J., C. S. Ponseca Jr., L. Kunneman et al. 2014. Thermally activated exciton dissociation and recombination control the carrier dynamics in organolead halide perovskite. *J. Phys. Chem. Lett.* 5; 2189–2194.

Snaith, H. J. 2013. Perovskites: The emergence of a new era for low-cost, high-efficiency solar cells. *J. Phys. Chem. Lett.* 4; 3623–3630.

Starr, D. E., G. Sadoughi, E. Handick, R. G. Wilks, J. H. Alsmeier, L. Köhler, M. Gorgoi, H. J. Snaith, and M. Bär 2015. Direct observation of an inhomogeneous chlorine distribution in $CH_3NH_3PbI_{3-x}Cl_x$ layers: Surface depletion and interface enrichment. *Energy Environ. Sci.* 8; 1609–1615.

Stoumpos, C. C., C. D. Malliakas, and M. G. Kanatzidis. 2013. Semiconducting tin and lead iodide perovskites with organic cations: Phase transitions, high mobilities, and near-infrared photoluminescent properties. *Inorg. Chem.* 52; 9019–9038.

Stranks, S. D., G. E. Eperon, G. Grancini, C. Menelaou, M. J. P. Alcocer, T. Leijtens, L. M. Herz, A. Petrozza, and H. J. Snaith. 2013. Electron-hole diffusion lengths exceeding 1 micrometer in an organometal trihalide perovskite absorber. *Science* 342; 341–344.

Umari, P., E. Mosconi, and F. De Angelis. 2014. Relativistic GW calculations on $CH_3NH_3PbI_3$ and $CH_3NH_3SnI_3$ perovskites for solar cell applications. *Sci. Rep.* 4; 4467.

Umebayashi, T., K. Asai, T. Kondo, and A. Nakao. 2003. Electronic structures of lead iodide based low-dimensional crystals. *Phys. Rev. B* 67; 155405.

Vanderbilt, D. 1990. Soft self-consistent pseudopotentials in a generalized eigenvalue formalism. *Phys. Rev. B* 41; 7892–7895.

Wu, X., M. T. Trinh, D. Niesner, H. Zhu, Z. Norman, J. S. Owen, O. Yaffe, B. J. Kudisch, and X.-Y. Zhu. 2015. Trap states in lead iodide perovskites. *J. Am. Chem. Soc.* 137; 2089–2096.

Xing, G., N. Mathews, S. Sun, S. S. Lim, Y. M. Lam, M. Grätzel, S. Mhaisalkar, and T. C. Sum. 2013. Long-range balanced electron- and hole-transport lengths in organic-inorganic $CH_3NH_3PbI_3$. *Science* 342; 344–347.

Yamada, Y., T. Nakamura, M. Endo, A. Wakamiya, and Y. Kanemitsu. 2014. Photocarrier recombination dynamics in perovskite CH₃NH₃PbI₃ for solar cell applications. *J. Am. Chem. Soc.* 136; 11610–11613.

Yin, W.-J., T. Shi, and Y. Yan. 2014. Unique properties of halide perovskites as possible origins of the superior solar cell performance. *Adv. Mater.* 26; 4653–4658.

Yin, W.-J., T. Shi, and Y. Yan. 2014. Unusual defect physics in CH₃NH₃PbI₃ perovskite solar cell absorber. *Appl. Phys. Lett.* 104; 063903.

Yin, W. J., J. H. Yang, J. Kang, Y. Yan, and S.-H. Wei. 2014. Halide perovskite materials for solar cells: A theoretical review. *J. Mater. Chem. A* 3; 8926–8942.

# CHAPTER 6

# Computational High-Throughput Screening for Solar Energy Materials

Ivano E. Castelli, Kristian S. Thygesen,
and Karsten W. Jacobsen

## CONTENTS

## 6.1 INTRODUCTION

The design and development of new materials is required to meet the challenge of sustainable and environmental friendly energy production and storage. Computer simulations in the form of atomic-scale electronic structure calculations can be expected to play an important role in this, as demonstrated since the early part of this century. Most of such calculations are carried out in the framework of Density Functional Theory (DFT) (Hohenberg and Kohn 1964; Kohn and Sham 1965), which demonstrates a good compromise between computational speed and accuracy of the calculations. The ever-increasing computational power makes it possible today to study fairly complex structures with several hundreds or even thousands of atoms. A few examples of materials design studies include the search for novel materials for carbon capture and storage (Lin et al. 2012), batteries (Ceder et al. 1998), stable binary and ternary alloys (Curtarolo et al. 2012), transparent conductors (Hautier et al. 2013), dye-sensitized solar cells (Ørnsø et al. 2013), photovoltaics (d'Avezac et al. 2012; Hachmann et al. 2011), and water splitting materials (Castelli et al. 2012a, 2012b; Wu et al. 2013).

The calculations usually address only a small part of the actual design problem and are nowhere near direct simulations of the functioning material. An important issue is therefore to determine a few simple parameters, so-called descriptors or design metrics, that describe key properties for the material function and that at the same time can be computed at the atomistic level (Curtarolo et al. 2013). For example, the formation enthalpy is a possible descriptor for the stability of a compound, the band gap for its absorption properties, the binding energies of reaction intermediates for the catalytic activity (Rossmeisl et al. 2005), and so on.

In this chapter, we first present some descriptors useful for identifying novel semiconductors for light harvesting in a photovoltaic or photoelectrochemical cell, and afterwards review the application of a screening procedure to more than 20,000 inorganic perovskites in the cubic and layered phases. Furthermore, we study trends and propose a handful of materials for further investigation. Lastly, we report the trends for a smaller set of 300 hybrid halide perovskites. Our *ab initio* quantum mechanics simulations have been performed in the framework of DFT, using the GPAW code (Enkovaara et al. 2010; Mortensen et al. 2005) and the Atomic Simulation Environment (ASE) (Larsen et al. 2017). All our results are collected in the open source and open data Computational Materials Repository (CMR) database (CMR 2016).

## 6.2 IDENTIFICATION OF DESCRIPTORS

To harvest and convert solar light into electrical energy in photovoltaic (PV) cells or into chemical energy using photoelectrochemical (PEC) reactions require as the first step the absorption of light-creating electron–hole pairs. The absorption is performed by a semiconducting material that has to be optimized for efficient absorption and that allows for further transformation of the energy in the electron–hole pairs.

A list of some—but certainly not all—properties that a semiconducting material has to fulfill to be used in a PEC cell might include (1) chemical and structural stability, (2) good light absorption, (3) photogenerated charges at the correct potentials and with good mobility, and (4) low cost and nontoxicity. Stability, good light absorption, and low cost are also among the desired properties for a material to be used in a PV device. The PEC technology is more challenging than the one behind PV cells because the electron and hole transfer should induce chemical reactions. However, the benefit is that high-energy storage capacity is available in the chemical bonds of the produced fuels.

There are many other issues that have to be dealt with in a working PEC device. One of the most important ones is the catalysis of the chemical processes happening at the material surfaces. The light-absorbing materials should therefore either have good catalytic properties themselves or be able to interface both structurally and electronically to efficient catalysts (Walter et al. 2010). However, none of these issues shall be addressed here.

In the following we discuss the computational descriptors that are used in the screening studies. The descriptor for the stability is the heat of formation for the material or more specifically the calculated total energy difference between the material and competing structures and materials at zero temperature. The simplest measure is the standard heat of formation, which is calculated as the energy difference between the material at hand and its components in their standard states. However, we also go beyond this by considering phase separation not only into elemental materials but also into a pool of reference materials of varying compositions and structures. The pool includes more than 2000 structures from the ICSD (2016) and the Materials Project database (MP 2016). If the energy of the candidate material is above the calculated convex hull (plus a metastability region of 0.2 eV/atom) that defines the stability frontier, the material under investigation is considered unstable. The total energies are calculated using the revised Perdew, Burke, and Ernzerhof functional (RPBE) as the exchange-correlation functional in the generalized gradient approximation (GGA) (Hammer et al. 1999).

If the material is in contact with water, such as, for example, in a PEC device without protection layers of the light absorber, the stability against dissolution should also be considered. The stability in water can be calculated by means of Pourbaix diagrams. A Pourbaix diagram is a pH-potential phase diagram that contains both solid and dissolved species in the pool of reference systems. In our case, the energy of the solid phases are calculated by DFT as noted previously and the dissolution energies are obtained from experimental results (Castelli et al. 2014a; Persson et al. 2012). Although Pourbaix diagrams give information about the bulk thermodynamic stability in water of a compound, they do not provide details about the reaction kinetics and the possible role of surfaces and their passivation.

The absorption properties of a material can be described, to a first approximation, by the band gap, assuming that photons with energies above the gap are absorbed while lower energy photons are not. Within DFT, the simplest estimate of the band gap would be the calculated gap in the single-particle Kohn–Sham spectrum. However, this leads to severe underestimation of the gap because of the so-called derivative discontinuity (Godby et al. 1986) and the self-interaction error for semilocal exchange-correlation functionals (Perdew and Zunger 1981). Several methods have been proposed to take these effects into account, usually at a significantly higher computational cost. Hybrid functionals such as PBE0 or HSE06, which include a fraction of exact exchange, or many-body methods such as the GW approximation improve the estimation of the band gaps, but are considerably more computationally expensive than the semilocal functionals.

As a compromise we calculate the band gaps using an improved description of the so-called Gritsenko, van Leeuwen, van Lenthe, and Baerends (GLLB) functional (Gritsenko et al. 1995), called GLLB-SC (Kuisma et al. 2010), particularly suited for solids. This functional is an additional approximation to the Krieger–Li–Iafrate (KLI) approximation (Krieger et al. 1992a, 1992b) to the exact exchange optimized effective potential (EXX-OEP) (Talman and Shadwick 1976). As a rather unique feature, it includes an explicit evaluation of the derivative discontinuity.

We note that the gap discussed in the preceding text is the so-called quasi-particle (QP) or fundamental gap, which is the difference between the first ionization potentials (IP) and the electron affinity (EA). It can be directly measured by photoemission and inverse photoemission experiments and within DFT it is obtained computationally by adding the derivative discontinuity, $\Delta_{xc}$, to the Kohn–Sham (KS) gap: $E_{gap}^{QP} = IP - EA = E_{gap}^{KS} + \Delta_{xc}$. QP gaps do not include excitonic effects which can result in a lower optical band gap. However, we shall not go any further into this issue here.

The GLLB-SC calculated band gaps show an error of the order 0.5 eV when tested against non-self-consistent $G_0W_0$ values and experiments for single metal oxides (Castelli et al. 2012a), binary semiconductors (Hüser et al. 2013), and for oxo-perovskites with a band gap in the visible range (Castelli et al. 2013a). More recently, we have benchmarked the GLLB-SC against different levels of the GW approximation ($G_0W_0$, $GW_0$, and GW) (Shishkin and Kresse 2007) and the hybrid HSE06 (Heyd et al. 2003) for a set of 20 ternary and quaternary semiconductors (Castelli et al. 2015a). Among the considered methods, the GLLB-SC gives the best approximation to the eigenvalue-self-consistent GW, with a mean relative error of around 15%. We therefore expect that the GLLB-SC is accurate enough for high-throughput calculations because it provides reliable results at only a slight increase in computational cost relative to standard GGA calculations even for large crystal structures.

An appropriate band gap is certainly a requirement for achieving efficient visible light absorption; however, it does not take into account the actual strength of the absorption at different photon energies. The absorption might, for example, be limited by symmetry or by lack of spatial overlap between electronic states in the valence and conduction bands (VB and CB, respectively). Therefore in some cases we perform explicit calculation of the absorption spectrum within time-dependent DFT where both the character of the band gap (direct/indirect) and the strengths of the dipole-transition matrix elements are included (Yan et al. 2011). We note that in screening studies by Yu and Zunger (Yu and Zunger 2012) the band gap characters and the absorption strengths have been combined into a single descriptor or metric.

In a PEC device, the photogenerated charges have to be at the right potential with respect to the redox levels of water to be able to induce the water oxidation and the proton reduction reactions. We use a simple empirical approach (Butler and Ginley 1978; Xu and Schoonen 2000) where the positions of the electron and hole levels relative to vacuum for a material with $N$ atoms in the unit cell is estimated by first calculating the center of the gap as the geometrical average of the electronegativities, $X_i$, of the constituent elements, $i$, with the Mulliken's scale (Putz et al. 2005). The CB and VB edges with respect to the vacuum level, $E_{VB}$ and $E_{CB}$, are then obtained by adding and subtracting half of the gap, respectively: $E_{CB,VB} = (\Pi_i X_i)^{1/N} \pm E_{gap}/2$. More sophisticated models, which include the investigation of surface properties and of realistic electrochemical environments, can be used also but at increased computational cost (Stevanović et al. 2014; Wu et al. 2011).

Low cost and nontoxicity of a material are also desirable properties and can at least to some extent be estimated based on the constituent elements. The cost can, for example, further be related to the abundance of the element (Vesborg and Jaramillo 2012). However, in the following screening studies we do not exclude elements based on their abundance or cost.

## 6.2.1 Inorganic Perovskites

In the following we consider compounds in the cubic perovskite structure with the formula $ABO_3$ (space group $Pm\bar{3}m$). We select this structure because of the generally high stability and because of the large variety of properties and applications of materials with this structure (Ishihara 2005). The unit cell of the cubic perovskite is formed of five atoms (two inequivalent metals, a large 12-coordinated cation [A-ion], and a small 6-coordinated cation [B-ion] and three nonmetals) and it can accommodate most of the elements in the periodic table.

First we consider all the possible cubic perovskites obtained by combining 52 nonradioactive metals of the periodic table in the A and B sites. As nonmetals we start out with oxygen, because of the high stability of oxides.

Figure 6.1 shows the formation energies per atom and the band gaps for the 2704 oxides. Each square that corresponds to a particular oxide containing two metals is divided into

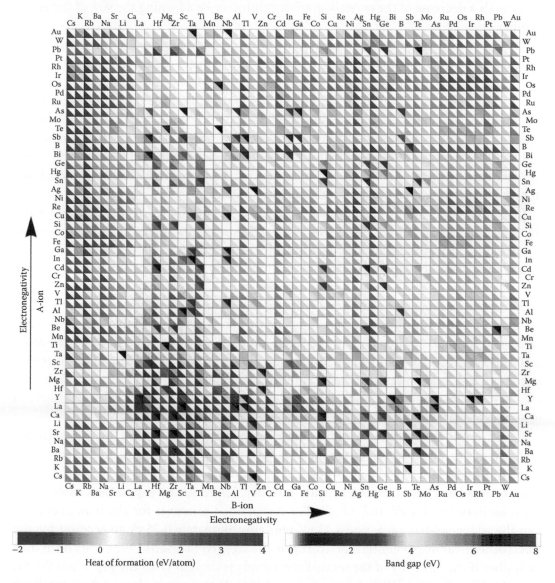

FIGURE 6.1 Calculated heats of formation per atom and band gaps of perovskite binary metal oxides. In each square the formation energy (lower, left triangle) and the band gap (upper, right triangle) of an oxide are shown. The interesting combinations for water splitting are indicated with red squares. The chemicals are sorted for increasing electronegativity. (Castelli, I. E., T. Olsen, S. Datta, D. D. Landis, S. Dahl, K. S. Thygesen, and K. W. Jacobsen. 2012a. Computational screening of perovskite metal oxides for optimal solar light capture. *Energy Environ. Sci.* 5; 5814–5819. Reproduced by permission of The Royal Society of Chemistry.)

two triangles where the lower left one indicates the stability (from red to blue with decreasing stability) and the upper right one, the band gap. It is apparent from the figure that there is a clear trend in the heat of formation where low electronegativity of the constituent metal atoms favors stability. The band gap does not show such a clear trend; however, there is a certain tendency that the band gaps increase with a decrease of the B-ion electronegativity or with a decrease of the crystallographic symmetry be adjusting the size of the A-ion (Aguiar et al. 2008).

The calculated data show that most of the semiconducting compounds have too large a band gap, and that the band edges are too deep with respect to the redox levels of water. We therefore also investigate other anions, in particular nitrogen because the N $2p$ levels are placed higher in energy than the O $2p$ levels and because the $p$-states dominate the VB edge, and smaller band gaps and in fact also a better matching with the redox levels of water can be obtained. However, the nitrides generally are less stable than the oxides.

Including nitrogen, sulfur, and fluorine as possible anions in the cubic perovskite structure, we have established a database of cubic perovskites with almost 19,000 combinations of metal atoms and anions (Castelli et al. 2012a, 2012b).

It is possible to "mine" the database to discover trends and correlations between the materials, for example, by applying clustering algorithms and constructing dendrograms (Castelli and Jacobsen 2014). The analysis shows what influences the stability of the compounds and their ability to form band gaps. Three fairly general rules are found to apply: (1) The number of electrons in the supercell has to be even. Without this a nonmagnetic material will have an electronic band crossing the Fermi level and therefore be metallic. (2) The sum of the possible oxidation numbers has to be zero. Again, otherwise a band gap cannot be formed. (3) The atom in the A-site should have a larger radius than the atom in the B-site. This rule is in spirit similar to rules based on the Goldschmidt tolerance factor, but turns out to be much better fulfilled. Applying these rules to, for example, layered perovskites leads to a significant speed-up in the identification of relevant materials (Castelli et al. 2013a).

Returning to the database of cubic perovskites, it can be taken as a starting point for identifying materials of relevance for water splitting by applying the following three screening criteria: (1) The heat of formation relative to the pool of reference systems must be less than 0.2 eV/atom to ensure stability. (2) The band gap in the visible range must be between 1.5 eV and 3.0 eV. The lower limit accounts for the bare energy to run the water-splitting reaction 1.23 eV plus approximately 0.25 eV to account for the electrochemical overpotentials of hydrogen and oxygen (Trasatti 1990). This limit may be higher if the splitting of the quasi-Fermi levels is smaller than the gap when the SC is under illumination (Weber and Dignam 1986). If the gap is above 3.0 eV too little of the solar spectrum is absorbed. Depending on the construction of the device, the photoactive material in a cell may be thin or thick and we therefore perform the search for either the direct or the indirect gap. (3) Without considering the overpotentials, the VB and CB edges should at least fulfill $VB_{edge} > 1.23$ V versus the Normal Hydrogen Electrode (NHE) and $CB_{edge} < 0$ V versus NHE for appropriate alignment of the band edges with the water redox levels.

Of the 19,000 compounds in the database only 20 fulfill these criteria: 10 oxides, 7 oxynitrides, and 3 oxyfluorides as shown in Figure 6.2. A further investigation of the crystal structure of these combinations shows that $SrSnO_3$ and $CaSnO_3$ undergo lattice distortions, and in their most stable structure they have gaps beyond the visible light absorption limit. $AgNbO_3$ and $BaSnO_3$ as well as the oxynitride series ($BaTaO_2N$, $SrTaO_2N$, $CaTaO_2N$, $LaTiO_2N$, and $LaTaON_2$) have been tested as water splitting materials. Owing to defect-assisted recombination, $BaSnO_3$ shows no activity (Zhang et al. 2007), while $AgNbO_3$ splits water in visible light in the presence of sacrificial reagents (Kato et al. 2002), and the oxynitrides perform well for hydrogen evolution (Yamasita et al. 2004). However, the oxynitrides may have stability issues in contact with water at high potentials, as also seen in Figure 6.2. To our knowledge, no other cubic perovskites that can split water in visible light have been identified. In addition to this set, we identify materials, such as $AgTaO_3$ and $SrTiO_3$, that can split water under UV light.

As mentioned earlier, the band gap is only a rough descriptor of the energy conversion efficiency. A more accurate estimation of the absorbed light can be obtained by considering the full absorption spectrum as calculated within time-dependent DFT. Details of

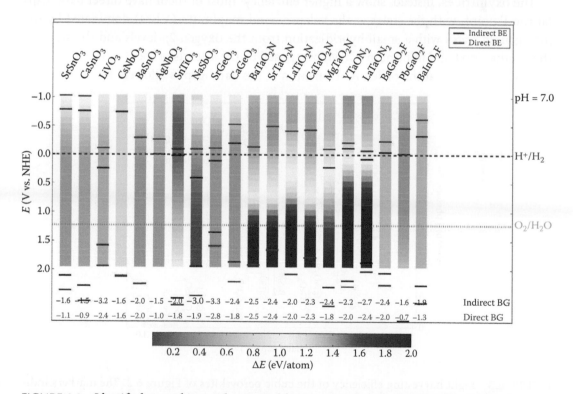

FIGURE 6.2    Identified perovskites with potential for one-photon water splitting. The stability in water of each material is estimated as the energy difference between the material and the most stable phases (solid and aqueous) in which it can separate in a potential range between –1 and 2 V and at pH = 7. The color scale runs from green (stable) to red (unstable compounds). The indirect and direct positions of the valence and conduction band edges (BE) are indicated in black and red as well as the indirect and direct band gap (BG) and the levels for hydrogen and oxygen evolution. Reproduced from CMR (CMR 2016).

the approach can be found in Yan et al. (2011). The results of such a calculation for the 20 perovskites identified in the preceding text are shown in Figure 6.3.

The absorption efficiencies are estimated as the ratio between the absorbed and the total amount of photons from the sun at AM1.5. In Figure 6.3, the efficiencies are plotted as a function of the possibly indirect band gap for a material thickness of 100 nm. The vertical lines go up to the result for infinite thickness corresponding to absorption of all light above the direct band gap if the matrix element at the band edge is nonvanishing. All the oxides and oxyfluorides have small efficiencies owing to at least two factors: they have indirect band gaps (from the *R*-point, with the wave functions usually located on the 2*p*-levels of oxygen, to the Γ-point, with wave functions localized on the *s*- or *d*-levels of the B-ion) and the absorption of photons for thin films therefore starts at a higher energy corresponding to at least the indirect band gap. Also, the dipole transition matrix elements are rather small owing to the lack of overlap between the wave functions at the band edges. The absorption at the onset is therefore not very intense. Only $LiVO_3$ has an efficiency that approaches 15% for a 100-nm-thick material. All the others are below 10%, and in all these cases, the band gap is also a fairly poor descriptor of the efficiencies.

The oxynitrides, instead, show a higher efficiency: most of them have direct band gaps (at the Γ-point, with the states at the valence band maximum (VBM) dominated by the nitrogen 2*p* states with a small hybridization from the oxygen 2*p*-levels and the states at the conduction band minimum (CBM) composed by the *d*-levels of the B-ion) with a more intense absorption at the offset. As transition at the band gap is allowed, these materials

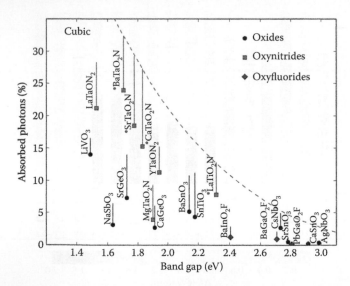

FIGURE 6.3   Light harvesting efficiency of the cubic perovskites of Figure 6.2. The markers indicate the efficiency of a 100-nm-thick material, while the vertical lines go up to the limit of infinite thickness. The green dashed line indicates the maximum theoretical efficiency as a function of the (possibly indirect) band gap. The materials marked with * have a direct transition at the band gap. (Castelli, I. E., K. S. Thygesen, and K. W. Jacobsen. 2015b. Calculated optical absorption of different perovskite phases. *J. Mater. Chem. A* 3; 12343–12349. Reproduced by permission of The Royal Society of Chemistry.)

have the maximum theoretical efficiency allowed by the gap, in the limit of infinite thickness (Castelli et al. 2015b).

In summary, the perovskite oxides and oxyfluorides are seen to possess indirect band gaps with fairly low absorption while some of the oxynitrides have both direct gap and more efficient absorption. The band edges of the oxynitrides are also generally positioned better for the water splitting reaction (Yamasita et al. 2004). However, as also pointed out previously, the oxynitrides are not so stable, which can be traced to breaking of the symmetry of the octahedra. The stability is particularly reduced in contact with water at high electron potential.

The computational approach discussed here has also been applied to the screening of different lower symmetry perovskites such as layered or double perovskites aimed at one- and two-photon water splitting or at identification of protecting shields against dissolution (Castelli et al. 2013a, 2013b, 2015b).

## 6.2.2 Hybrid Halide Perovskites

Over the past few years new hybrid halide perovskites have attracted immense interest as PV light absorbers starting with studies on $CH_3NH_3PbBr_3$ and $H_3NH_3PbI_3$ (Kojima et al. 2009) and followed by numerous other investigations of these and related systems (Christians et al. 2014; Edri et al. 2014; Jeon et al. 2013; Singh and Nagarjuna 2014).

The screening approach discussed in the previous section can also be applied to the class of hybrid halide perovskites. In the following we shall more specifically report investigations of perovskites obtained by combining cesium (Cs), methylammonium $\left(MA, CH_3NH_3^+\right)$, or formamidinium (FA, $^+HC(NH_2)_2$) as an A-ion; tin ($Sn^{2+}$) or lead ($Pb^{2+}$) as a B-ion; and chlorine (Cl$^-$), bromine (Br$^-$), iodine (I$^-$), or their combinations as anions. We also allow for four different symmetries (cubic, tetragonal, and two orthorhombic phases with space groups $Pm3m$, $P4/mbm$, $Pbnm$, and $Pnma$, respectively) for a total of 240 combinations (Castelli et al. 2014b) that have been fully relaxed using the PBESol exchange-correlation functional (Perdew et al. 2008).

Figure 6.4 shows the light absorption efficiencies for a selection of hybrid halide perovskites analogous to Figure 6.3 for the oxides, oxynitrides, and oxyfluorides. The superior absorption properties of the halide perovskites are immediately apparent. They exhibit a direct band gap transition with the states at the VBM located on the tin/lead $s$-states and the CBM on the tin/lead $p$-states. This leads to a strong overlap with a large dipole matrix elements.

Castelli et al. (2014b) investigated the stability of the perovskites in the four different phases. At low temperatures for all materials, the cubic phase is less stable than the lower symmetry phases (tetragonal and both of the orthorombic phases) that come into play depending on the detailed composition. However, the energy difference between the different phases turns out to be small, typically below 0.1 eV per formula unit. Mixing the halide atoms typically costs energy, except for the compound $FASnBr_2I$, which turns out to be marginally more stable than the corresponding bromide and iodide. However, again the energy differences are small and the combined system can be expected to be metastable considering also that the entropy of mixing is not taken into account (Usanmaz et al. 2016).

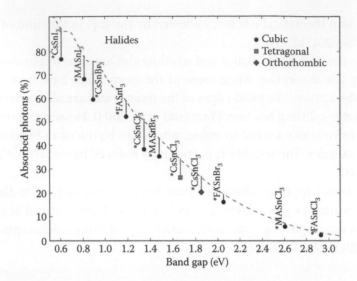

FIGURE 6.4 Light harvesting efficiency for hybrid halide perovskites in different symmetries (cubic: dots; tetragonal: square; and orthorhombic: diamond). (Castelli, I. E., K. S. Thygesen, and K. W. Jacobsen. 2015b. Calculated optical absorption of different perovskite phases. *J. Mater. Chem. A* 3; 12343–12349. Reproduced by permission of The Royal Society of Chemistry.)

In fact, both Cs and MA mixed compounds have already been synthesized (Scaife et al. 1974; Singh and Nagarjuna 2014).

The high absorption efficiency with direct gaps and strong dipole matrix elements independent of the detailed composition (Figure 6.4) make it of interest to study the variation of the band gap in the whole database of 240 compounds. Owing to the orbitals involved in the transition at the band gap, the spin–orbit effect plays an important role in the calculation of the band gaps, in particular in the lead-based perovskites for which the effect is larger (Amat et al. 2014; Umari et al. 2014). Test calculations on a limited set of systems show that the spin–orbit correction, $\Delta_{soc}$, is rather independent of the A-ion and anion configuration but depends sensitively on the character of the B-ion. Based on the calculations on the test systems we therefore introduce an *ad hoc* reduction of the band gap of 0.25 ± 0.05 eV for the Sn perovskites and of 1.02 ± 0.06 eV for the Pb systems. Focusing on the optical band gap we also need to take into account the exciton energies, which based on calculations using the Bethe–Salpeter equation are found to lead to a reduction of the gap in a range from 0.11 to 0.15 eV for all the cubic systems formed by Cs with $I_3$, $Br_3$, or $Cl_3$. We thus reduce the calculated gaps by $\Delta_{e-h} = 0.13$ eV to take this effect into account. Summarizing, the optical gap is therefore obtained starting from the GLLB-SC Kohn–Sham band gap and adding the derivative discontinuity contribution, which depends on the composition and phase considered, and subtracting the spin–orbit and electron–hole interaction corrections. The agreement between the calculated and measured gaps for the cases in which experimental studies have been carried out is very good, with a mean absolute error of 0.2 eV (Castelli et al. 2014b).

In Figure 6.5, we report the calculated optical band gaps of the 240 systems. Each colored square represents a particular phase and composition. The band gaps span over a region between 0.5 and 5 eV. We can identify several trends, summarized in the figure by red arrows, which can be helpful to design novel materials with gaps in a desired region. We note: (1) The band gaps increase as the size of the A-cation is decreased. The FA molecule is larger than MA, which is again larger than the Cs ion. The reason behind this effect is that a larger cation leads to a larger lattice constant, which again leads to a down-shift of the Sn or Pb s-states at the VBM (Borriello et al. 2008; Li et al. 2015). (2) The Pb systems have a larger gap than the Sn systems. (3) The band gaps increase with increased electronegativity of the anions. This is clearly seen in Figure 6.5, where the anions are listed according to this parameter. This effect is probably caused by the more electronegative anions pulling

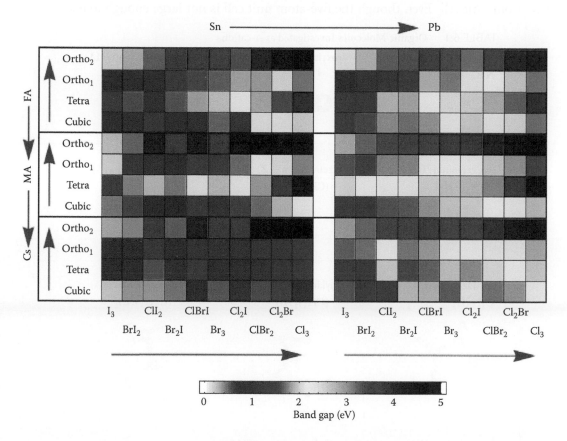

FIGURE 6.5 The 240 calculated gaps. Each square corresponds to a different phase and composition. The phase and the A-ion are indicated by the labels on the left vertical axis. The B-ion and the anion composition are shown on the top and bottom of the horizontal axis, respectively. The anion combinations are sorted according to increased geometrical mean of the electronegativity. The increase of the gaps with respect to a change in the phase and composition is followed by red arrows. (Reprinted from Castelli, I. E., J. M. García-Lastra, K. S. Thygesen, and K. W. Jacobsen. 2014b. Band gap calculations and trends of organometal halide perovskites. *APL Mater.* 2; 081514–081517; used in accordance with the Creative Commons Attribution (CC BY) license. (https://creativecommons.org/licenses/by/4.0/).)

down the extended Sn or Pb *s*-states more than the more localized *p*-states. (4) Symmetry breaking leads generally to increased band gaps so that the cubic phase exhibits smaller gaps than the other phases. In particular, the orthorhombic phase exhibits higher gaps than the other phases.

The character of the A-cation may play an important role for the functionality of a hybrid halide perovskite as a PV material. It has, for example, been suggested that the spontaneous electric polarization of A-cation molecules is of key importance (Frost et al. 2014). In the following, we therefore investigate the properties of perovskites with different common nitrogen-based cations (Table 6.1) together with Sn (or Pb) and I in the cubic phase. For each structure we generate 10 different configurations in which the molecule is randomly rotated. Each structure is then allowed to relax, in both lattice and atomic positions, in the five-atom unit cell. Even though the five-atom unit cell is not large enough to encompass

TABLE 6.1   Organic Molecules Investigated as A-cations

| Ammonium (A) | Hydroxylammonium (HA) | Methylammonium (MA) |
|---|---|---|
| $NH_4^+$ | $H_3NOH^+$ | $(CH_3)NH_3^+$ |

| Hydrazinium (HZ) | Azetidinium (AZ) | Formamidinium (FA) |
|---|---|---|
| $H_3N^+ - NH_2^+$ | $(CH_2)_3 NH_2^+$ | $NH_2(CH)^+NH_2$ |

| Imidazolium (IA) | Dimethylammonium (DMA) | Ethylammonium (EA) |
|---|---|---|
| $C_3N_2H_5^+$ | $(CH_3)_2 NH_2^+$ | $(C_2H_5)NH_3^+$ |

| Guanidinium (GA) | Tetramethylammonium (TMA) |
|---|---|
| $^+C(NH)_3^+$ | $(CH_3)_4 N^+$ |

*Source:*  Kieslich, G. et al., *Chem Sci.* 5; 4712–4715, 2014.

all the possible distortions, it gives an estimate of the coupling between the molecule orientation, volume and lattice distortions, and their effects on the magnitude of the band gap.

In Figure 6.6, the calculated band gaps for the systems $MSnI_3$ (or $MPbI_3$, where M is one of the organic molecules) are plotted as a function of the pseudo-cubic lattice parameter, defined as the cubic root of the unit cell volume. The error bars along the x-axis indicate the average noncubicity, defined as the average of $\frac{1}{3}\sum_{i=x,y,z}\left|a_i - a_{pc}\right|$, where $a_i$ is the lattice parameter along the ith direction and $a_{pc}$ is the pseudo-cubic lattice parameter. The error bars on the gap axis represent the variation of the gap depending on the orientation of the molecules.

The Sn-based systems show an almost linear relation between the pseudo-cubic lattice parameter and the band gap, and the molecules are clearly ordered according to size. This confirms the gap-volume trend discussed earlier, and opens up for significant band gap modification by molecular replacement. The correlation is less clear for the Pb-based systems, but again the molecules are roughly ordered by size.

For the smaller molecules the lattice distortions are seen to be less pronounced than for the larger ones, pointing to a—maybe not unexpected—stronger coupling between lattice and molecular orientation for the large molecules. Molecular replacements can therefore be expected not only to affect the band gap but also to play a major role for the stabilization of the different symmetry-breaking phases.

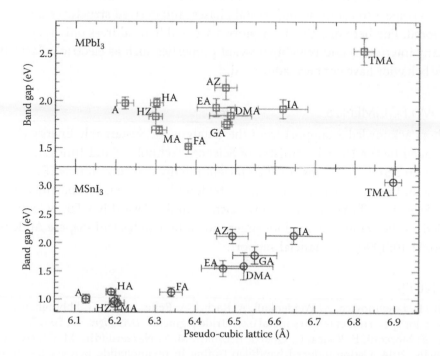

FIGURE 6.6 Calculated band gaps as a function of the pseudo-cubic lattice parameters for the set of organic molecules of Table 6.1. The error bars indicate the average noncubicity (x-axis) and the variation in the gap (y-axis) depending on the different orientations of the organic molecules.

## 6.3 CONCLUSIONS AND PERSPECTIVES

The identification and design of new light-absorbing materials for use in PV or PEC devices is traditionally based on a combination of trial and error and of sophisticated chemical/physical intuition by its practitioners. In this chapter we have tried to illustrate how screening with *ab initio* quantum mechanical calculations can be used as a supplementary tool in this endeavor.

The screenings necessarily have to be based on computable descriptors that address key issues in the material performance. Even though such descriptors may be quite simple and address only part of the problem, they can still be useful in reducing the number of possible candidate materials for a particular application. As an example, consider the database discussed earlier with 19,000 perovskite oxides, oxynitrides, and oxyfluorides. Considering only the band gap, heat of formation, and level alignment, the number of candidate materials for one-photon water splitting was reduced to only 20—a much more reasonable number of materials for further investigation. Explicit calculation of the absorption efficiency could then reveal that the oxynitrides had superior absorption properties compared to the oxides.

The calculations can also be used to discover or quantify trends. For example, the band gaps of the hybrid halide peroskites showed a clear correlation with the electronegativity of the anions, underpinning the possibility of band gap tuning by anion substitution. Likewise the band gaps are seen to increase systematically with the size of the molecules used as A-cations, opening up for enlargement of the band gaps of Sn-based perovskites, bringing them into a range of higher relevance to water splitting.

The computational exploration of perovskites as PV or PEC materials is far from complete. With respect to composition, crystal phases, and derived structures much territory is unexplored. Furthermore, the descriptors discussed here address only very basic issues while many interesting and possibly relevant properties such as electric polarization and magnetic behavior have not been addressed.

## ACKNOWLEDGMENTS

The authors acknowledge support from the Catalysis for Sustainable Energy (CASE) initiative funded by the Danish Ministry of Science, Technology and Innovation and from the Center on Nanostructuring for the Efficient Energy Conversion (CNEEC) at Stanford University, an Energy Frontier Research Center founded by the US Department of Energy, Office of Science, Office of Basic Energy Sciences under Award No. DE-SC0001060. KST acknowledges the Danish Council for Independent Research's DFF-Sapere Aude program (grant no. 11-1051390) for financial support.

## REFERENCES

Aguiar, R., D. Logvinovich, A. Weidenkaff, A. Rachel, A. Reller, and S. G. Ebbinghaus. 2008. The vast color spectrum of ternary metal oxynitride pigments. *Dyes Pigm.* 76; 70–75.

Amat, A., E. Mosconi, E. Ronca, C. Quarti, P. Umari, M. K. Nazeeruddin, M. Grätzel, and F. De Angelis. 2014. Cation-induced band-gap tuning in organohalide perovskites: Interplay of spin-orbit coupling and octahedra tilting. *Nano Lett.* 14; 3608–3616.

Borriello, I., G. Cantele, and D. Ninno. 2008. *Ab initio* investigation of hybrid organic-inorganic perovskites based on tin halides. *Phys. Rev. B* 77; 235214–235219.

Butler, M. A., and D. S. Ginley. 1978. Prediction of flatband potentials at semiconductor-electrolyte interfaces from atomic electronegativities. *J. Electrochem. Soc.* 125; 228–232.

Castelli, I. E., and K. W. Jacobsen. 2014. Designing rules and probabilistic weighting for fast materials discovery in the perovskite structure. *Model. Simul. Mater. Sci. Eng.* 22; 055007–055014.

Castelli, I. E., T. Olsen, S. Datta, D. D. Landis, S. Dahl, K. S. Thygesen, and K. W. Jacobsen. 2012a. Computational screening of perovskite metal oxides for optimal solar light capture. *Energy Environ. Sci.* 5; 5814–5819.

Castelli, I. E., D. D. Landis, K. S. Thygesen, S. Dahl, I. Chorkendorff, T. F. Jaramillo, and K. W. Jacobsen. 2012b. New cubic perovskites for one- and two-photon water splitting using the computational materials repository. *Energy Environ. Sci.* 5; 9034–9043.

Castelli, I. E., J. M. García-Lastra, F. Hüser, K. S. Thygesen, and K. W. Jacobsen. 2013a. Stability and band gaps of layered perovskites for one- and two-photon water splitting. *New J. Phys.* 15; 105026, 15pp.

Castelli, I. E., K. S. Thygesen, and K. W. Jacobsen. 2013b. Band gap engineering of double perovskites for one- and two-photon water splitting. *Mater. Res. Soc. Symp. Proc.* 1523; 18–23.

Castelli, I. E., K. S. Thygesen, and K. W. Jacobsen. 2014a. Calculated Pourbaix diagrams of cubic perovskites for water splitting: Stability against corrosion. *Top. Catal.* 57; 265–272.

Castelli, I. E., J. M. García-Lastra, K. S. Thygesen, and K. W. Jacobsen. 2014b. Band gap calculations and trends of organometal halide perovskites. *APL Mater.* 2; 081514–081517.

Castelli, I. E., F. Hüser, M. Pandey, H. Li, K. S. Thygesen, B. Seger, A. Jain, K. A. Persson, G. Ceder, and K. W. Jacobsen. 2015a. New light-harvesting materials using accurate and efficient band-gap calculations. *Adv. Energy Mater.* 5; 1400915–1400917.

Castelli, I. E., K. S. Thygesen, and K. W. Jacobsen. 2015b. Calculated optical absorption of different perovskite phases. *J. Mater. Chem. A* 3; 12343–12349.

Ceder, G., Y.-M. Chiang, D. R. Sadoway, M. K. Aydinol, Y.-I. Jang, and B. Huang. 1998. Identification of cathode materials for lithium batteries guided by first-principles calculations. *Nature* 392; 694–696.

Christians, J. A., R. C. M. Fung, and P. V. Kamat. 2014. An inorganic hole conductor for organo-lead halide perovskite solar cells. Improved hole conductivity with copper iodide. *J. Am. Chem. Soc.* 136; 758–764.

CMR (Computational Materials Repository). 2016. https://cmr.fysik.dtu.dk/ (Accessed February, 2016).

Curtarolo, S., W. Setyawan, G. L. W. Hart, M. Jahnatek, R. V. Chepulskii, R. H. Taylor, S. Wang, J. Xue, K. Yang, O. Levy, M. J. Mehl. 2012. AFLOW: An automatic framework for high-throughput materials discovery. *Comp. Mater. Sci.* 58; 218–226.

Curtarolo, S., G. L. W. Hart, M. Buongiorno Nardelli, N. Mingo, S. Sanvito, and O. Levy. 2013. The high-throughput highway to computational. *Nat. Mater.* 12; 191–201.

d'Avezac, M., J.-W. Luo, T. Chanier, and A. Zunger. 2012. Genetic-algorithm discovery of a direct-gap and optically allowed superstructure from indirect-gap Si and Ge semiconductors. *Phys. Rev. Lett.* 108; 027401–027405.

Edri, E., S. Kirmayer, M. Kulbak, G. Hodes, and D. Cahen. 2014. Chloride inclusion and hole transport material doping to improve methyl ammonium lead bromide perovskite-based high open-circuit voltage solar cells. *J. Phys. Chem. Lett.* 5; 429–433.

Enkovaara, J., C. Rostgaard, J. J. Mortensen, J. Chen, M. Dułak, L. Ferrighi, J. Gavnholt, C. Glinsvad, V. Haikola, H. A. Hansen, H. H. Kristoffersen, 2010. Electronic structure calculations with GPAW: A real-space implementation of the projector augmented-wave method. *J. Phys.: Condens. Matter* 22; 253202–253224.

Frost, J. M., K. T. Butler, F. Brivio, C. H. Hendon, M. van Schilfgaarde, and A. Walsh. 2014. Atomistic origins of high-performance in hybrid halide perovskite solar cells. *Nano Lett.* 14; 2584–2590.

Godby, R. W., M. Schlüter, and L. J. Sham. 1986. Accurate exchange-correlation potential for silicon and its discontinuity on addition of an electron. *Phys. Rev. Lett.* 56; 2415–2418.

Gritsenko, O., R. van Leeuwen, E. van Lenthe, and E. J. Baerends. 1995. Self-consistent approximation to the Kohn-Sham exchange potential. *Phys. Rev. A* 51; 1944–1954.

Hachmann, J., R. Olivares-Amaya, S. Atahan-Evrenk, C. Amador-Bedolla, R. S. Sánchez-Carrera, A. Gold-Parker, L. Vogt, A. M. Brockway, and A. Aspuru-Guzik. 2011. The Harvard clean energy project: Large-scale computational screening and design of organic photovoltaics on the world community grid. *J. Phys. Chem. Lett.* 2; 2241–2251.

Hammer, B., L. B. Hansen, and J. K. Nørskov. 1999. Improved adsorption energetics within density-functional theory using revised Perdew-Burke-Ernzerhof functionals. *Phys. Rev. B* 59; 7413–7421.

Hautier, G., A. Miglio, G. Ceder, G.-M. Rignanese, and X. Gonze. 2013. Identification and design principles of low hole effective mass *p*-type transparent conducting oxides. *Nat. Commun.* 4; 2292–2297.

Heyd, J., G. E. Scuseria, and M. Ernzerhof. 2003. Hybrid functionals based on a screened coulomb potential. *J. Chem. Phys.* 118; 8207–8215.

Hohenberg, P., and W. Kohn. 1964. Inhomogeneous electron Gas. *Phys. Rev.* 136; 864–871.

Hüser, F., T. Olsen, and K. S. Thygesen. 2013. Quasiparticle GW calculations for solids, molecules, and two-dimensional materials. *Phys. Rev. B* 87; 235132, 14pp.

ICSD. 2016. http://www.fiz-karlsruhe.de/icsd_web.html (Accessed February, 2016).

Ishihara, T. (ed.). 2005. *Perovskite oxide for solid oxide fuel cell*. New York: Springer Science+Business Media.

Jeon, N. J., J. Lee, J. H. Noh, M. K. Nazeeruddin, M. Grätzel, and S. I. Seok. 2013. Efficient inorganic-organic hybrid perovskite solar cells based on pyrene arylamine derivatives as hole-transporting materials. *J. Am. Chem. Soc.* 135; 19087–19090.

Kato, H., H. Kobayashi, and A. Kudo. 2002. Role of $Ag^+$ in the band structures and photocatalytic properties of $AgMO_3$ (M: Ta and Nb) with the perovskite structure. *J. Phys. Chem. B* 106; 12441–12447.

Kieslich, G., S. Sun, and A. K. Cheetham. 2014. Solid-state principles applied to organic–inorganic perovskites: New tricks for an old dog. *Chem. Sci.* 5; 4712–4715.

Kohn, W., and L. J. Sham. 1965. Self-consistent equations including exchange and correlation effects. *Phys. Rev.* 140; 1133–1138.

Kojima, A., K. Teshima, Y. Shirai, and T. Miyasaka. 2009. Organometal halide perovskites as visible-light sensitizers for photovoltaic cells. *J. Am. Chem. Soc.* 131; 6050–6051.

Krieger, J. B., Y. Li, and G. J. Iafrate. 1992a. Construction and application of an accurate local spin-polarized Kohn-Sham potential with integer discontinuity: Exchange-only theory. *Phys. Rev. A* 45; 101–126.

Krieger, J. B., Y. Li, and G. J. Iafrate. 1992b. Systematic approximations to the optimized effective potential: Application to orbital-density-functional theory. *Phys. Rev. A* 46; 5453–5458.

Kuisma, M., J. Ojanen, J. Enkovaara, and T. T. Rantala. 2010. Kohn-Sham potential with discontinuity for band gap materials. *Phys. Rev. B* 82; 115106–115107.

Larsen, A., J. Mortensen, J. Blomqvist, I. Castelli, R. Christensen, M. Dulak, J. Friis, M. Groves, B. Hammer, C. Hargus, E. Hermes, P. Jennings, P. Jensen, J. Kermode, J. Kitchin, E. Kolsbjerg, J. Kubal, K. Kaasbjerg, S. Lysgaard, J. Maronsson, T. Maxson, T. Olsen, L. Pastewka, A. Peterson, C. Rostgaard, J. Schiøtz, O. Schütt, M. Strange, K. Thygesen, T. Vegge, L. Vilhelmsen, M. Walter, Z. Zeng, and K. W. Jacobsen. 2017. The atomic simulation environment—A Python library for working with atoms. *J. Phys.: Condens. Matter* (accepted). https://doi.org/10.1088/1361-648X/aa680e.

Li, H., I. E. Castelli, K. S. Thygesen, and K. W. Jacobsen. 2015. Strain sensitivity of band gaps of Sn-containing semiconductors. *Phys. Rev. B* 91; 045204–045206.

Lin, L.-C., A. H. Berger, R. L. Martin, J. Kim, J. A. Swisher, K. Jariwala, C. H. Rycroft, A. S. Bhown, M. W. Deem, M. Haranczyk, and B. Smit. 2012. In silico screening of carbon-capture materials. *Nat. Mater.* 11; 633–641.

Mortensen, J. J., L. B. Hansen, and K. W. Jacobsen. 2005. Real-space grid implementation of the projector augmented wave method. *Phys. Rev. B* 71; 035109–035111.

MP. 2016. Materials Project—A Materials Genome Approach. http://materialsproject.org/ (Accessed February, 2016).

Ørnsø, K. B., J. M. Garcia-Lastra, and K. S. Thygesen. 2013. Computational screening of functionalized zinc-porphyrins for dye sensitized solar cells. *Phys. Chem. Chem. Phys.* 15; 19478–19486.

Perdew, J. P., A. Ruzsinszky, G. I. Csonka, O. A. Vydrov, G. E. Scuseria, L. A. Constantin, X. Zhou, and K. Burke. 2008. Restoring the density-gradient expansion for exchange in solids and surfaces. *Phys. Rev. Lett.* 100; 136406, 4pp.

Perdew, J. P., and A. Zunger. 1981. Self-interaction correction to density-functional approximations for many-electron systems. *Phys. Rev. B* 23; 5048–5079.

Persson, K. A., B. Waldwick, P. Lazic, and G. Ceder. 2012. Prediction of solid-aqueous equilibria: Scheme to combine first-principles calculations of solids with experimental aqueous states. *Phys. Rev. B* 85; 235438, 12pp.

Putz, M. V., N. Russo, and E. Sicilia. 2005. About the Mulliken electronegativity in DFT. *Theor. Chem. Acc.* 114; 38–45.

Rossmeisl, J. A. Logadottir, and J. K. Nørskov. 2005. Electrolysis of water on (oxidized) metal surfaces. *Chem. Phys.* 319; 178–184.

Scaife, D. E., P. F. Weller, and W. G. Fisher. 1974. Crystal preparation and properties of cesium tin (II) trihalides. *J. Solid State Chem.* 9; 308–314.

Shishkin, M., and G. Kresse. 2007. Self-consistent GW calculations for semiconductors and insulators. *Phys. Rev. B* 75; 235102–235109.

Singh, S. P., and P. Nagarjuna. 2014. Organometal halide perovskites as useful materials in sensitized solar cells. *Dalton Trans.* 43; 5247–5251.

Stevanović, V., S. Lany, D. S. Ginley, W. Tumas, and A. Zunger. 2014. Assessing capability of semiconductors to split water using ionization potentials and electron affinities only. *Phys. Chem. Chem. Phys.* 16; 3706–3714.

Talman, J. D., and W. F. Shadwick. 1976. Optimized effective atomic central potential. *Phys. Rev. A* 14; 36–40.

Trasatti, S. 1990. Surface chemistry of oxides and electrocatalysis. *Croat. Chem. Acta* 63; 313–329.

Umari, P., E. Mosconi, and F. De Angelis. 2014. Relativistic GW calculations on $CH_3NH_3PbI_3$ and $CH_3NH_3SnI_3$ perovskites for solar cell applications. *Sci. Rep.* 4; 4467, 7pp.

Usanmaz, D., P. Nath, J. J. Plata, G. L. W. Hart, I. Takeuchi, M. Buongiorno Nardelli, M. Fornari, and S. Curtarolo. 2016. First principles thermodynamical modeling of the binodal and spinodal curves in lead chalcogenides. *Phys. Chem. Chem. Phys.* 18; 5005–5011.

Vesborg, P. C. K., and T. F. Jaramillo. 2012. Addressing the terawatt challenge: Scalability in the supply of chemical elements for renewable energy. *RSC Adv.* 2; 7933–7947.

Walter, M. G., E. L. Warren, J. R. McKone, S. W. Boettcher, Q. Mi, E. A. Santori, and N. S. Lewis. 2010. Solar water splitting cells. *Chem. Rev.* 110; 6446–6473.

Weber, M. F., and M. J. Dignam. 1986. Splitting water with semiconducting photoelectrodes-efficiency considerations. *Int. J. Hydrogen Energy* 11; 225–232.

Wu, Y., M. Chan, and G. Ceder. 2011. Prediction of semiconductor band edge positions in aqueous environments from first principles. *Phys. Rev. B* 83; 235301–235307.

Wu, Y., P. Lazic, G. Hautier, K. Persson, and G. Ceder. 2013. First principles high throughput screening of oxynitrides for water-splitting photocatalysts. *Energy Environ. Sci.* 6; 157–168.

Xu, Y., and M. A. A. Schoonen. 2000. The absolute energy positions of conduction and valence bands of selected semiconducting minerals. *Am. Mineral.* 85; 543–556.

Yamasita, D., T. Takata, M. Hara, J. N. Kondo, and K. Domen. 2004. Recent progress of visible-light-driven heterogeneous photocatalysts for overall water splitting. *Solid State Ionics* 172; 591–595. *Proceedings of the Fifteenth International Symposium on the Reactivity of Solids.*

Yan, J., J. J. Mortensen, K. W. Jacobsen, and K. S. Thygesen. 2011. Linear density response function in the projector augmented wave method: Applications to solids, surfaces, and interfaces. *Phys. Rev. B* 83; 245122, 10pp.

Yu, L., and A. Zunger. 2012. Identification of potential photovoltaic absorbers based on first-principles spectroscopic screening of materials. *Phys. Rev. Lett.* 108; 068701–068705.

Zhang, W., J. Tang, and J. Ye. 2007. Structural, photocatalytic, and photophysical properties of perovskite MSnO$_3$ (M = Ca, Sr, and Ba) photocatalysts. *J. Mater. Res.* 22; 1859–1871.

# CHAPTER 7

# Organic–Inorganic Halide Perovskite Quasi-Particle Nature Analysis via the Interplay among Classic Solid-State Concepts, Density Functional, and Many-Body Perturbation Theory

Jacky Even, Giacomo Giorgi, Claudine Katan,
Hiroki Kawai, and Koichi Yamashita

## CONTENTS

## 7.1 QUASI-PARTICLE BEHAVIOR IN 2D/3D ORGANIC–INORGANIC HALIDE PEROVSKITES

### 7.1.1 Excitons

For the sake of clarity and coherence, this chapter devoted to the nature of quasi-particles in organic–inorganic halide perovskites will concentrate on lead-based perovskites. The exceptional light emission properties of layered hybrid organic-inorganic perovskite (HOP) thin films or crystal structures have been well documented in the literature since the early 1980s (Ishihara et al. 1990; Koutselas et al. 1996; Mitzi et al. 1995). Intense light emission is indeed observed from helium to room temperature. Light absorption experiments have unambiguously revealed the excitonic character of the optical absorption at the electronic band gap edge. Extremely large exciton binding energies, approximately 300 meV (see Figure 7.1 and Ishihara et al. 1990), have been measured as well as bound state resonances for the 1s up to 4s exciton states or for the biexciton and triexciton transitions (Even et al. 2012; Kato et al. 2003; Muljarov et al. 1995; Shimizu et al. 2006). Conversely, until recently little was known about excitons in 3D perovskites. The 50 meV binding energy recorded at 4 K in the early stages has only recently been shown to correspond to a large overestimation of the room temperature value (Even et al. 2014a; Hirasawa et al. 1994).

The excitonic properties can be tailored mainly by two means: (1) by chemical engineering of the organic moiety and (2) by changing the thickness of the inorganic layer. For instance, a series of layered 2D/3D structures of general formula $(R)_2(CH_3NH_3)_{n-1}Pb_nI_{3n+1}$, where R indicates a large size organic cation and $n$ the number of lead iodide layers in the inorganic sheet, have been investigated experimentally for $n = 1-4$. Noteworthy, the

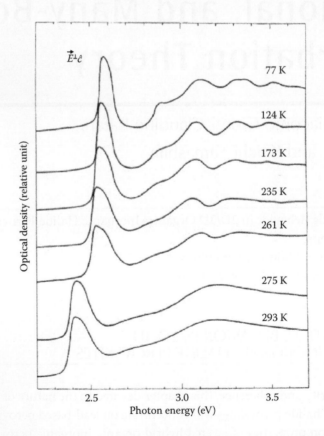

FIGURE 7.1 Optical density spectra in a cleaved thin crystal of $(C_{10}H_{21}NH_3)_2PbI_4$ at several temperatures. (Reprinted with permission from Ishihara, T., J. Takahashi, and T. Goto. 1990. Optical properties due to electronic transitions in two-dimensional semiconductors $(C_nH_{2n+1}NH_3)_2PbI_4$. *Phys. Rev. B* 42; 11099–11107. Copyright 1990 by the American Physical Society.)

bulk 3D compound $CH_3NH_3PbI_3$ corresponds to $n = \infty$. Structural phase transitions of the layered hybrid crystal, which affect both the 2D inorganic lattice and the organic barrier, may lead to the switching of the exciton resonances and Peierls-like transitions. The energy shifts of the exciton lines are mostly related to distortions of the inorganic sublattice, that is, rotations of the metal–halide octahedra that are coupled to ordering of the organic cations. These structural changes essentially renormalize the monoelectronic state energies without affecting much the oscillator strength of the optical transitions related to the bound exciton state. The situation is very different for 3D hybrid perovskites: structural phase transitions lead to both renormalization of the monoelectronic states and screening of the exciton resonances (Even et al. 2014a).

Unfortunately, an accurate theoretical description of the monoelectronic states in layered hybrid perovskite crystals is involved. These materials are very large systems for atomistic approaches based on Density Functional Theory (DFT), which may face difficulties related to limited computational resources (Even et al. 2012, 2014b). Furthermore, while numerous self-assembled hybrid perovskites have been studied as thin films, fewer crystallographic structures are known precisely. In fact, growth of monocrystals for x-ray or neutron diffraction may be difficult and the degradation of the lattice under irradiation or illumination can be a serious issue during experimental characterization. The flexibility of the molecules used to synthesize layered hybrid perovskite materials often promotes growth of self-assembled thin film, while introducing lattice disorder. For instance, with long alkyl chains, structural phase transitions with an order–disorder character have been reported at low temperature (Billing and Lemmerer 2007). For ordered structures having a reasonable number of atoms, for example, $(pFC_6H_5C_2H_4NH_3)_2PbI_4$, $(C_5H_{11}NH_3)_2PbI_4$, or $(C_2H_4INH_3)_2PbI_4$, DFT computations are feasible (Even et al. 2012; Pedesseau et al. 2014). Moreover, it has been shown that a fast assessment of electronic properties and monoelectronic states, close to the band gap, can be achieved by replacing the molecular cation by a $Cs^+$ located at the position of the N atom. In most cases, this trick allows mimicking the ionic interactions due to the organic cations. Once spin–orbit coupling (SOC) is considered, the DFT electronic structure reveals a direct band-gap character in agreement with the observed luminescence at room temperature (Even et al. 2012, 2013). We stress that SOC plays a major role for the effective mass and degeneracy of the conduction band edge electronic state. However, DFT+SOC calculations underestimate the fundamental transition energy, as a result of an inherent limitation of DFT, which is a ground state theory. This limit can be overcome by implementing many-body effects, for example, GW self-energy corrections for conduction and valence band states close to the band gap. Such calculations are already involved for crystals with smaller unit cells, namely 3D hybrid perovskites (Brivio et al. 2014; Even et al. 2014c; Umari et al. 2014), and are currently beyond available computational resources for layered hybrid perovskites. Meanwhile, the overall picture catch by DFT+SOC calculations on layered hybrids holds: they can be viewed as pure 2D systems with electronic states of specific symmetries. They provide detailed information on Bloch states and selection rules that can lead to further investigations using semi-empirical approaches (e.g., in k·p theory, tight-binding approximation) (Even et al. 2012, 2014b). Such fundamental studies are important to understand the optical properties as well as the anisotropic electronic transport. Good in-plane carrier transport allowed a nice demonstration of thin-film transistors, but similar vertical transport has not yet been reported.

For layered hybrid perovskites, most of the experimental optical absorption spectra evidence very large exciton binding energies, in the 250–350 meV range (see Figure 7.1). The enhanced exciton binding energy, as compared to that of 3D hybrid perovskites, is related to dielectric confinement and image-charge effects. These effects stem from the dielectric mismatch between the inorganic and organic layers. In layered hybrid perovskites, qualitative analyses of the excitons rely on the 2D Wannier exciton picture (Muljarov et al. 1995). It provides a reasonable account of experimental results and can be extended to understand the influence of halogen alloying. Unfortunately, unlike many of the conventional semiconductors, a quantitative analysis based on the Bethe–Salpeter equation (BSE) starting from monoelectronic states computed at the DFT+GW level is still beyond reach (Even et al. 2014c). Thus, alternative empirical approaches have been implemented to investigate the monoelectronic states and the enhancement of the exciton resonances due to dielectric confinement.

In the 1990s, David Mitzi introduced the quantum well (QW) concept that affords a schematic/qualitative picture: the layered HOP is approximated as being built from two bulk semiconductors A and B (similar to those sketched on Figure 7.2a), with inorganic sheets (A) alternating with organic layers (B) of much larger band gaps (Even et al. 2014b; Mitzi et al. 2001). However, this qualitative picture of a type I QW-like heterostructure leads to inadequate analysis of quantum confinement, especially when it is combined with an effective mass approach. As a matter of fact, slowly varying envelope functions $F_i(r)$ are required to write the electronic wavefunctions of the heterostructure:

$$\psi(r) = \sum_i F_i(r) U_i(r) \tag{7.1}$$

where $U_i(r)$ are periodic and rapidly oscillating basis functions. The Bloch functions of the bulk materials $U_{A,i}(r)$ and $U_{B,i}(r)$ are assumed to differ only slightly at high symmetry points of the Brillouin zone:

$$U_i(r) \approx U_{A,i}(r) \approx U_{B,i}(r) \tag{7.2}$$

This condition is obviously not fulfilled in layered hybrid perovskites. Moreover, this fully empirical approach predicts a superlattice effect and a strong vertical electronic coupling that are neither confirmed by DFT, nor evidenced through carrier transport measurements. Alternatively, defining the whole layered hybrid perovskite as a composite material allows quantitatively evaluation of confinement potentials. It leads to a rigorous composite approach to afford the valence band-lineup between inorganic and organic layers (see Figure 7.2 and Even et al. 2014b).

To analyze the confinement effect, the band gap of all-inorganic and hybrid perovskite colloidal slabs, as well as layered hybrid perovskite structures, can be written as a function of the number of layers $n$:

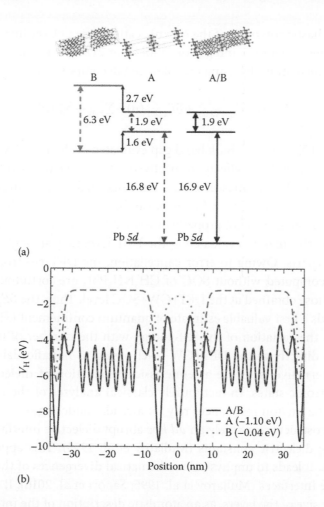

(a)

(b)

FIGURE 7.2 (a) Schematic representation of the CB and VB band alignment considering the $(C_{10}H_{21}NH_3)_2PbI_4$ layered hybrid perovskite (A/B) as a composite structure made of A and B modeled by $(Na)_2PbI_4$ and $C_{10}H_{21}CH_3$, respectively. (b) Computed potential profiles for the real layered material (A/B, straight line) and A (dashed line) and B (dotted line) bulk-like materials. Band alignment and match of potential profiles of A/B are obtained thanks to downward shifts of the data computed for A and B amounting to 1.1 and 0.04 eV, respectively. Alignment of Pb 5d orbitals between A/B and A also require a 1.1 eV shift. (Reprinted with permission from Even, J., L. Pedesseau, and C. Katan. 2014b. Understanding quantum confinement of charge carriers in layered 2D hybrid perovskites. *Chem. Phys. Chem.* 15; 3733–3741. Copyright © 2014 John Wiley & Sons.)

$$E_g(n) = E_{g;bulk} + \delta E_g(n) \qquad (7.3)$$

where $E_{g;bulk}$ is the band gap of the bulk core material (e.g., $CH_3NH_3PbX_3$, X = I, Br, or Cl) and $\delta E_g(n)$ stems from quantum and dielectric confinement (Even et al. 2014b; Sapori et al. 2016). Proper evaluation by DFT of the band gap, for instance $E_{g;bulk}$, requires the inclusion of self-energy corrections. For $CH_3NH_3PbI_3$ several values of $E_{g;bulk}$ are available at the GW+SOC level. Noteworthy, room temperature contributions related to electron–phonon coupling and stochastic rotations of the organic cations are not yet included. The various contributions to the electronic band gap of colloidal or 2D/3D structures can be estimated from theoretical calculations by using the following decomposition:

$$E_g(n) \approx E_{g;DFT,bulk} + \delta E_{g,DFT}(n) + \Sigma_{bulk} + \delta\Sigma(n) \tag{7.4}$$

where $E_{g;DFT,bulk}$ and $\Sigma_{bulk}$ are the bulk band gap evaluated at the DFT level and self-energy corrections due to many-body effects, respectively. Whenever the band gap is not badly underestimated and effective masses of the bulk materials are accurately described by plain DFT, the effect of quantum confinement can be estimated from $\delta E_{g,DFT}(n)$. As stated previously, this contribution is not properly described by a purely empirical approach. The band gap and effective masses evaluated at the DFT level may strongly differ from their DFT+GW counterparts. Owing to error cancellation, the DFT electronic band gap and effective masses computed without SOC of $CH_3NH_3PbI_3$ are fortuitously in reasonable agreement with those obtained at the DFT+GW+SOC level. Thus, the $\delta E_{g,DFT}(n)$ calculated without SOC yields a first valuable estimate of quantum confinement (Figure 7.3).

Unfortunately, the variation of the self-energy with the number of layers, $\delta\Sigma(n)$, cannot be evaluated directly at the DFT+GW level owing to nonaffordable computational resources. An alternative is to use a semiclassical evaluation of dielectric confinement on the monoelectronic states. In fact, purely classical analyses of the effect of dielectric confinement on the exciton resonances rely on a crude modeling of the dielectric profile through the heterostructure, namely an *ad hoc* abrupt dielectric interface separating layers with bulk-like dielectric constants (Muljarov et al. 1995). This approach has several severe limitations. It leads to unphysical mathematical divergences of the self-energy profile $\delta\Sigma(z, n)$ at the interfaces (Muljarov et al. 1995; Sapori et al. 2016). It is also difficult to exactly define the size of the layers, as an atomistic description of the interfaces is lacking. Finally, often, the bulk dielectric constants of the bulk core materials $CH_3NH_3PbX_3$ itself

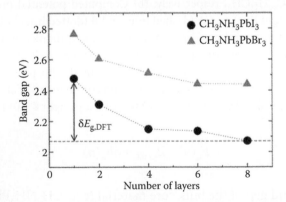

FIGURE 7.3 Band gaps with respect to the slab thickness computed for slabs of $CH_3NH_3PbX_3$ (X = I, Br). (Sapori, D., M. Kepenekian, L. Pedesseau, C. Katan, and J. Even. 2016. Quantum confinement and dielectric profiles of colloidal nanoplatelets of halide inorganic and hybrid organic–inorganic perovskites. *Nanoscale* 8; 6369–6378. Reproduced by permission of The Royal Society of Chemistry.)

is considered for the inorganic sheet, even for an ultrathin ($n = 1$) inorganic layer. Using DFT to describe dielectric profiles of layered systems has been shown to improve accuracy, yet still at a reasonable computational cost (Sapori et al. 2016). This approach takes advantage of the *ab initio* description of the nanoplatelets or 2D/3D heterostructures and in particular of interfaces. It shows that in real 2D hybrid perovskites, a single standalone layer derived from the corresponding 3D $CH_3NH_3PbX_3$ material cannot solely approximate the inorganic layer. It allows accounting for various effects such as substitution of halide atoms, phonon contributions to the dielectric constant, or comparison between all-inorganic and hybrid colloidal platelets. The dramatic effect of phonons is especially evident for the inorganic parts of the hybrid materials. The semiclassical evaluation of $\delta\Sigma(n)$, both for the conduction band (CB) and the valence band (VB) states, can be obtained by computing the following integrals:

$$\delta\Sigma_{CB(VB)}(n) \approx \int \delta\Sigma(z,n)\rho_{CB(VB)}(z,n)dz \qquad (7.5)$$

where CB (VB) is the electronic density profile for the CB (VB) state. This value can be estimated at the center of the slab as shown in Figure 7.4.

We underline that it is not straightforward to identify the various contributions to the band gap that is actually derived from the optical absorption of colloidal or 2D/3D hybrid structures. Because the optical band gap corresponds to the electronic band gap minus the exciton binding energy, it undergoes a smoother variation than the band gap predicted using monoelectronic states. When the thickness decreases, the exciton binding energy increases and partially compensates the gap increase related to effects of quantum and dielectric confinement on the monoelectronic states.

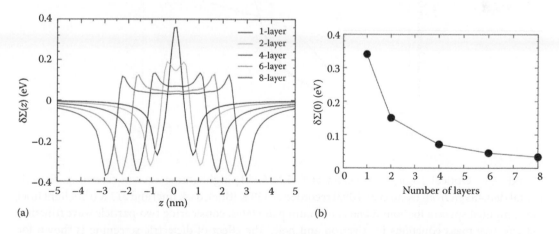

FIGURE 7.4    (a) Self-energy profile $\delta\Sigma(z)$ for slabs of $CH_3NH_3PbI_3$. (b) Self-energy taken at the slab center $\delta\Sigma(0)$. (Sapori, D., M. Kepenekian, L. Pedesseau, C. Katan, and J. Even. 2016. Quantum confinement and dielectric profiles of colloidal nanoplatelets of halide inorganic and hybrid organic–inorganic perovskites. *Nanoscale* 8; 6369–6378. Reproduced by permission of The Royal Society of Chemistry.)

At first approximation, the exciton binding energy $E_X(n)$ of all-inorganic and hybrid perovskite colloidal slabs, as well as layered hybrid perovskite structures, can be expressed in terms of the number of layers $n$ as

$$E_X(n) = E_{X,bulk}(n) + \delta E_X(n), \tag{7.6}$$

where $E_{X,bulk}(n)$ is the exciton binding energy of the bulk core material, for example $CH_3NH_3PbI_3$, and $\delta E_X(n)$ stems from quantum and dielectric confinement. $E_{X,bulk}(n)$ is an important parameter for photovoltaic devices operating at room temperature, especially for the ratio of free carrier to exciton populations. It is expected to strongly affect the charge transport through the structures. Besides, a careful reassessment of old experimental absorption spectra of $CH_3NH_3PbI_3$ revealed the importance of exciton screening at high temperature. In fact, simulation of the optical absorption spectra using an expression accounting both for bound and continuum pair states clearly evidences that the generally accepted value for $CH_3NH_3PbI_3$ [~50 meV, measured at 4 K (Hirasawa et al. 1994)] is inappropriate for describing the real exciton resonance at room temperature as shown in Figure 7.5 (Even et al. 2014a; Ishihara 1994).

Excitonic effects in 3D hybrid perovskites can be accounted for using BSE starting from the monoelectronic states calculated at the DFT level. For instance, when taking the pseudo-cubic high-temperature reference phase of $CH_3NH_3PbI_3$, enhancement of absorption at the band gap can clearly be evidenced (Figure 7.6 and Even et al. 2014c). Noteworthy,

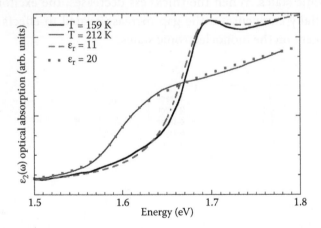

FIGURE 7.5  Optical absorption spectra of $CH_3NH_3PbI_3$ highlighting exciton screening. Experimental data taken from (Kato et al. 2003) recorded at 159 K (thick dark line) and 212 K (thin light line) and computed spectra for bound and continuum pair states, considering two-particle wave function and effective mass equations for electron and hole. The effect of dielectric screening is shown for $\varepsilon_{eff} = 11$ (dashed line) and 20 (dotted line) and leads to a good fit of the experimental spectra below (159 K) and above $T_c$ (212 K), respectively. (Reprinted with permission from Even, J., L. Pedesseau, and C. Katan. 2014a. Analysis of multi-valley and multi-bandgap absorption and enhancement of free carriers related to exciton screening in hybrid perovskites. *J. Phys. Chem. C* 118; 11566–11572. Copyright © 2014 American Chemical Society.)

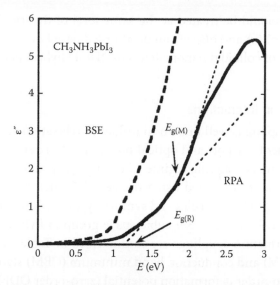

FIGURE 7.6 Comparison of the dielectric constant variation of $CH_3NH_3PbI_3$ in the cubic phase with (dotted line, DFT+BSE calculation) and without (straight line, DFT+RPA calculation) the excitonic interaction. Onsets of optical transitions at R and M points of the Brillouin zone are indicated. (Reprinted with permission from Even, J., L. Pedesseau, J.-M. Jancu, and C. Katan. 2014c. DFT and k · p modelling of the phase transitions of lead and tin halide perovskites for photovoltaic cells. *Phys. Status Solidi RRL* 8; 31–35. Copyright © 2014 John Wiley & Sons.)

in the perturbative BSE/DFT approach, screening of the electron–hole interaction due to atomic motion is not taken into account. Thus, such a description is more suited to low-temperature phases where these motions are expected to be frozen.

The discrepancy with high-temperature experimental results stems from additional exciton screening, related to interactions between electrons and polar phonons, leading to an increased effective dielectric constant. Analysis of the structural phase transition of $CH_3NH_3PbI_3$ near 160 K further shows that besides a renormalization of the optical band gap related to distortion of the inorganic lattice, the thermally activated disorder of the polar cation's orientations associated with the phase transition may further screen the excitonic interaction (Even et al. 2014a). Multiple exciton lines in the 25–75 K temperature range have also been evidenced on a single crystal (Fang et al. 2015). The helium temperature exciton line was attributed to a bound exciton with a long lifetime, possibly arising from a strong coupling with the organic cations.

In the available literature, $\delta E_X(n)$ is mainly attributed to dielectric confinement effects. In a pioneer contribution, Muljarov et al. calculated the binding energies of excitons in layered hybrid perovskites (Muljarov et al. 1995). Their method, based on a classical resolution of BSE, was able to account for dielectric superlattice effects but, unfortunately, it relied on parameters fitted to experimental data. Moreover, to avoid unphysical divergences related to abrupt dielectric interfaces, a transitional layer at the interface between organic and inorganic parts had to be considered. Considering a more realistic dielectric profiles affords significant improvement for evaluating $\delta E_X(n)$. In fact, both monoelectronic and excitonic pair states should be treated at the same level of theory. Currently, semiclassical estimates of both $\delta\Sigma(n)$ and $\delta E_X(n)$ can be afforded based on a preliminary step where the dielectric profiles of colloidal nanoplatelets or 2D/3D hybrid perovskite are computed using DFT.

### 7.1.2 Electron–Phonon Interactions

Electron–phonon coupling is well known to influence relaxation and broadening mechanisms as well as carrier transport and optical properties. Carrier scattering processes can be investigated in the weak coupling regime assuming independent particles and the scattering probabilities can be computed using the Fermi Golden rule. Group theory allows inferring corresponding selection rules and symmetry allowed couplings (Even 2015; Even et al. 2016a). For the reference $Pm$-$3m$ cubic space group of hybrid perovskites, the possible intravalley deformation couplings are summarized in Table 7.1. None of the valence band maximum (VBM) and conduction band minimum (CBM) states can couple to optical phonons via a zero-order deformation potential (zero-order ODP). Next, acoustic phonons and electrons can also interact via a deformation potential, but the coupling always vanishes at zero order. Because of SOC, only local volumetric strain $\left(\Gamma_1^+\right)$ is expected to couple to VBM and CBM states and influence the first-order acoustic deformation potential (ADP). Molecular degrees of freedom, which can be streamlined based on the concept of pseudo-spins (PS), may also yield an additional deformation potential mechanism for electrons in hybrid perovskites via elastic PS (EPS), similarly to deformation potential for phonons. The dynamical/stochastic configurations of elastic multipoles can show up

TABLE 7.1    Possible Intravalley Deformation Couplings in the Reference $Pm$-$3m$ Cubic Phase of Hybrid Perovskites

| | VBM $R_1^+$ SOC = 0 | VBM $R_6^+$ SOC ≠ 0 | CBM $R_4^-$ SOC = 0 | CBM $R_6^-$ SOC ≠ 0 |
|---|---|---|---|---|
| Strain $\Gamma_1^+$ | ✖ | ✖ | ✖ | ✖ |
| Strain $\Gamma_3^+$ | – | – | ✖ | – |
| Strain $\Gamma_5^+$ | – | – | ✖ | – |
| Optical Phonons | – | – | – | – |
| Organic Cation elastic PS (**A**) | ✖ | ✖ | ✖ | ✖ |
| Organic Cation elastic PS (**B**) | ✖ | ✖ | ✖ | ✖ |
| Organic cation elastic PS (**C**) | ✖ | ✖ | ✖ | ✖ |

*Note:* Crosses indicate couplings between CBM and VBM band edge electronic states at the R point of the Pm-3m Brillouin zone, and acoustic phonons (first-order ADP, strains), optical phonons (zero-order ODP) or elastic pseudospin (PS) at $\Gamma$ point. Elastic PS are defined for the three scenarios **A**, **B**, and **C** where the molecular axis points toward a cubic cell facet, a halogen atom, or the center of an octahedron, respectively (Even 2015, 2016a).

along different scenarios (Table 7.1 and Even 2015; Even et al. 2016a). All of them can induce crystal potential fluctuations that can change the electron density. Regarding polar couplings of charge carrier, piezoelectric electron–phonon coupling (PZA) vanishes as a result of vanishing piezoelectric tensor components in *Pm-3m* crystal structures. The Frölich interaction between the delocalized CBM and VBM states and polar optical phonons (FOP) is most probably the dominant scattering process in $CH_3NH_3PbI_3$ as a result of dielectric increment between the high- and medium-frequency ranges related to $\Gamma_4^-$ longitudinal optical modes. Dynamical and static disordered configurations of the molecular cations may induce additional polar PS (PPS) excitations, which can couple to the charge carrier depending on the strength of the molecular electric dipole and the size of the polar domains.

Next, in standard semiconductors, one of the most relevant mechanisms of intraband hot-carrier relaxation is thermalization coupled with lattice vibrations. As the *slow hot-hole cooling* process is currently among the most controversial and debated feature of 3D $CH_3NH_3PbI_3$ (Xing et al. 2013), we discuss this process in more detail. The UV–VIS absorbance spectrum analysis shows two main absorption peaks: one at 760 nm and another at 480 nm; this latter is characterized by an impressively long lifetime (~0.4 ps), a possible fingerprint of the mentioned slow hot-hole cooling in the VB.

In detail, while the peak at 760 nm is undoubtedly associated with the VBM → CBM direct excitation, the attribution of the peak at 480 nm still remains unclear. The results of Xing (Xing et al. 2013) show that the two states are composed of different VB states but the same CB state, assigning the 480 nm peak to the transition from a band situated approximately 1 eV below VBM (VB2) to the CBM. To complete the scenario, early theoretical studies by some of us (Even et al. 2014a) have suggested the state at 480 nm as composed of multi-band gap absorption, not only VB2 → CBM. There is thus still a need to make the process clear and accordingly give a conclusive attribution to the 480-nm peak state.

Full first-principles calculations on electron-phonon (*e–ph*) interactions, that is, one of the most relevant mechanisms of intraband hot-carrier relaxation, have been introduced combining DFT, Density Functional Perturbation Theory (DFPT), and many-body perturbation theory (MBPT) (Cannuccia and Marini 2011; Kawai et al. 2014; Marini 2008). According to such a scheme, $CH_3NH_3PbI_3$ and $CsPbI_3$, and also the charged semiconductor network, $PbI_3^-$ (Giorgi et al. 2014), have been here analyzed.

Concerning the theoretical details of the MBPT, in the coupled electron–nuclei system, the Hamiltonian, $\hat{H}$, is constituted of three components, where $\hat{H}_0$ represents the electronic Hamiltonian corresponding to the case in which the atoms are frozen at their equilibrium position, $\mathbf{R}_0$, while $\hat{H}_1$ and $\hat{H}_2$ are the first and second terms of the Taylor expansion of $\hat{H}_0$ with the atomic positions $\{\mathbf{R}\}$, respectively. In the present DFT-based analysis, the electron–electron correlations were treated at the mean field level. $\hat{H}_0$ is approximated to the Kohn–Sham (KS) Hamiltonian:

$$\hat{H}_0 \approx \sum_j \left[ \hat{h}(\mathbf{r}_j) \right] \text{ with } \hat{h}(\mathbf{r}_j) = \frac{1}{2}\frac{\partial^2}{\partial r^2} + \hat{V}_{\text{scf}}(\mathbf{r},\mathbf{R})\bigg|_{R=R_0},$$

where the electronic effective potential, $\hat{V}_{scf}$ is defined as $\hat{V}_{scf} = \hat{V}_{ion} + \hat{V}_{Hartree} + \hat{V}_{XC}$. MBPT enables us to calculate the two self-energies, the "Fan," $\sum_{ik}^{Fan}(\omega,T)$, and the "Debye-Waller (DW)," $\sum_{ik}^{DW}(T)$, associated with $\hat{H}_1$ and $\hat{H}_2$, respectively (Cannuccia and Marini 2013). The quasi-particle (QP) energy of the $i$th band at the point $k$ in the Brillouin zone, $E_{nk}(T)$, that derives from such self-energies is

$$E_{ik}(T) = \varepsilon_{ik}^{KS} + Z_{ik}(T)\left[\sum_{ik}^{Fan}\left(\varepsilon_{ik}^{KS},T\right) + \sum_{ik}^{DW}(T)\right] \tag{7.7}$$

with $Z_{ik}(T) = \left(1 - \left(\left(\partial \sum_{ik}^{Fan}(\omega,T)\right)/\partial\omega\right)\Big|_{\omega=\varepsilon_{ik}^{KS}}\right)^{-1}$ representing the renormalization factor and $\varepsilon_{ik}^{KS}$ the KS energy of the electronic state at $(i', \mathbf{k-q})$. The lifetime of the QP, $\tau_{ik}(T)$, is related to the imaginary part of QP energy. Since the $\sum_{ik}^{DW}(T)$ is real, only the $\sum_{ik}^{Fan}(\omega,T)$ determines $\tau_{ik}(T)$. $\sum_{ik}^{Fan}(\omega,T)$ is represented as follows.

$$\sum_{ik}^{Fan}(\omega,T) = \sum_{i',q\lambda} \frac{\left|g_{ii'k}^{q\lambda}\right|^2}{N_q}\left[\frac{N_{q\lambda}(T)+1-f_{i'\mathbf{k-q}}}{\omega - \varepsilon_{i'\mathbf{k-q}}^{KS} - \omega_{q\lambda} - i0^+} + \frac{N_{q\lambda}(T)-f_{i'\mathbf{k-q}}}{\omega - \varepsilon_{i'\mathbf{k-q}}^{KS} - \omega_{q\lambda} - i0^+}\right] \tag{7.8}$$

In Equation 7.8 $\omega_{q\lambda}$ represents the phonon energy relative to the mode $\lambda$ and the transferred momentum $\mathbf{q}$. The term $N_{q\lambda}(T)$ is the Bose–Einstein distribution function of the phonon mode $(\lambda, \mathbf{q})$ at temperature $T$, and $f_{i'\mathbf{k-q}}$ is the occupation number of the bare electronic state at $(i', \mathbf{k-q})$. $N_q$ is the number of $\mathbf{q}$-point in the simulation and $0^+$ is a damping parameter. $g_{ii'k}^{q\lambda}$ is the $e$-$ph$ coupling matrix element defined as

$$g_{ii'k}^{q\lambda} = \sum_{scf}(2M_s\omega_{q\lambda})^{-1/2}e^{iq\tau_i}\left\langle ik\left|\frac{\partial\hat{V}_{scf}(\mathbf{r},\mathbf{R})}{\partial\mathbf{R}_{scf}}\right|i'\mathbf{k-q}\right\rangle\xi_\alpha(q\lambda|s) \tag{7.9}$$

with $M_s$ representing the mass of the atom occupying the $\tau_s$ position in the unit cell. The phonon polarization vectors are represented by the term $\xi_\alpha(q\lambda|s)$. Phonons and the derivatives of $\hat{V}_{scf}$ in Equation 7.9 were calculated using DFPT. According to the on-mass-shell approximation, (i.e., $Z_{ik}(T) = 1$ in Equation 7.7) that we initially adopted, and focus on the case at $T = 0$ K, then $\tau_{nk}$ is

$$\tau_{ik}(T=0K) = \left[\frac{2}{N_q}\sum_{i'q\lambda}\left|g_{ii'k}^{q\lambda}\right|^2\delta\left(\varepsilon_{ik}^{KS} - \varepsilon_{i'\mathbf{k-q}}^{KS} \pm \omega_{q\lambda}\right)\right]^{-1} \tag{7.10}$$

where an upper (lower) sign of $\omega_{q\lambda}$ is assigned when a state $i'$ belongs to valence (conduction) band.

The optimized structures (Giannozzi et al. 2009; Kawai et al. 2015) of the $CH_3NH_3PbI_3$, $CsPbI_3$, and $PbI_3^-$ systems are shown in Figure 7.7, with $CH_3NH_3PbI_3$, at variance with $CsPbI_3$ and $PbI_3^-$, showing a noticeable distorted structure from the ideal cubic one. The volume of the semiconductor network $PbI_3^-$ increases with respect to the case of $CH_3NH_3PbI_3$ because of the absence of the attractive Coulomb interactions between aminic hydrogen and iodine atoms (Giorgi et al. 2014; Mosconi et al. 2013).

From the density of states (DOS) analysis, shown in Figure 7.8, the edges of both $CH_3NH_3PbI_3$ and $CsPbI_3$ are characterized by large band dispersions, resulting in light effective masses of both the electrons and holes, confirming the ambipolar nature of $CH_3NH_3PbI_3$ (Xing et al. 2013).

$PbI_3^-$, whose DOSs are reported in Figure 7.8c, shows, at variance with the two previous cases, a reduced dispersion in the VBM, induced by the absence of the methylammonium group [reduction of the Pb $6s$ and I $5p$ orbital antibonding interaction (Giorgi et al. 2014; Huang and Lambrecht 2013)]. Accordingly, the DOSs in the VBM region of $CH_3NH_3PbI_3$ and $CsPbI_3$ will appear smooth, while that of $PbI_3^-$ shows a noticeably high peak. We will see below how this difference has a marked impact on the carrier lifetimes.

The phonons and $e$–$ph$ coupling matrix (Equation 7.10) were calculated using the DFPT scheme by PHonon code in the Quantum ESPRESSO (Giannozzi et al. 2009). For both $CsPbI_3$ and $CH_3NH_3PbI_3$ we have considered the cubic polymorph, because of the computational burden associated with the large size of the room temperature stable polymorph, which is orthorhombic for $CsPbI_3$ ($o$-$CsPbI_3$, $Pnma$, $Z = 4$, 20 atoms) and tetragonal for $CH_3NH_3PbI_3$ ($t$-$CH_3NH_3PbI_3$, $I4cm$, $Z = 4$, 48 atoms). In this way we can also compare

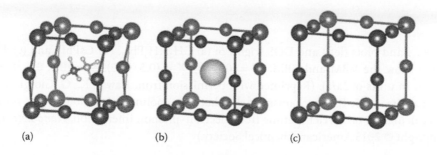

(a)                    (b)                    (c)

FIGURE 7.7  Optimized structures of (a) $CH_3NH_3PbI_3$, (b) $CsPbI_3$, and (c) $PbI_3^-$. Large and small atoms in the network represent Pb and I, respectively. Inside the network (a) dark (light) larger atoms are C (N) while small white ones are H, respectively. In (b) the large atom inside the cage is Cs. (Reprinted with permission from Kawai, K., G. Giorgi, A. Marini, and K. Yamashita. 2015. The mechanism of slow hot-hole cooling in lead-iodide perovskite: First-principles calculation on carrier lifetime from electron–phonon interaction. *Nano Lett.* 15; 3103–3108. Copyright © 2015 American Chemical Society.)

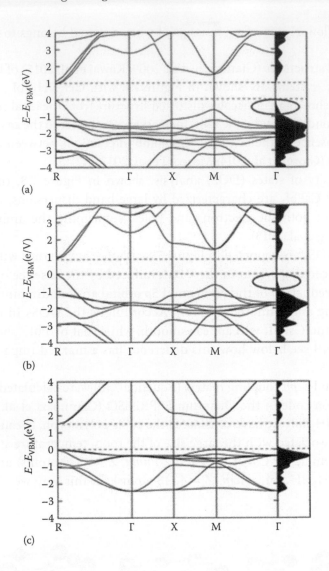

FIGURE 7.8   Band plots (left) and DOS (right) of (a) $CH_3NH_3PbI_3$, (b) $CsPbI_3$, and (c) $PbI_3^-$. The dotted lines show the VBM and CBM. [$\Gamma$ = (0, 0, 0), X = (0.5, 0, 0), M = (0.5, 0.5, 0), and R = (0.5, 0.5, 0.5) in units of $2\pi/a$]. (Reprinted with permission from Kawai, K., G. Giorgi, A. Marini, and K. Yamashita. 2015. The mechanism of slow hot-hole cooling in lead-iodide perovskite: First-principles calculation on carrier lifetime from electron–phonon interaction. *Nano Lett.* 15; 3103–3108. Copyright © 2015 American Chemical Society.)

systems with similar properties but with the drawback that the metastable cubic structures result in some imaginary phonon modes for the three species under analysis.

The resulting phonon bands are shown in Figure 7.9 for the case of $CH_3NH_3PbI_3$. Notably, imaginary modes do not affect the mechanism of slow hot-hole cooling and the carrier lifetimes. In detail, the phonon bands whose wavenumbers are higher than 200 cm$^{-1}$ refer to methylammonium vibration. These contributions are decoupled from those with lower wavenumber associated with the lead iodide semiconductor network (Papavassiliou 1997);

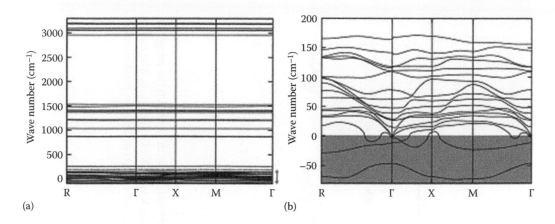

(a)          (b)

FIGURE 7.9 (a) Phonon band of $CH_3NH_3PbI_3$. Imaginary modes are represented as negative values in the shaded areas. (b) The phonon band in low wavenumber region is indicated by an arrow in (a). (Reprinted with permission from Kawai, K., G. Giorgi, A. Marini, and K. Yamashita. 2015. The mechanism of slow hot-hole cooling in lead-iodide perovskite: First-principles calculation on carrier lifetime from electron–phonon interaction. *Nano Lett.* 15; 3103–3108. Copyright © 2015 American Chemical Society.)

thus the band dispersion is not remarkable. To better observe the character of phonons with low wavenumbers, such a region is indicated by an arrow in Figure 7.9a and magnified in Figure 7.9b.

The lifetimes $\tau$ calculated with and without the imaginary modes are plotted in Figure 7.10 along with the KS energies.* The fact that only the optical phonon modes strongly couple with the VBs makes the impact of these imaginary modes marginal in the three investigated systems.

Figure 7.10 also reveals that the $\tau$ in the VB of $CH_3NH_3PbI_3$ ($CsPbI_3$) begins to increase at $E–E_{VBM} \sim -0.3$ eV ($E–E_{VBM} \sim -0.6$ eV) (long lifetime region). Because of the enhancement in the lifetime, the relaxation of the hot holes in the two perovskites, the organic–inorganic and the fully inorganic, will be suppressed in the range $E–E_{VBM}$ between −0.3 and −0.6 eV, confirming the available experimental results for $CH_3NH_3PbI_3$ (Xing et al. 2013).

In Figure 7.10 the inverses of the DOSs are also plotted, showing the general correlation between the lifetime and the DOSs: in particular, the lifetime increases as the DOS becomes smaller (Bernardi et al. 2014; Lautenschlager et al. 1986; Restrepo et al. 2009), and smaller DOS offers a reduced number of carrier relaxation paths leading to longer carrier lifetimes and vice versa. As a result, the long-lifetime region in both the organic–inorganic mixed perovskite and the fully inorganic one can be attributed to the small DOS region shown in Figure 7.8, long-lifetime that still from Figure 7.8 is absent in the case of the charged semiconductor network, $PbI_3^-$. Accordingly, we clearly predict that the slow hot-hole cooling is universally observed in $APbI_3$, as long as the characteristic small DOS is presented in their VBs. Figure 7.11a and b show the generalized Eliashberg functions

---

* MBPT calculations performed to obtain the self-energies and the lifetimes are performed with the Yambo code (Marini et al. 2009). For further details see Kawai et al. (2015).

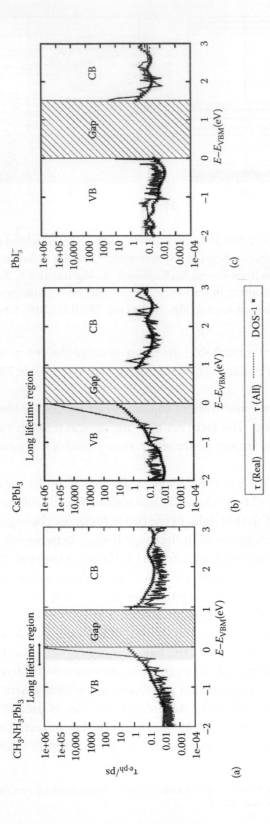

FIGURE 7.10 Hot-carrier lifetimes of (a) CH₃NH₃PbI₃, (b) CsPbI₃, and (c) PbI₃⁻. The blue (red) dotted lines are lifetimes with (without) the imaginary modes. Black crosses are the inverse of the DOSs shown in Figure 7.8. (Partially adapted with permission from Kawai, K., K. G. Giorgi, A. Marini, and K. Yamashita. 2015. The mechanism of slow hot-hole cooling in lead-iodide perovskite: First-principles calculation on carrier lifetime from electron–phonon interaction. *Nano Lett.* 15; 3103–3108. Copyright © 2015 American Chemical Society.)

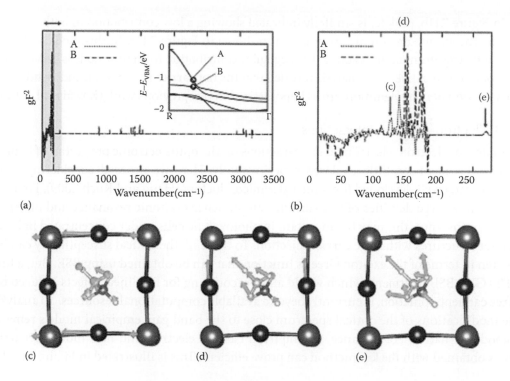

FIGURE 7.11   (a) Generalized Eliashberg functions of $CH_3NH_3PbI_3$ for the two valence states A and B. Their positions are indicated in the inset band plot. (b) Generalized Eliashberg functions in small wave number region indicated by an arrow in (a). (c), (d), (e) Atomic motions of the phonon modes for $(\lambda, q) = (14, (0.5, 0.5, 0))$, $(\lambda, q) = (17, (0.5, 0.5, 0))$, $(\lambda, q) = (19, (0.5, 0.5, 0))$ respectively. As in Figure 7.7, large and small atoms in the network represent Pb and I, respectively. Inside the network dark (light) larger atoms are C (N) while small white ones are H, respectively.

of the two valence states A and B in the inset band plot. The analysis of such functions is extremely useful to explain the contribution of phonons to the carrier relaxation for each electronic state. Figure 7.11a clearly confirms that only phonon modes in a low wavenumber range below 200 cm$^{-1}$ have a dominant contribution.

Similarly as in Figure 7.9, the low wavenumber region is indicated by an arrow in Figure 7.11a and magnified in Figure 7.11b. In state A, the highest peaks (c) were in the range from 110 to 120 cm$^{-1}$. In state B, on the other hand, the highest peaks (d) were in the range from 140 to 170 cm$^{-1}$. The representative atomic motions in these peaks are shown in Figure 7.11c and 7.11d. In both phonon modes, methylammonium vibrations are due to the light atomic masses of H, C, and N. However, the Pb–I network is similarly active. The difference between mode (c) and (d) is the motion of lead: it is indeed active in mode (c) but inactive in mode (d). This trend indicates that the carrier relaxation on the electronic state A is governed by motions of both Pb and I while the relaxation on the state B is governed by only I motion. This difference stems from the fact that state A is associated with the antibonding coupling between the Pb 6s and I 5p orbitals, but state B is composed mainly of I 5p orbital. This result is exactly the same as CsPbI$_3$ (Kawai et al. 2015).

In Figure 7.11b, peak (e) is similarly indicated showing a low contribution in both states A and B. The atomic motion of this peak is also shown in Figure 7.11e. In this mode, only the part of methylammonium is active. The negligible contribution from this mode is because the part of $PbI_3$ is inactive. Our analysis revealed that the hot carrier relaxation is independent of the vibration of methylammonium as hypothesized in our previous work (Kawai et al. 2015).

### 7.1.3 Electron–Electron Interactions

The effect of electron–electron ($e$–$e$) interactions on the optoelectronic properties of hybrid perovskites has been little explored experimentally. Here we proceed considering concepts already well developed for standard semiconductors (Haug and Koch 2009). First we explore how large densities of free carriers can influence excitonic resonance and material gain. For instance, this may be important in photovoltaic cells under concentrated light or in laser structures with large carrier injection. To this end, the optical susceptibility can be written in terms of the exciton Green's function that can be obtained using BSE. But, a full DFT+GW+BSE treatment, which would allow accounting for nonlinear effects induced by a free carrier population, is currently beyond available computational resources. To analyze the modifications of the optical spectrum close to the band gap, empirical models remain a good alternative. For instance, an empirical basis of electron and hole monoelectronic states obtained with the **k·p** method can prove efficient. This is illustrated in Figure 7.12. In

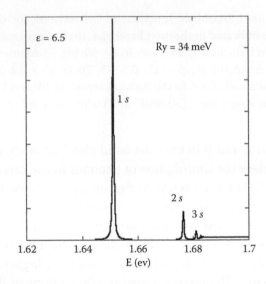

FIGURE 7.12 Optical absorption of $CH_3NH_3PbI_3$ computed using the BSE and the exciton Green's function in the linear regime at low temperature. An empirical basis of electron and hole monoelectronic states obtained with the **k·p** method close to the R point of the Brillouin zone is used to compute the exciton Green's function. The $e$–$h$ Coulomb interaction is computed with an effective dielectric constant of 6.5. $1s$–$3s$ exciton resonances are shown. The electronic band gap is set to 1.685 eV. (Reprinted with permission from Even J., Boyer-Richard S., Carignano M., Pedesseau L., Jancu, J.-M., and Katan C. 2016b. Theoretical insights into hybrid perovskites for photovoltaic applications. *Proc. SPIE* 9742, Physics and Simulation of Optoelectronic Devices XXIV, 97421A.)

the linear regime, the BSE contains the dipole of the *e–h* optical transition at *R* as a source term and an *e–h* Coulomb interaction, which singularity deserves a specific numerical procedure.

In the nonlinear regime, the BSE equation is strongly modified: (1) the *e–h* Coulomb interaction is screened, which is usually represented within the plasmon-pole approximation; (2) the oscillator strength is reduced by the phase-space population due to free carriers; and (3) self-energy contributions for both electrons and holes lead to band gap renormalization and damping of the exciton resonances.

As illustrated in Figure 7.13, for moderate free carrier concentration of about $10^{17}$ cm$^{-3}$, the screening of the exciton resonance is the most important phenomenon. For larger concentrations $5 \times 10^{17}$ cm$^{-3}$/$10^{18}$ cm$^{-3}$, phase space filling effects start to dominate until population inversion is reached for $2 \times 10^{18}$ cm$^{-3}$. For low carrier concentrations, the band gap renormalization is partially compensated by the screening of the exciton resonance. For large concentrations the band gap is shifted to lower values (about 1.65 eV for $2 \times 10^{18}$ cm$^{-3}$).

Light emission close to the band gap is also affected by the presence of large densities of charge carriers in the material. The third-order nonradiative Auger process yields a nonradiative channel in competition with radiative recombination. Before gauging such Auger effects in hybrid perovskites, let us recall what is well known in the field of III–V

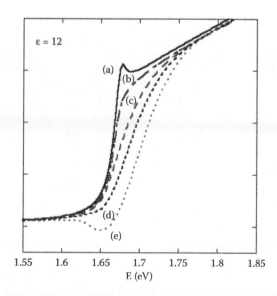

FIGURE 7.13   Optical absorption of $CH_3NH_3PbI_3$ computed using the BSE and the exciton Green's function in the nonlinear regime at $T = 160$ K. The concentration of free carriers amounts to (a) 0; (b) $10^{17}$ cm$^{-3}$; (c) $5.10^{17}$ cm$^{-3}$; (d) $10^{18}$ cm$^{-3}$; and (e) $2.10^{18}$ cm$^{-3}$. The band gap in the linear regime is set to 1.685 eV. (Reprinted with permission from Even J., Boyer-Richard S., Carignano M., Pedesseau L., Jancu, J.-M., and Katan C. 2016b. Theoretical insights into hybrid perovskites for photovoltaic applications. *Proc. SPIE* 9742, Physics and Simulation of Optoelectronic Devices XXIV, 97421A.)

semiconductors (Figure 7.14a and Even et al. 2015). An electron in the CB may recombine with a heavy hole (HH) in the VB, either by transferring energy and momenta to an electronic transition in the CB (CCCH or CHCC process) or to a hole jumping from the HH band to the spin–orbit split-off (SO) band (CHHS or CHSH process). Among the numerous other possible processes, a third CHLH or CHHL process, not represented here, is usually also considered. This is weaker than the CHHS process, except when the spin–orbit splitting energy is larger than the band gap. As electronic states away from the band gap are involved, the full simulation requires repeated summations over the particle momenta and a careful description of the electronic band structure. The dependence on the electronic band gap, effective masses, and carrier densities can be estimated using parabolic approximations for the electronic dispersions:

$$R_{\text{Auger, CCCH}} \propto n^2 p e^{-(E_T - E_{\text{gap}})/kT}, \tag{7.11}$$

$$R_{\text{Auger, CHHS}} \propto n p^2 e^{-(E_T - E_{\text{gap}})/kT}, \tag{7.12}$$

where

$$E_{T,\text{CCCH}} = \frac{2m_C + m_{HH}}{m_C + m_{HH}} E_{\text{gap}}, \tag{7.13}$$

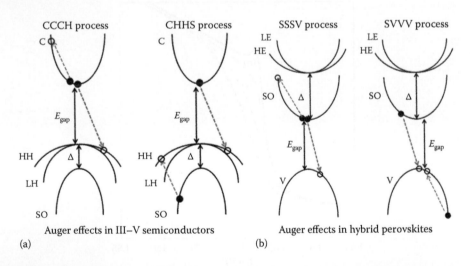

CCCH process     CHHS process     SSSV process     SVVV process

Auger effects in III–V semiconductors      Auger effects in hybrid perovskites

(a)            (b)

FIGURE 7.14 Dominant Auger processes in (a) IIIV semiconductors and (b) hybrid perovskites. (a) HH, LH, and SO correspond to heavy hole, light hole, and hole spin–orbit split-off states of the VB in III–V semiconductors. (b) HE, LE, and SO correspond to heavy electron, light electron, and electron spin–orbit split-off states of the CB in hybrid perovskites. (Reprinted with permission from Even, J., L. Pedesseau, C. Katan, M. Kepenekian, J.-S. Lauret, D. Sapori, and E. Deleporte. 2015. Solid-state physics perspective on hybrid perovskite semiconductors. *J. Phys. Chem. C* 119; 10161–10177. Copyright © 2015 American Chemical Society.)

$$E_{\text{T,CHHS}} = \frac{2m_{\text{HH}} + m_{\text{C}}}{2m_{\text{HH}} + m_{\text{C}} - m_{\text{SO}}}(E_{\text{gap}} - \Delta). \tag{7.14}$$

In III–V semiconductors, the Auger effect decreases exponentially as the band gap increases. Moreover, when SOC and the band gap are of the same order of magnitude, the CHHS process prevails in most cases. However, in n-type III–V semiconductors, the CCCH process is dominant through the $n^2p$ term. For large band gap materials such as GaAs, where Auger processes are very weak, phonon-assisted Auger processes (not described here) give the most important contribution.

The electronic band structure of hybrid perovskites has two important differences with respect to that of III–V semiconductors: (1) the band structure shows reverse band ordering, with SOC on CB, and (2) the effect of SOC is significantly larger than that occurring in III–V semiconductors. Based on the method developed for III–V semiconductors, we can investigate the two most important Auger processes occurring in hybrid perovskites as follows (Figure 7.14b and Even et al. 2015):

$$R_{\text{Auger,SSSV}} \propto n^2 p e^{-(E_{\text{T}} - E_{\text{gap}})/kT} \tag{7.15}$$

$$R_{\text{Auger,SVVV}} \propto np^2 e^{-(E_{\text{T}} - E_{\text{gap}})/kT}, \tag{7.16}$$

where

$$E_{\text{T,SSSV}} = \frac{2m_{\text{S}} + m_{\text{V}}}{m_{\text{S}} + m_{\text{V}}} E_{\text{gap}}, \tag{7.17}$$

$$E_{\text{T,SVVV}} = \frac{2m_{\text{V}} + m_{\text{S}}}{m_{\text{S}} + m_{\text{V}}} E_{\text{gap}}. \tag{7.18}$$

An electron in the split-off state (S) of CB may recombine with a hole (V) in the VB, either by transferring energy and momenta to an electronic transition in the split-off state of the CB (SSSV process) or to a hole in the valence band (SVVV process). Using effective masses of 0.12 and 0.15 for the split-off CB and for the VB, respectively, we deduce that the SVVV process slightly dominates over the SSSV process, especially when considering the natural p-doping of hybrid perovskites. However, the overall effect is expected to be small, like in large band gap III–V semiconductors such as GaAs. A process involving the SOC, and heavy electrons (HE) or light electrons (LE) in the CB (HSSV or LSSV processes), would require a more detailed analysis, as HE and LE electronic states lying at high energy in the CB are hybridized with molecular states.

## REFERENCES

Bernardi, M., D. Vigil-Fowler, J. Lischner, J. B. Neaton, and S. G. Louie. 2014. *Ab initio* study of hot carriers in the first picosecond after sunlight absorption in silicon. *Phys. Rev. Lett.* 112; 257402–257405.

Billing, D. G., and A. Lemmerer. 2007. Synthesis, characterization and phase transitions in the inorganic–organic layered perovskite-type hybrids [(C$_n$H$_{2n+1}$NH$_3$)$_2$PbI$_4$], n = 4, 5 and 6. *Acta Cryst. B* 63; 735–747.

Brivio, F., K. T. Butler, A. Walsh, and M. van Schilfgaarde. 2014. Relativistic quasiparticle self-consistent electronic structure of hybrid halide perovskite photovoltaic absorbers. *Phys. Rev. B* 89; 155204.

Cannuccia, E., and A. Marini. 2011. Effect of the quantum zero-point atomic motion on the optical and electronic properties of diamond and trans-polyacetylene. *Phys. Rev. Lett.* 107; 255501.

Cannuccia, E., and A. Marini. 2013. Ab-initio study of the effects induced by the electron-phonon scattering in carbon based nanostructures. *Phys. Rev. B* arXiv:1304.0072 [cond-mat.mtrl-sci].

Even, J. 2015. Pedestrian guide to symmetry properties of the reference cubic structure of 3D all-inorganic and hybrid perovskites. *J. Phys. Chem. Lett.* 6; 2238–2242.

Even, J., L. Pedesseau, M.-A. Dupertuis, J.-M. Jancu, and C. Katan. 2012. Electronic model for self-assembled hybrid organic/perovskite semiconductors: Reverse band edge electronic states ordering and spin-orbit coupling. *Phys. Rev. B* 86; 205301.

Even, J., L. Pedesseau, J.-M. Jancu, and C. Katan. 2013. Importance of spin–orbit coupling in hybrid organic/inorganic perovskites for photovoltaic applications. *J. Phys. Chem. Lett.* 4; 2999–3005.

Even, J., L. Pedesseau, and C. Katan. 2014a. Analysis of multi-valley and multi-bandgap absorption and enhancement of free carriers related to exciton screening in hybrid perovskites. *J. Phys. Chem. C* 118; 11566–11572.

Even, J., L. Pedesseau, and C. Katan. 2014b. Understanding quantum confinement of charge carriers in layered 2D hybrid perovskites. *Chem. Phys. Chem.* 15; 3733–3741.

Even, J., L. Pedesseau, J.-M. Jancu, and C. Katan. 2014c. DFT and k · p modelling of the phase transitions of lead and tin halide perovskites for photovoltaic cells. *Phys. Status Solidi RRL* 8; 31–35.

Even, J., L. Pedesseau, C. Katan, M. Kepenekian, J.-S. Lauret, D. Sapori, and E. Deleporte. 2015. Solid-state physics perspective on hybrid perovskite semiconductors. *J. Phys. Chem. C* 119; 10161–10177.

Even, J., M. Carignano, and C. Katan. 2016a. Molecular disorder and translation/rotation coupling in the plastic crystal phase of hybrid perovskites. *Nanoscale* 8; 6222–6236.

Even, J., S. Boyer-Richard, M. Carignano, L. Pedesseau, J.-M. Jancu, and C. Katan 2016b. Theoretical insights into hybrid perovskites for photovoltaic applications. *Proc. SPIE* 9742, Physics and Simulation of Optoelectronic Devices XXIV, 97421A.

Fang, H.-H., R. Raissa, M. Abdu-Aguye, S. Adjokatse, G. R. Blake, J. Even, and M. A. Loi. 2015. Photophysics of organic–inorganic hybrid lead iodide perovskite single crystals. *Adv. Func. Mater.* 25; 2378–2385.

Giannozzi, P., S. Baroni, N. Bonini, M. Calandra, R. Car, C. Cavazzoni, D. Ceresoli, G. L. Chiarotti, M. Cococcioni, I. Dabo, A. Dal Corso et al. 2009. Ab-initio study of the effects induced by the electron-phonon scattering in carbon based nanostructures *J. Phys.: Condens. Matter* 21; 395502–395519.

Giorgi, G., J.-I. Fujisawa, H. Segawa, and K. Yamashita. 2014. Small photocarrier effective masses featuring ambipolar transport in methylammonium lead iodide perovskite: A density functional analysis. *J. Phys. Chem. C* 118; 12176–12183.

Haug, H., and S. W. Koch. 2009. *Quantum theory of the optical and electronic properties of semiconductors* (5th ed.). Hackensack, NJ: World Scientific.

Hirasawa, M., T. Ishihara, T. Goto, K. Uchida, and N. Miura. 1994. Magnetoabsorption of the lowest exciton in perovskite-type compound (CH$_3$NH$_3$)PbI$_3$. *Physica B* 201; 427–430.

Huang, L., and W. R. L. Lambrecht. 2013. Electronic band structure, phonons, and exciton binding energies of halide perovskites $CsSnCl_3$, $CsSnBr_3$, and $CsSnI_3$. *Phys. Rev. B* 88; 165203–165212.

Ishihara, T. 1994. Optical properties of PbI-based perovskite structures. *J. Luminescence* 60; 269–274.

Ishihara, T., J. Takahashi, and T. Goto. 1990. Optical properties due to electronic transitions in two-dimensional semiconductors $(C_nH_{2n+1}NH_3)_2PbI4$. *Phys. Rev. B* 42; 11099–11107.

Kato, Y., D. Ichii, K. Ohashi, H. Kunugita, K. Ema, K. Tanaka, T. Takahashi, and T. Kondo. 2003. Extremely large binding energy of biexcitons in an organic–inorganic quantum-well material $(C_4H_9NH_3)_2PbBr_4$. *Solid State Comm.* 128; 15–18.

Kawai, K., G. Giorgi, A. Marini, and K. Yamashita. 2015. The mechanism of slow hot-hole cooling in lead-iodide perovskite: First-principles calculation on carrier lifetime from electron–phonon interaction. *Nano Lett.* 15; 3103–3108.

Kawai, H., K. Yamashita, E. Cannuccia, and A. Marini. 2014. Electron-electron and electron-phonon correlation effects on the finite-temperature electronic and optical properties of zinc-blende GaN. *Phys. Rev. B* 89; 085202.

Koutselas, B., L. Ducasse, and G. C. Papavassiliou. 1996. Electronic properties of three- and low-dimensional semiconducting materials with Pb halide and Sn halide units. *J. Phys. Condens. Matter.* 8; 1217–1227.

Lautenschlager, P., P. Allen, and M. Cardona. 1986. Phonon-induced lifetime broadenings of electronic states and critical points in Si and Ge. *Phys. Rev. B* 33; 5501–5511.

Marini, A. 2008. *Ab initio* finite-temperature excitons. *Phys. Rev. Lett.* 101; 06405.

Marini, A., C. Hogan, C., M. Grüning, and D. Varsano, D. 2009. Yambo: An ab initio tool for excited state calculations. *Comput. Phys. Commun.* 180; 1392–1403.

Mitzi, D. B., K. Chondroudis, and C. R. Kagan. 2001. Organic–inorganic electronics. *IBM J. Res. Dev.* 45; 29–46.

Mitzi, D. B., S. Wang, C. A. Field, C.A. Chess, and A. M. Guloy. 1995. Conducting layered organic–inorganic halides containing [110]-oriented perovskite sheets. *Science* 267; 1473–1476.

Mosconi, E., A. Amat, Md. K. Nazeeruddin, M. Grätzel, and F. De Angelis. 2013. First-principles modeling of mixed halide organometal perovskites for photovoltaic applications. *J. Phys. Chem. C* 117; 13902–13913.

Muljarov, E. A., S. G. Tikhodeev, N. A. Gippius, and T. Ishihara. 1995. Excitons in self-organized semiconductor/insulator superlattices: PbI-based perovskite compounds. *Phys. Rev. B* 51; 14370–14378.

Papavassiliou, G. C. 1997. Three-and low-dimensional inorganic semiconductors. *Prog. Solid State Chem.* 25; 125–270.Pedesseau, L., J.-M. Jancu, A. Rolland, E. Deleporte, C. Katan, and J. Even. 2014. Electronic properties of 2D and 3D hybrid organic/inorganic perovskites for optoelectronic and photovoltaic applications. *Opt. Quant. Electron.* 46; 1225–1232.

Restrepo, O. D., K. Varga, and S. T. Pantelides. 2009. First-principles calculations of electron mobilities in silicon: Phonon and Coulomb scattering. *Appl. Phys. Lett.* 94; 212103.

Sapori, D., M. Kepenekian, L. Pedesseau, C. Katan, and J. Even. 2016. Quantum confinement and dielectric profiles of colloidal nanoplatelets of halide inorganic and hybrid organic–inorganic perovskites. *Nanoscale* 8; 6369–6378.

Shimizu, M., J. Fujisawa, and T. Ishihara. 2006. Photoluminescence of the inorganic-organic layered semiconductor $(C_6H_5C_2H_4NH_3)_2PbI_4$: Observation of triexciton formation. 2006. *Phys. Rev. B* 74; 155206.

Umari, P., E. Mosconi, and F. De Angelis. 2014. Relativistic GW calculations on $CH_3NH_3PbI_3$ and $CH_3NH_3SnI_3$ perovskites for solar cell applications. *Sci. Rep.* 4; 4467.

Xing, G., N. Mathews, S. Sun, S. S. Lim, Y. M. Lam, M. Grätzel, S. Mhaisalkar, and T. C. Sum. 2013. Long-range balanced electron- and hole-transport lengths in organic–inorganic $CH_3NH_3PbI_3$. *Science* 342; 344–347.

# Index

Printed and bound by CPI Group (UK) Ltd, Croydon, CR0 4YY

01/11/2024

01782603-0004